Probabilistic Methods in
the Theory of Structures

Probabilistic Methods in the Theory of Structures

Isaac Elishakoff
Department of Aeronautical Engineering
Technion, Israel Institute of Technology, Haifa

A Wiley-Interscience Publication

JOHN WILEY & SONS

New York · Chichester · Brisbane · Toronto · Singapore

Library of Congress Cataloging in Publication Data:

Elishakoff, Isaac.
  Probabilistic methods in the theory of structures.

  Includes indexes.
  1. Structures, Theory of.   2. Probabilities.
I. Title.

TA646.E44   1983   624.1'71   82-13470
ISBN 0-471-87572-4

Printed in the United States of America

10 9 8 7 6 5 4 3 2 1

To the Blessed Memory of My Parents

Preface

This book is written both to serve as a first-course text on probabilistic methods in the theory of structures and to provide a more advanced treatment of random vibration and buckling. It is intended in particular for the student in aeronautical engineering, mechanical engineering, or theoretical and applied mechanics, and may also be used by practicing engineers and research workers as a reference. In fact, it combines the features of a textbook and a monograph.

Probability theory and random functions are playing an ever more prominent role in structural mechanics due to the growing realization that many mechanical phenomena can be satisfactorily described by probabilistic means only. In the last 25 years, much work has been done and many studies have been published on this subject. However, despite significant advances, the probabilistic approach to the theory of structures has not yet found its proper place in engineering education.

Chapter 1 introduces the role of probabilistic methods in the theory of structures. Chapters 2 through 4 deal exclusively with elements of the theory of probability for a single random variable. This apparent preoccupation with the single random variable stems from my own feeling that it would be unfair to offer the reader a mere taste of the theory of probability and then immediately confront him or her with a wide range of applications. Chapter 5 is devoted to the reliability of structures described by a single random variable. Chapter 6 discusses elements of the theory of probability of two or more random variables, while Chapter 7 examines the reliability of such multivariable structures. Chapter 8 introduces the theory of random functions. Chapter 9 deals with random vibration of single- and multidegree-of-freedom structures, and Chapter 10 with random vibration of continuous systems. These chapters concentrate on the role of modal cross correlations in random vibration analysis, usually overlooked in literature, as well as treat point-driven structures, and random vibration and flutter. These chapters constitute, among others, a prerequisite to study the fatigue life of structures—a topic which is outside the scope of this book. The reader interested in this subject is referred to other sources where it is adequately treated. Finally, Chapter 11 is devoted to the Monte Carlo method for treating problems incapable of exact solution. Special emphasis is placed on buckling of nonlinear structures where random imperfections may be responsible for drastic reduction of the buckling loads.

Ample examples are included in the book, because it is my experience that much of the material in question may be taught most effectively by this means. An additional purpose of the examples is to examine the validity of some widely accepted simplifying assumptions concerning the probabilistic nature of the output quantities and to observe the errors that these assumptions may cause. Numerous exercises are provided with each chapter, to deepen the reader's grasp of the subject and widen his or her perspectives.

The material in Chapters 1 to 5, together with Sections 11.1–11.3, are suitable for a one-semester, first-level course at the junior or senior level. Prerequisite courses for this part are calculus, differential equations, and mechanics of solids. For departments whose curriculum requires a course in the theory of probability, the first four chapters may be rapidly recapitulated, in which case the one-semester course may include also Chapters 6 and 7, as well as Sections 11.4 and 11.5. The material in Chapters 8 through 11 is open to both the analytically minded senior and the graduate student and may form an advanced course on random vibration and buckling. The additional prerequisite for this part is knowledge of matrix theory and the basics of vibration and buckling of structures, although the necessary material is reviewed at the beginning of each chapter.

It is my agreeable duty to thank the Department of Aerospace Engineering of Delft University of Technology for their invitation to present a series of lectures (from which this text grew) to their students and scientific staff during my sabbatical leave in the academic year 1979–1980—an experience of endless Dutch courtesy and good will. My sincere thanks are due to the Dean, Prof. Ir. Jaap A. van Ghesel Grothe, and to Professor of Aircraft Structures Dr. Johann Arbocz for their constant encouragement and help. Appreciation is expressed to the staff members and the students of Delft, and especially to Ir. Johannes van Geer, Ir. Willie Koppens, and Ir. Kees Venselaar for their able assistance in a number of calculations and constructive suggestions concerning the lecture notes. I acknowledge the help of Ir. J. K. Vrijling of Delft in writing Sec. 4.18. I also thank the Department of Aeronautical Engineering, Technion-Israel Institute of Technology, in whose encouraging atmosphere I was able to bring this work to completion. I am also most indebted to Eliezer Goldberg of Technion, for his kind help in editing the text, to Marijke Schillemans and Dvora Zirkin for typing much of the original manuscript, to Alice Aronson and Bernice Hirsch for typing Chapter 9, and to Willem Spee and Irith Nizan for preparing the drawings.

ISAAC ELISHAKOFF

Technion City, Haifa
January 1983

Contents

Probabilistic Methods in
the Theory of Structures

chapter 1

Introduction

For an adequate description of structural behavior, probabilistic methods must be resorted to. Properly speaking, an element of probability is embodied even in the deterministic approach, which claims to "simplify" the structure by eliminating all aspects of uncertainty. Under the deterministic approach, external loading and the properties of the structure are represented as though they were fully determined, and available (often highly sophisticated) tools yield, with sufficient accuracy, the strains and stresses in systems with complex configurations. At the same time, these stresses are compared with allowable ones obtained by dividing their ultimate levels by a "safety factor," so as to yield a level below that of failure, a practice that recognizes the uncertain, and random, features of the stress distribution in the material. This is how a probabilistic consideration is admitted "via the back door"; indeed, the safety factor has often been referred to as the "ignorance factor."

The quality of "randomness" is characteristic both of loads borne by structures and of the properties of the structure themselves. No two structures, even if they have been produced by the same manufacturing process, have identical properties. Thin-walled structures are often sensitive to imperfections—deviations from their prescribed geometry—in the sense that the buckling load of an imperfect structure may be lower than that of its ideal counterpart by several percentage decades. The shape and magnitude of these initial imperfections vary widely from case to case, since differences are inherent in any manufacturing process, which is itself subject (by its very nature) to a large number of random influences. These and other examples clearly indicate that it is impossible to investigate structural behavior without resorting to probabilistic methods.

The need for a probabilistic approach does not obviate the classical treatment of the behavior of an ideal structure with given properties, subjected to given loading. In fact, the solution to a deterministic problem may very often prove useful in a probabilistic setting. For example, assume that the properties of a structure are fully determined, while the external forces or moments are random. We begin by constructing explicit equations of motion (or equilibrium) in terms of these forces and moments, which are then used as input in determining the probabilistic characteristics of the response (output). Where the exact relationship between input and output is unavailable, or its application proves too cumbersome, statistical simulation (such as with the Monte Carlo method) is the logical remedy, now that high-speed digital computers are so readily available. The first step of this method consists in simulating the random variable; the second step is numerical solution of the problem for each realization of the random variable; the third and last is statistical analysis (computation of the characteristics of the output by averaging over the ensemble). Thus, one of the cornerstones of the Monte Carlo method is the solution of a deterministic problem.

The deterministic and probabilistic approaches to design differ in principle. Deterministic design is based on total "discounting" of the contingency of failure. The designer is trained in the doctrine that with the relevant quantities properly chosen, admissible levels would never be exceeded; it is postulated that, as it were, the structure is immune to failure and will survive indefinitely. This approach dates back to antiquity, when design analysis and control were unknown and everything centered on the personal responsibility of the artisan. Its earliest written record is probably Hammurabi's Code, according to which, if a house collapses and the householder is killed the builder is liable to the death penalty.

Deterministic design has now reached a very high level of sophistication, and modern computation techniques make it possible to determine stresses, strains, and displacements in highly complex structures. However, problems of structural design always involve an element of uncertainty, unpredictability, or randomness: No matter how much is known about the phenomenon, the behavior of a structure is incapable of precise prediction. In these circumstances there always exists some likelihood of failure, that is, of an unfavorable state of the structure setting in. Even with safety factors—empirical reserve margins—failures did and still do occur. There can in principle be no "never-fail" structure; it is a question only of a higher or lower probability of failure. Accordingly, probabilistic design is concerned with the probability of failures or, preferably, of nonfailure performance, the probability that the structure will realize the function assigned to it—in other words, with *reliability*. The *McGraw-Hill Dictionary of Scientific and Technical Terms* gives the following definition of this basic concept: "Reliability—the probability that a component part, equipment, or system will satisfactorily perform its intended function under given circumstances, such as environmental conditions, limitations as to operating time, and frequency and thoroughness of maintenance, for a specified period of time." The reliability approach was initiated by Maier and Khozialov and carried on by Freudenthal, Johnson, Pugsley, Rzhanitsyn, Shinozuka, Streletskii, Tye, and Weibull. The contributions of Ang and Tang, Augusti, Barrata and Casciati, Benjamin and Cornell, Bolotin, Ferry Borges

Fig. 1.1. Section of Hammurabi's stela at high magnification (photoassembly by courtesy of J. Kogan).

and Castanheta, Ditlevsen, Haugen, Kogan, Lind, Moses, Murzewski, Rackwitz, Rosenblueth, Schuëller, and Veneziano should also be mentioned.

The development of high-power rocket jet engines and supersonic transport since the 1950s has brought out new problems of mechanical and structural vibration, namely the response of panel-like structures to aerodynamic noise and to a turbulent boundary layer, with the attendant aspects of acoustic fatigue and interior noise, all of which are incapable of deterministic solution. The probabilistic methods for these and other problems are embodied in a new discipline called "Random Vibration," dealt with by numerous research centers and their offshoots, which have come into being throughout the world in the last 20 years. Of these, the teams of Caughey (Caltech), Crandall (M.I.T.), Lin (University of Illinois, Urbana-Champaign), and Shinozuka (Columbia University) in the United States; of Bolotin (Moscow Energetics Institute) and Pal'mov (Leningrad Polytechnic) in the Soviet Union; of Clarkson (Southampton) and Robson (Glasgow) in the United Kingdom, of Ariaratnam (Waterloo) in Canada; and of Parkus (Vienna) in Austria are perhaps the most well-known.

The probabilistic approach proved extremely useful in analysis of flexible buildings subjected to earthquakes (Cornell; Newmark and Rosenblueth; Vanmarcke) or wind (Cermak); offshore structures subjected to random wave loading (BOSS conference); ships in rough seas (Ekimov; Price and Bishop); structures undergoing fatigue failure (Bogdanoff, Freudenthal, Gumbel, Payne, Weibull); structures subjected to environmental temperatures (Heller); structurally inhomogeneous media (Beran, Kröner, Lomakin, Shermergor, Volkov); stability of stochastic systems (Khas'minskii, Kozin, Kushner); probabilistic identification of structures (Hart, Ibrahim, Masri and Caughey); and other fascinating problems.

GENERAL REFERENCES*

Ang, A. H.-S., and Tang, W. H., *Probability Concepts in Engineering Planning and Design*, Vol. 1, *Basic Principles*, John Wiley & Sons, New York, 1975.

Ariaratnam, S. T., "Dynamic Stability of Column under Random Loading," in G. Herrmann, Ed., *Dynamic Stability of Structures*, Pergamon, New York, 1967, pp. 255–266.

_____, "Stability of Mechanical Systems under Stochastic Parametric Excitations," in R. F. Curtain, Ed., *Lecture Notes in Mathematics*, No. 294 Springer-Verlag, Berlin, 1972, pp. 291–302.

Augusti, G., Barrata, A., and Casciati, F., *Probabilistic Methods in Structural Engineering*, Chapman and Hall, London, in press.

Benjamin, J. R., and Cornell, C. A., *Probability, Statistics and Decision for Civil Engineers*, McGraw-Hill, New York, 1970.

Beran, M. J., *Statistical Continuum Theories*, (*Monographs in Statistical Physics and Thermodynamics*, Vol. 9), Wiley-Interscience, New York, 1968.

Bogdanoff, J. L., "A New Cumulative Damage Model," Part 1, *ASME J. Appl. Mech.*, **100**, 246–250 (1978).

Bolotin, V. V., *Statistical Methods in Structural Mechanics*, State Publ. House for Building, Architecture and Building Materials, Moscow, 1961 (2nd ed., 1965, translated into English by S. Aroni, Holden-Day, San Francisco, 1969).

_____, *Application of the Methods of the Theory of Probability and the Theory of Reliability to Analysis of Structures*, State Publ. House for Buildings, Moscow, 1971 (English translation: FTD-MT-24-771-73, Foreign Technol. Div., Wright Patterson AFB, Ohio, 1974).

_____, *Random Vibrations of Elastic Bodies*, "Nauka" Publ. House, Moscow, 1979.

_____, "Reliability of Structures," in J. F. Besseling, and A. M. A. Van der Heijden, Eds., *Trends in Solid Mechanics*, (Proc. Symp. Dedicated to the 65th Birthday of W. T. Koiter), Delft Univ. Press, Sijthoff and Noordhoff Intern. Publ., 1979, pp. 79–91.

BOSS 1976, Proceedings of an International Conference on the Behavior of Offshore Structures, Norwegian Inst. Technol., Trondheim, 1976.

---

*Many highly interesting studies simply could not be mentioned here, since the subject is much too vast. Since a complete bibliography on probabilistic methods in mechanics could fill by itself a hefty volume, I have confined myself mostly to books and reviews, so as to give some idea of what has been done. A list of cited references and of recommended further reading is given at the end of each chapter.

Caughey, T. K., "Response of a Nonlinear String to Random Loading," *ASME J. Appl. Mech.*, **26**, 341–349 (1959).

_____, "Derivation and Application of the Fokker-Planck Equation to Discrete Nonlinear Dynamic Systems Subjected to White Random Excitation," *J. Acoust. Soc. Am.*, **35** (11), 1683–1692 (1963).

_____, "Equivalent Linearization Techniques," *ibid.*, **35** (11), 1706–1711 (1963).

_____, "Nonlinear Theory of Random Vibrations," *Advan. Appl. Mech.*, **11**, 209–253 (1971).

Cermak, J. E., "Applications of Fluid Mechanics to Wind Engineering—A Freeman Scholar Lecture," *J. Fluids Eng.*, **97**, 9–38 (1975).

Clarkson, B. L., "Stresses in Skin Panels Subjected to Random Acoustic Loading," *J. Roy. Aeronaut. Soc.*, **72**, 1000–1010 (1968).

_____, and Ford, R. D., "The Response of a Typical Aircraft Structure to Jet Noise," *ibid.*, **66**, 31–40 (1962).

_____, and Mead, D. J., "High Frequency Vibration of Aircraft Structures," *J. Sound Vibration*, **28** (3), 487–504 (1973).

_____, Ed., *Stochastic Problems in Dynamics*, Pitman, London, 1977.

Cornell, C. A., "Probabilistic Analysis of Damage to Structures under Seismic Loads," in D. A. Howells et al., Eds., *Dynamic Waves in Civil Engineering*, John Wiley & Sons, London, 1971, Chap. 27.

Crandall, S. H., Ed., *Random Vibration*, Vol. 1, Technology Press, Cambridge, MA, 1958; Vol. 2, M.I.T. Press, Cambridge, MA, 1963.

_____, and Mark, W. D., *Random Vibration in Mechanical Systems*, Academic Press, New York, 1963.

_____, *Wide-band Random Vibration of Structures* (*Proc. Seventh U.S. Nat. Congr. Appl. Mech.*), ASME, New York, 1974, pp. 131–138.

_____, *Random Vibration of Vehicles and Structures* (*Proc. Seventh Can. Congr. Appl. Mech.*), Sherbrooke, Ont., 1979, pp. 1–12.

Ditlevsen, O., *Uncertainty Modeling*, McGraw-Hill, New York, 1981.

Ekimov, V. V., *Probabilistic Methods in the Structural Mechanics of Ships*, "Sudostroenie" Pub. House, Leningrad, 1966.

Ferry Borges, J., and Castanheta, M., *Structural Safety*, 2nd ed., National Civil Eng. Lab., Lisbon, Portugal, 1971.

Freudenthal, A. M., "Safety of Structures," *Trans. ASCE*, **112**, 125–180 (1947).

_____, "Safety and Probability of Structural Failure," *Trans. ASCE*, **121**, 1337–1375 (1956), Proc. Paper 2843.

_____, and Gumbel, E. J., "Physical and Statistical Aspects of Fatigue," *Advanc. Appl. Mech.*, **4**, 117–158 (1956).

_____, "Statistical Approach to Brittle Fracture," in H. Liebowitz, Ed., *Fracture—An Advanced Treatise*, Vol. 2, Academic Press, New York, 1968, pp. 592–621.

Gumbel, E. J., *Statistics of Extremes*, Columbia Univ. Press, New York, 1958.

Hart, G. C., ed., Dynamic Response of Structures: Instrumentation, Testing Methods and System Identification, ASCE/EMD Specialty Conference, UCLA, 1976.

Haugen, E. B., *Probabilistic Approaches to Design*, John Wiley & Sons, London, 1968.

_____, *Probabilistic Mechanical Design*, John Wiley & Sons, New York, 1980.

Heller, R. A., "Temperature Response of an Initially Thick Slab to Random Surface Temperatures," *Mechanics Research Communications*, **3**, No. 5, 379–385 (1976).

_____, "Thermal Stresses as a Narrow–Band Random Load," *J. Eng. Mech. Div.*, ASCE, EM5, No. 12450, 787–805 (1976).

Holand, I., Kalvie, D., Moe, G., and Sigbjörnsson R., Eds., *Safety of Structures under Dynamic Loading*, papers presented at Intern. Res. Seminar, June 1977, Norweg. Inst. Technol., Tapir Press, Trondheim, 1978.

Ibrahim, S. R., "Random Decrement Technique for Modal Identification of Structures," *J. Spacecraft and Rockets*, **14**, 696–700 (1977).

_____, "Modal Confidence Factor in Vibration Testing," *ibid*, **15**, 313–316 (1978).

Johnson, A. I., *Strength, Safety and Economical Dimensions of Structures*, National Swedish Institute for Building Research, Document D7, 1971 (1st ed., 1953).

Khas'minskii, R. Z., *Stability of Systems of Differential Equations with Random Parametric Excitation*, "Nauka" Publ. House, Moscow, 1969.

Khozialov, N. F., "Safety Factors," *Building Ind.*, **10**, 840–844 (1929).

Kogan, J., *Crane Design*, John Wiley & Sons, Jerusalem, 1976.

Kozin, F., "A Survey of Stability of Stochastic Systems," *Automatica*, **5**, 95–112, (1969).

_____, "Stability of Linear Stochastic Systems," in R. F. Curtain, Ed., *Lecture Notes in Mathematics*, No. 294, Springer-Verlag, New York, 1972, pp. 186–229.

Kröner, E., *Statistical Continuum Mechanics*, Intern. Centre for Mechanical Sciences, Udine, Italy, Course No. 92, Springer-Verlag, Vienna, 1973.

Kushner, H., *Stochastic Stability and Control*, Academic Press, New York, 1967.

Lin, Y. K., *Probabilistic Theory of Structural Dynamics*, McGraw-Hill, New York, 1967.

_____, "Response of Linear and Nonlinear Continuous Structures Subject to Random Excitation and the Problem of High-level Excursions," in A. M. Freudenthal, Ed., *International Conference in Structural Safety and Reliability*, Pergamon Press, 1972, pp. 117–130.

_____, "Random Vibrations of Periodic and Almost Periodic Structures," *Mechanics Today*, **3**, 93–125 (1976).

_____, "Structural Response under Turbulent Flow Excitations," in H. Parkus, Ed., *Random Excitation of Structures by Earthquakes and Atmospheric Turbulence*, Springer-Verlag, Vienna, 1977, pp. 238–307.

Lind, N., Ed., "Structural Reliability and Codifiel Design," SM Study No. 2, Solid Mech. Div., Univ. Waterloo Press, 1970.

_____, "Mechanics, Reliability and Society," *Proc. Seventh Can. Cong. Appl. Mech.*, Sherbrooke, *1979*, pp. 13–23.

Lomakin, V. A., *Statistical Problems of the Mechanics of Solid Deformable Bodies*, "Nauka" Publ. House, Moscow, 1970.

Maier, M., *Die Sicherheit der Bauwerke und Ihre Berechnung nach Grenzkräften anstatt nach Zulässigen Spannungen*, Springer-Verlag, Berlin, 1926.

Masri, S. F., and Caughey, T. K., "A Nonparametric Identification Technique for Nonlinear Dynamic Problems," *ASME J. Appl. Mech.*, **46**, 433–447 (1979).

Moses, F., "Design for Reliability Concepts and Applications," in R. H. Gallagher and D. C. Zienkiewicz, Eds., *Optimum Structural Design*, John Wiley & Sons, New York, 1973, pp. 241–265.

Murzewski, J., *Bezpieczenstwo Konstrukeji Budowlanych*, "Arkady" Publ. House, Warsaw, 1970.

Newmark, N. M., and Rosenblueth, E., *Fundamentals of Earthquake Engineering*, Prentice-Hall, Englewood Cliffs, NJ, 1971.

Pal'mov, V. A., "Thin Shells Acted by Broadband Random Loads," *PMM J. Appl. Math. Mech.*, **29**, 905–913 (1965).

_____, *Vibrations of Elasto-Plastic Bodies*, "Nauka" Publ. House, Moscow, 1978.

Parkus, H., *Random Processes in Mechanical Sciences*, Intern. Centre Mech. Sci., Udine, Italy, Course No. 9, Springer-Verlag, Vienna, 1969.

_____, Ed., *Random Excitation of Structures by Earthquakes and Atmospheric Turbulence*, Intern. Centre Mech. Sci., Udine, Italy, Course No. 225, Springer-Verlag, Vienna, 1977.

Payne, A. O., "The Fatigue of Aircraft Structures," in H. Liebowitz, Ed., *Progress in Fatigue and Fracture*, Pergamon Press, Oxford, 1976, pp. 157–203.

Pugsley, A., "A Philosophy of Airplane Strength Factors," *Brit. A.R.C. R & M*, **1906**, 1942.

_____, *The Safety of Structures*, Edward Arnold Publ., London, 1966.

Price, W. G., and Bishop, R. E. D., *Probabilistic Theory of Ship Dynamics*, Chapman and Hall, London, 1974.

Rackwitz, R., "Non-Linear Combination for Extreme Loadings," in Technical University of Munich, Rept. No. 29, 1978.

Robson, J. D., *An Introduction to Random Vibration*, Edinburgh at the University Press, 1963.

_____, Dodds, C. J., Macvean, D. B., and Paling, V. R., *Random Vibrations*, Intern. Centre Mech. Sci., Udine, Italy, Course No. 115, Springer-Verlag, Vienna, 1971.

Rosenblueth, E., "The Role of Research and Education in Risk Control of Structures," in T. Moan and M. Shinozuka, Eds., *Structural Safety and Reliability*, Elsevier, Amsterdam, 1981, pp. 1–18.

Rzhanitsyn, A. R., *Calculation of Structures with Materials Plasticity Properties Taken into Account*, State Publ. House for Buildings, Moscow, 1954.

_____, *Theory of Reliability Analysis in Building Constructions*, State Publ. House for Buildings, Moscow, 1978.

Shermergor, T. D., *Theory of Elasticity of Micro-Inhomogeneous Media*, "Nauka" Publ. House, Moscow, 1977.

Shinozuka, M., "Safety, Safety Factors and Reliability of Mechanical Systems," *Proc. 1st Symp. Eng. Appl. Random Function Theory Probability*, Purdue Univ., Lafayette, IN., 1962, J. L. Bogdanoff and F. Kozin, Eds., John Wiley & Sons, New York, 1963, pp. 130–162.

_____, "Methods of Safety and Reliability Analysis," in A. M. Freudenthal, Ed., *International Conference in Structural Safety and Reliability*, Pergamon Press, New York, 1972, pp. 11–45.

_____, "Time and Space Domain Analysis on Structural Reliability Assessment," in H. Kupfer, M. Shinozuka, and G. I. Schuëller, Eds., *Second Intern. Conf. Structural Safety and Reliability* Werner-Verlag, Düsseldorf, 1977, pp. 9–28.

_____, "Application of Digital Simulation of Gaussian Random Processes," in H. Parkus, Ed., *Random Excitation of Structures by Earthquakes and Atmospheric Turbulence*, Springer-Verlag, Vienna, 1977, pp. 201–237.

Schuëller, G. I., *Einführung in die Sicherheit und Zuverlässigkeit von Tragwerken*, W. Ernst und Sohn, Berlin, in press.

Streletskii, N. S., *Foundations of Statistical Account of Factor of Safety of Structural Strength*, State Publ. House for Buildings, Moscow, 1947.

Tye, W., "Factors of Safety—or of Habit?," *J. Roy. Aeronaut. Soc.* **48**, 487–494 (1944).

_____, "Basic Safety Concepts," *Aeronaut. J.*, **81**, 271–275 (1977).

Vanmarcke, E. H., "Structural Response to Earthquakes," in C. Lomnitz and E. Rosenblueth, Eds., *Seismic Risk and Engineering Decisions*, Elsevier, Amsterdam, 1976, Chap. 8, pp. 287–337.

_____, "Seismic Safety Assessment," in H. Parkus, Ed., *Random Excitation of Structures by Earthquakes and Atmospheric Turbulence*, Intern. Centre Mech. Sci., Udine, Italy, Course No. 225, Springer-Verlag, Vienna, 1977.

Veneziano, D., "Synopsis of some Recent Research on Reliability Formats", in I. Holand, Ed., "Safety of Structures, under Dynamic Loading", Norwegian Institute of Technology (Trondheim) Press, 1978, pp. 595–605.

Volkov, S. D., *Statistical Strength Theory* (translated from Russian), Gordon and Breach, New York, 1962.

Weibull, W., "A Statistical Theory of the Strength of Materials," *Proc. Roy. Swedish Inst. Eng. Res., Stockholm*, **151**, 1–45 (1939).

_____, "Scatter of Fatigue Life and Fatigue Strength of Aircraft Structural Materials and Parts," *Proc. Intern. Conf. Fatigue Aircraft Structures*, Columbia Univ., New York, 1956.

chapter **2**

Probability Axioms

2.1 RANDOM EVENT

We will associate mechanical phenomena with a complex of conditions under which they may proceed, assuming that this complex is realizable (or rather reproducible, at least conceptually) an arbitrarily large number of times in essentially identical circumstances, with an observation or a measurement taken at each such realization. Such a process of observation or measurement will be referred to as a trial or an *experiment*. In this sense an experiment may consist in checking whether stresses in a structure exceed some specified value, or in determining the profile of imperfections of its surface, or else (in modern supersonic aircraft) in determining the noise level. We define an *event* as an outcome, or a collection of outcomes, of a given experiment (a positive or a negative conclusion; readings of the scanning mechanism; the final result of a highly complex calculation). The outcome of a deterministic phenomenon is totally predictable and is, or can be, known in advance: Deterministic phenomena are either *certain* or *impossible*, depending on whether, inevitably, they do or do not occur in the course of the given experiment.

For example, consider a perfectly elastic beam with symmetric uniform cross section (section modulus S), subject to given constraints under a given transverse load resulting in a maximal bending moment M_{max} ("complex of conditions," Fig. 2.1a and 2.1b). The maximum bending stress, according to the theory of strength of materials, is then given by

$$\sigma_{max} = \frac{M_{max}}{S} \tag{2.1}$$

Another example is a perfectly cylindrical shell made of perfectly elastic material, with radius R, length l, thickness h, Young's modulus E, and Poisson's ratio ν, under uniform axial compression with ends simply supported

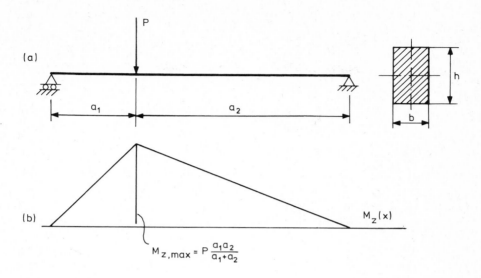

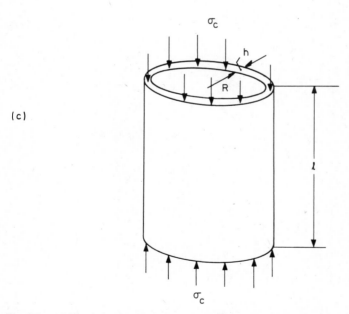

Fig. 2.1. (a) Elastic beam simply supported at its ends. (b) Bending moment diagram. (c) Elastic cylindrical shell under uniform axial compression.

("complex of conditions," Fig. 2.1c). For shells that are not too short, the buckling stress is given by

$$\sigma_c = \frac{Eh}{R\sqrt{3(1 - \nu^2)}} \tag{2.2}$$

If the conditions specified in both examples are realized, the maximal stress in the beam and the buckling stress in the shell will be determined by Eqs. (2.1) and (2.2), respectively.

The statement of impossibility of some event under a given complex of conditions reduces readily to one of certainty of the opposite event. An event that is neither certain nor impossible is referred to as *random*, signifying that it may or may not occur under given essentially identical conditions, in other words, that the outcome of the experiment is not known in advance, before it has taken place.

Consider as an example, in more detail, a cylindrical shell manufactured by electroplating from pure copper, and tested on a controlled end-displacement-type compression testing machine ("complex of conditions," Fig. 2.2) of which

Fig. 2.2. Compression testing machine for buckling tests on thin-walled structures. (Designed by W. D. Verduyn, Delft University of Technology.)

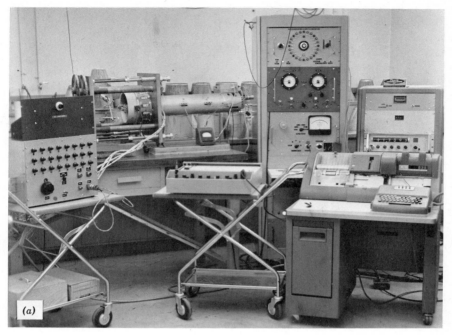

(a)

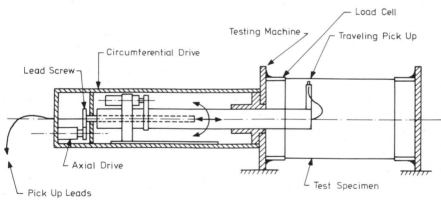

(b)

Fig. 2.3. (a) Testing machine and data acquisition equipment. (b) Cylindrical shell testing configuration. (Courtesy of J. Arbocz)

a suitable stock may be visualized as available. Due to the very nature of the manufacturing process, each realization of the shell will have a different initial shape that cannot be predicted in advance. The imperfections (deviations of the initial shape from the ideal circular cylinder), amounting to a fraction of the wall thickness, can be picked up and recorded by the special experimental setup (Fig. 2.3) developed by Arbocz at the California Institute of Technology. The scanning device, moving in both the axial and circumferential directions, yields a complete surface map of the shell. As illustrated by the examples in Figs. 2.4 and 2.5, the two shells produced by the same manufacturing process have totally different imperfection profiles, and it is intuitively obvious that even when tested on the same machine, they would generally have different buckling loads. These differ considerably from the classical buckling load of the perfect cylindrical shells as per Eq. (2.2): 0.736 σ_c for shell A9, and 0.673 σ_c for shell A12.*

2.2 SAMPLE SPACE

Although not absolutely essential, some mathematical preliminaries are given below.

The axiomatic foundation of the theory of probability was laid by Kolmogorov, according to whom the primary notion is not the random event, but rather the *sample space*. When an experiment or phenomenon gives rise to one of a totality of mutually exclusive events, we denote it by the Greek letter ω and refer to it as an *elementary event*, an *elementary outcome*, or finally a *sample point*. The totality of all possible sample points are denoted by Ω and referred to as the sample space, of which the sample points are the elements. A sample point is *indivisible*, in that it embodies no distinguishable outcomes.

Sample spaces are usually classified according to the number of elements they contain. If such a space contains a *countable* number of elements, that is, a finite number or a denumerable infinity, so that its elements can be put in one-to-one correspondence with positive integers, it is referred to as *discrete*; otherwise, it is said to be *continuous*.

We will claim that event A is *associated with the experiment* (or *the elementary outcome ω*) if, according to each elementary outcome ω, we know precisely that A does or does not take place. Denote by the same letter A the totality (or set) of all ω's as a result of which A takes place. Obviously, A takes place when and only when one of the ω's does; in other words, instead of speaking of A, we may speak of an event of an elementary outcome ω (which belongs to A) happening. Events are thus simple subsets of the sample space Ω.

---

*This effect of small imperfections considerably reducing the buckling load of a cylindrical shell was pointed out by Koiter in his pioneering doctoral thesis, an achievement, in terms of its impact on the course of the general theory of structural stability, that any scientist, conceited or modest, should dream of.

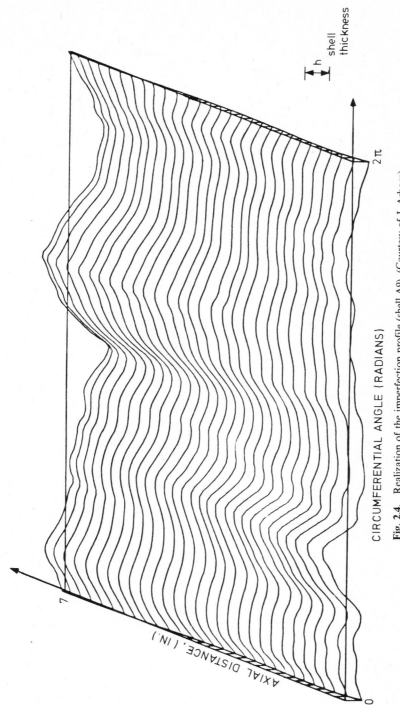

CIRCUMFERENTIAL ANGLE (RADIANS)

Fig. 2.4. Realization of the imperfection profile (shell A9). (Courtesy of J. Arbocz)

13

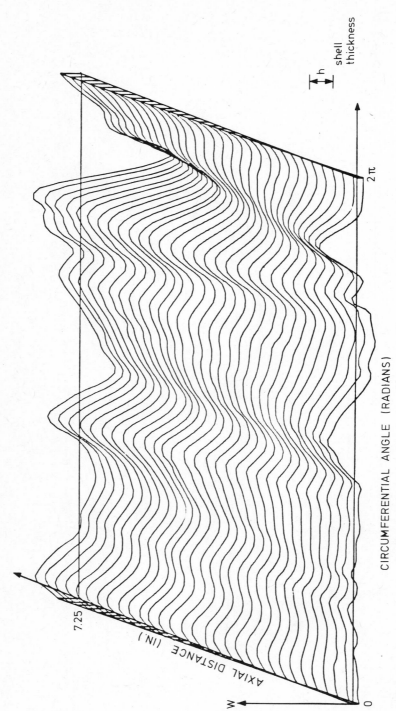

CIRCUMFERENTIAL ANGLE (RADIANS)

Fig. 2.5. Realization of the imperfection profile (shell A12). (Courtesy of J. Arbocz)

A certain or a sure event A, which takes place as a result of any outcome of the experiment under consideration, is formally identified with the whole sample space Ω, while an impossible event (denoted by $\varnothing$) is treated as an empty set, not containing any of the ω's.

If ω is an elementary outcome belonging to [included in] A, we write $\omega \in A$; if ω is not an element of A, we write $\omega \notin A$.

Two events A and B are said to be *equal*, $A = B$, if and only if [denoted *iff*] every element of A is also an element of B and vice versa—every element of B is also an element of A; that is, iff $A \subset B$, $B \subset A$. (Read: "A is contained in B" and "B is contained in A.")

Events A and B are referred to as *mutually exclusive* (or *disjoint*) if they have no sample points in common, that is, if they cannot take place simultaneously.

The *union* or *sum* of two events A_1 and A_2 is defined as an event A signifying realization of at least one of the events A_1, A_2:

$$A = A_1 \cup A_2 \quad (A = A_1 + A_2)$$

where $\cup$ is the special symbol of union. That is, A's elements are all the elements of A_1, or A_2, or both. A union of multiple events $A_1, A_2, \ldots, A_n$ is defined in an analogous manner and denoted by $A = \cup_{k=1}^{n} A_k$ or $A = \sum_{k=1}^{n} A_k$.

Sample spaces and events are conveniently represented by Venn diagrams. The sample space Ω is represented by a rectangle, whereas the events are represented by a region (or part of one), within the rectangle. Many relationships involving events can be demonstrated by this means (see Fig. 2.6).

The *intersection* or *product* of two events A_1 and A_2 is an event A, signifying realization of both A_1 and A_2: $A = A_1 \cap A_2$ (or $A_1 A_2$), where $\cap$ is the special symbol of intersection. The product of multiple events $A_1, A_2, \ldots, A_n$ is defined in an analogous manner and denoted by $A = \cap_{k=1}^{n} A_k$ (or $A = \prod_{k=1}^{n} A_k$).

The *difference* A of events A_1 and A_2 is an event signifying realization of A_1 but not of A_2: $A = A_1 \setminus A_2$ (or $A = A_1 - A_2$). The *complement* of an event A with respect to the sample space Ω, denoted by $\bar{A}$ or A^c, is an event signifying that A does not take place: $\bar{A} = \Omega \setminus A$ or $\bar{A} = \Omega - A$.

It is readily shown that

$$\text{If } A = A_1 + A_2, \quad \text{then } \bar{A} = \bar{A_1}\bar{A_2}$$

and

$$\text{If } A = A_1 A_2, \quad \text{then } \bar{A} = \bar{A_1} + \bar{A_2}$$

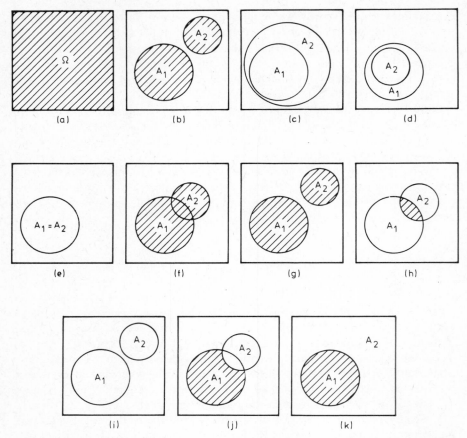

Fig. 2.6. (a) Ω shaded. (b) A_1 and A_2 are mutually exclusive events. (c) A_1 is contained in A_2. (d) A_2 is contained in A_1. (e) A_1 and A_2 are equal. (f) Sum $A_1 + A_2$ shaded. (g) Sum $A_1 + A_2$ shaded. (h) Product A_1A_2 shaded. (i) Product A_1A_2 of mutually exclusive events is an impossible event. (j) Difference $A_1 - A_2$ shaded. (k) A_1 is a complement of A_2.

also,

$$\text{If } A_1 \subset A_2, \text{ then } \overline{A}_1 \supset \overline{A}_2$$

the symbol $\supset$ signifying "contains." These are known as De Morgan's laws, signifying that if there is some link between given events, then the link obtained from the original one by transfer to the complementary events, by formally replacing the symbols of union $\cup$, intersection $\cap$, and inclusion $\subset$ by $\cap$, $\cup$, and $\supset$, respectively, are likewise valid.

Observe also that

$$\overline{\Omega} = \varnothing, \qquad \overline{\varnothing} = \Omega$$

and therefore the interlinkage will also be preserved if the following formal substitution is resorted to in addition to the above

$$\Omega \to \varnothing, \qquad \varnothing \to \Omega$$

A collection of events $A_1, A_2, \ldots, A_n$ is said to *partition* the sample space Ω iff they are pairwise mutually exclusive and their sum equals the sample space, that is,

$$A_i A_j = \varnothing, \quad i \neq j$$

and

$$\sum_{i=1}^{n} A_i = \Omega$$

2.3   PROBABILITY AXIOMS

Consider now the discrete sample space Ω, denoting as before the set of all possible outcomes of a random experiment. We now formulate axioms, defining the concept of probability.

Axiom 1 ("Nonnegativity" Axiom)

To each event A, there can be assigned a nonnegative real number $P(A) \geqslant 0$ called its probability.

Axiom 2 ("Normalization" Axiom)

The probability of a certain event equals unity:

$$P(\Omega) = 1$$

Axiom 3 ("Additivity" Axiom)

If $A_1, A_2, A_3, \ldots$ is a countable sequence of mutually exclusive events of Ω, then

$$P(A_1 + A_2 + A_3 + \cdots) = P(A_1) + P(A_2) + P(A_3) + \cdots$$

From these axioms, the following conclusions can be drawn immediately:

1. From the obvious equality,

$$\Omega = \Omega + \varnothing$$

and Axiom 3, we conclude

$$P(\Omega) = P(\Omega) + P(\varnothing)$$

so that

$$P(\varnothing) = 0$$

that is, the probability of the impossible event is zero.

2. For any event A,

$$P(\bar{A}) = 1 - P(A) \tag{2.3}$$

To prove this, we note that an event and its complement are mutually exclusive:

$$A\bar{A} = \varnothing$$

and the sum of an event and its complement represents the sample space

$$A + \bar{A} = \Omega$$

Then,

$$P(A + \bar{A}) = P(\Omega)$$

By Axiom 3 we have

$$P(A + \bar{A}) = P(A) + P(\bar{A})$$

On the other hand, by Axiom 2 we have

$$P(A + \bar{A}) = P(\Omega) = 1$$

which together yield

$$P(A) + P(\bar{A}) = 1 \quad \text{or} \quad P(\bar{A}) = 1 - P(A)$$

3. For any pair of events A_1 and A_2 in a sample space Ω

$$P(A_1 - A_2) = P(A_1) - P(A_1 A_2)$$

$$P(A_2 - A_1) = P(A_2) - P(A_1 A_2)$$

For proof, we note that each of the events A_1 and A_2 can be represented as

$$A_1 = (A_1 - A_2) + (A_1 A_2)$$

$$A_2 = (A_2 - A_1) + (A_1 A_2)$$

where the events $A_1 - A_2$, $A_1 A_2$, and $A_2 - A_1$ are mutually exclusive. Then by Axiom 3 we have

$$P(A_1) = P(A_1 - A_2) + P(A_1 A_2)$$

$$P(A_2) = P(A_2 - A_1) + P(A_1 A_2)$$

Furthermore, we conclude that if $A_1 \subset A_2$, then $A_1 A_2 = A_1$ and

$$P(A_1) = P(A_2) - P(A_2 - A_1) \leqslant P(A_2)$$

that is, if event A_1 is contained in A_2, then

$$P(A_1) \leqslant P(A_2)$$

4. The sum $A_1 + A_2$ of events A_1 and A_2 can be represented as the sum of the following mutually exclusive events

$$A_1 + A_2 = (A_1 - A_2) + (A_2 - A_1) + (A_1 A_2)$$

and therefore

$$
\begin{aligned}
P(A_1 + A_2) &= P(A_1 - A_2) + P(A_2 - A_1) + P(A_1 A_2) \\
&= [P(A_1) - P(A_1 A_2)] \\
&\quad + [P(A_2) - P(A_1 A_2)] + P(A_1 A_2) \\
&= P(A_1) + P(A_2) - P(A_1 A_2)
\end{aligned}
\tag{2.4}
$$

If A_1 and A_2 are mutually exclusive events, that is, if $A_1 A_2 = \varnothing$, then $P(A_1 A_2) = 0$ and we are back with Axiom 3.

Due to the nonnegativity of $P(A_1 A_2)$, we also conclude from (2.4) that

$$P(A_1 + A_2) \leqslant P(A_1) + P(A_2)$$

5. Let $A_1, A_2, \ldots, A_n$ be events in sample space Ω. We seek to calculate the probability of their sum. Denote

$$P_1 = \sum_{i=1}^{n} P(A_i), \quad P_2 = \sum_{i<j}^{n}\sum^{n} P(A_i A_j), \quad P_3 = \sum_{i<j<k}^{n}\sum^{n}\sum^{n} P(A_i A_j A_k), \ldots$$

$$\tag{2.5}$$

where $1 \leqslant i < j < k < \cdots \leqslant n$. The following formula is valid for the probability $P(\Sigma_{i=1}^{n} A_i)$ of the sum $\Sigma_{i=1}^{n} A_i$ of events $A_1, A_2, \ldots, A_n$:

$$P\left(\sum_{i=1}^{n} A_i\right) = P_1 - P_2 + P_3 - P_4 + \cdots + (-1)^{n-1} P_n \qquad (2.6)$$

We will prove this formula by induction. For $n = 2$, it is identical with Eq. (2.4). Assume that it is valid for the sum of any $n - 1$ events, so that

$$P\left(\sum_{i=2}^{n} A_i\right) = \sum_{i=2}^{n} P(A_i) - \sum_{2 \leqslant i < j}^{n} \sum^{n} P(A_i A_j) + \sum_{2 \leqslant i < j < k}^{n} \sum^{n} \sum^{n} P(A_i A_j A_k) - \cdots$$

We then have

$$P\left(\sum_{i=2}^{n} A_1 A_i\right) = \sum_{i=2}^{n} P(A_1 A_i) - \sum_{2 \leqslant i < j}^{n} \sum^{n} P(A_1 A_i A_j)$$

$$+ \sum_{2 \leqslant i < j < k}^{n} \sum^{n} \sum^{n} P(A_1 A_i A_j A_k) - \cdots$$

Now we use (2.4) to yield

$$P\left(\sum_{i=1}^{n} A_i\right) = P(A_1) + P\left(\sum_{i=2}^{n} A_i\right) - P\left(\sum_{i=2}^{n} A_1 A_i\right)$$

$$= \sum_{i=1}^{n} P(A_i) - \sum_{1 \leqslant i < j}^{n} \sum^{n} P(A_i A_j) + \sum_{1 \leqslant i < j < k}^{n} \sum^{n} \sum^{n} P(A_i A_j A_k) - \cdots$$

$$= P_1 - P_2 + P_3 - \cdots \qquad \text{Q.E.D.}$$

Example 2.1

When an ordinary die (a regular hexahedron, that is, a cube, marked on each face with one to six spots) is thrown, there are six possible outcomes. Any one of the six numbered faces may land upwards. The sample space is thus

$$\Omega = \{1, 2, 3, 4, 5, 6\}$$

These six outcomes are mutually exclusive, since two or more faces cannot turn up simultaneously. The event of 7 as the outcome of a throw is an impossible event: $A = \varnothing$; $P(A) = 0$.

Example 2.2

If a pair of ordinary dice are thrown, there are 36 possible outcomes:

$$\Omega = \{(i, j): \text{ the numbers of spots, } i \text{ and } j \text{ integers, } 1 \text{ to } 6\}$$

Describe the following three events

A: "The sum of the spots on the upward-landing faces is an even number"
A_1: "The outcome of a throw of each die is an even number"
A_2: "The outcome of a throw of each die is an odd number"

A represents the union of mutually exclusive events A_1 and A_2. ($A = A_1 + A_2$). $A_1 = A - A_2$, $A_2 = A - A_1$. The complement to A is $\overline{A}$: "The sum of spots on the upward-landing faces is an odd number". The complement to A_1 is $\overline{A_1}$: "An odd number will appear on the upward-landing face of at least one die" and the complement to A_2 is $\overline{A_2}$: "An even number will appear on the upward-landing face of at least one die."

Significantly, the probability axioms do not advise us how to assign probabilities to events; they merely impose restrictions on how this can be done. For instance, in the example on throwing a die, the following are two possible ways in which we can assign probabilities:

1. $P(B_1) = P(B_2) = P(B_3) = P(B_4) = P(B_5) = P(B_6) = \frac{1}{6}$ (for a "fair" die)
2. $P(B_1) = P(B_3) = P(B_5) = \frac{1}{5}$; $P(B_2) = P(B_4) = P(B_6) = \frac{2}{15}$ (for a "loaded" die),

so that all probability axioms are satisfied. Here B_i is an event of the number of spots on the upward-landing face equalling i.

2.4   EQUIPROBABLE EVENTS

Denote by $E_1, E_2, \ldots, E_n$ the elementary outcomes of the sample space Ω. Assume that they partition the sample space such that

$$\Omega = E_1 + E_2 + \cdots + E_n \quad \text{and} \quad E_i E_j = \varnothing, \quad i \neq j$$

That is, they "cover" the sample space and are pairwise disjoint.

Consider the particular case where all outcomes are equiprobable: $P(E_i) = p$, $i = 1, 2, \ldots, n$. Then, according to Axiom 3,

$$P(E_1 + E_2 + \cdots + E_n) = P(E_1) + P(E_2) + \cdots + P(E_n) = np$$

On the other hand, according to Axiom 2,

$$P(E_1 + E_2 + \cdots + E_n) = P(\Omega) = 1$$

Comparison of the last two equalities yields

$$p = \frac{1}{n}.$$

Consider now an event A such that

$$A = E_1 + E_2 + \cdots + E_m, \quad m \leqslant n$$

then

$$P(A) = P(E_1 + E_2 + \cdots + E_m) = pm = \frac{m}{n}$$

Consequently, if a random experiment can result in *mutually exclusive* and *equiprobable* outcomes, and if in m of these outcomes the event A occurs, then $P(A)$ is given by the fraction

$$P(A) = \frac{m}{n} \tag{2.7}$$

and these m outcomes are "favorable" to A.

The above cannot be used as such to define probability, since the procedure would be circular. The classical definition of probability (Laplace, 1812), is, however, very similar to (2.7). It states:

If a random experiment can result in n *mutually exclusive* and *equally likely* outcomes, of which m are favorable to A, then the probability of A equals the ratio of these favorables to the total number of outcomes n.

This definition reduces the notion of probability to that of "equilikelihood." Axiom 1 is satisfied, since the fraction m/n cannot be negative, and so are Axiom 2, since n outcomes are favorable to the sample space and $P(\Omega) = n/n = 1$, and Axiom 3. Assume that m_1 elementary events are favorable to an event A_1 and m_2 elementary events to an event A_2. A_1 and A_2 are mutually exclusive, the events E_i favorable to one of them are different from those favorable to the other. Thus there are $m_1 + m_2$ events E_i favorable to one of the events A_1 or A_2, that is, favorable to the event $A_1 + A_2 = A$. Consequently,

$$P(A) = \frac{m_1 + m_2}{n} = \frac{m_1}{n} + \frac{m_2}{n} = P(A_1) + P(A_2) \quad \text{Q.E.D.}$$

Equation (2.7) for the probability of an event composed of equiprobable events has many useful applications wherever symmetry considerations are involved. The probabilities of a homogeneous, balanced die, properly thrown, turning up each face are equal, $P(E_i) = \frac{1}{6}$. The probabilities of an "honest" coin, properly tossed, turning up *heads* or *tails* are the same, $P(E_i) = \frac{1}{2}$, and the probability of any card being drawn from a properly shuffled deck is $P(E_i) = \frac{1}{52}$. In these cases, 6, 2, and 52 are, respectively, the numbers of outcomes.

Example 2.3

The number of heads turned up in one toss of a pair of coins equals 2, 1, or 0, and we seek the probability of each event. The three events are not equiprobable, although they partition the sample space. However, in order to invoke the classical definition of probability the sample space must be represented as the sum of elementary, equiprobable events. We first tabulate the possible pairs:

First Coin	Second Coin
heads	heads
heads	tails
tails	heads
tails	tails

The four outcomes are natural equiprobables, being mutually exclusive and covering the sample space. Considering only the number of heads in each pair, we invoke the classical definition of probability, obtaining:

Probability of two "heads" equals $\frac{1}{4}$
Probability of one "heads" equals $\frac{2}{4} = \frac{1}{2}$
Probability of no "heads" equals $\frac{1}{4}$

2.5 PROBABILITY AND RELATIVE FREQUENCY

Consider a sequence of n identical experiments in each of which occurrence or nonoccurrence of some event A is recorded. A natural characteristic of A appears to be the *relative frequency* of its occurrence, defined as the ratio of its occurrences to the total number of trials. Denote by $\hat{P}(A)$ the relative frequency of A. We have

$$\hat{P}(A) = \frac{n(A)}{n} \tag{2.8}$$

where $n(A)$ is the number of occurrences of event A in n trials.

Note that the relative frequency is bounded between zero and unity:

$$0 \leqslant \frac{n(A)}{n} \leqslant 1$$

since the number of times $n(A)$ of the event A occurring in n trials is bounded between zero and n.

If A_1 and A_2 are mutually exclusive, and if in n experiments A_1 occurred $n(A_1)$ times and A_2 occurred $n(A_2)$ times, then the union $A_1 + A_2$ occurred $n(A_1) + n(A_2)$ times, and its relative frequency is given by

$$\hat{P}(A_1 + A_2) = \frac{1}{n}\left[n(A_1) + n(A_2)\right]$$

However, the relative frequencies of A_1 and A_2 are $n(A_1)/n$ and $n(A_2)/n$, respectively, and the last equation can be rewritten as

$$\hat{P}(A_1 + A_2) = \hat{P}(A_1) + \hat{P}(A_2)$$

Past experience has shown remarkable conformity, which imparted a deep significance to the probability notion. It turned out that in different series of experiments the corresponding relative frequencies $n(A)/n$ practically coincide at large values of n and are concentrated in the vicinity of some number. For example, if a die is made of a homogeneous material and represents a perfect cube (an "honest" die), then the relative frequencies 1, 2, 3, 4, 5, or 6 turning up oscillate in the vicinity of $\frac{1}{6}$.

Table 2.1 lists the relative frequencies of a simple toss of a coin turning up tails in experiments (total number 10,000) conducted in discrete series of 100 and 1000, respectively. It is seen that the relative frequencies $n(A)/n$ in the "1000" series differ surprisingly little from the probability $P(A) = \frac{1}{2}$ (the relative frequency in series of 10,000 experiments is 0.4979). This stability of the relative frequency could be interpreted as a manifestation of an objective property of the random event, namely existence of a definite degree of its possibility.

Formally, Eq. (2.8) should be understood in the following way:

$$P(A) = \lim_{n \to \infty} \frac{n(A)}{n} \qquad (2.9)$$

Realization of an infinite number of trials is only feasible conceptually, whereas in a physical experiment the number n may be large but always remains finite. Accordingly, definition (2.9) refers to the existence of a limit.

TABLE 2.1

Relative Frequencies in Series of 100 Experiments										Relative Frequency in Series of 1000 Experiments
0.54	0.46	0.53	0.55	0.46	0.54	0.41	0.48	0.51	0.53	0.501
0.48	0.46	0.40	0.55	0.49	0.49	0.48	0.54	0.53	0.45	0.485
0.43	0.52	0.58	0.51	0.51	0.50	0.52	0.50	0.53	0.49	0.509
0.58	0.60	0.54	0.55	0.50	0.48	0.47	0.57	0.52	0.55	0.536
0.48	0.51	0.51	0.49	0.44	0.52	0.50	0.46	0.53	0.41	0.485
0.49	0.50	0.45	0.52	0.52	0.48	0.47	0.47	0.47	0.51	0.488
0.45	0.47	0.41	0.51	0.49	0.59	0.60	0.55	0.53	0.50	0.500
0.53	0.52	0.46	0.52	0.44	0.51	0.48	0.51	0.46	0.54	0.497
0.45	0.47	0.46	0.52	0.47	0.48	0.59	0.57	0.45	0.48	0.494
0.47	0.41	0.51	0.59	0.51	0.52	0.55	0.39	0.41	0.48	0.484

For large n, however, (2.9), or more properly (2.8), may be used as an estimate of probability.

Although Kolmogorov's axiomatic method is superior, the relative frequency definition (due to von Mises) is suitable for physical applications and is by no means incompatible with Kolmogorov's axiomatics. In these circumstances, results obtained in terms of relative frequency are often generalized to the appropriate probabilities.

2.6 CONDITIONAL PROBABILITY

When analyzing some phenomenon, the observer is often concerned as to how occurrence of an event A is influenced by that of another event B. The simplest modes of intercorrelation of such a pair of events are (1) occurrence of B *necessarily results* in that of A, or, on the contrary, (2) occurrence of B *eliminates* that of A. In the theory of probability, this intercorrelation is characterized by the *conditional probability* $P(A|B)$ of event A, it being known that B (whose own probability is positive) actually took place:

$$P(A|B) = \frac{P(AB)}{P(B)} \tag{2.10}$$

We shall illustrate this on the example of an experiment with a finite number of equiprobable outcomes ω. Let n be the total number of outcomes; $n(B)$, the number favorable to B; $n(AB)$, the number favorable to both A and B. Then,

$$P(B) = \frac{n(B)}{n}, \qquad P(AB) = \frac{n(AB)}{n}$$

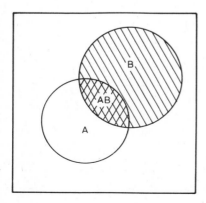

Fig. 2.7. Conditional probability: Probability of event A given that the event B has taken place: $P(A|B) = P(AB)/P(B)$.

so that the conditional probability is

$$P(A|B) = \frac{n(AB)}{n(B)} = \frac{n(AB)/n}{n(B)/n} = \frac{P(AB)}{P(B)} \qquad (2.11)$$

where $n(B)$ is the number of all elementary outcomes ω when B occurs and $n(AB)$, those favorable to A. Recalling (2.8), Eq. (2.11) determines the probability of A under new conditions, which arise when B occurs.

Conditional probability retains all the features of ordinary probability. Axiom 1 is satisfied in an obvious manner, since for each of the events A and B the nonnegative function $P(A|B)$ is defined according to (2.10). If A equals B, then according to the definition,

$$P(B|B) = \frac{P(BB)}{P(B)} = \frac{P(B)}{P(B)} = 1$$

and, therefore

$$0 \leqslant P(A|B) \leqslant 1$$

If the occurrence of B eliminates that of A, then $P(AB) = 0$ and therefore $P(A|B) = 0$. If occurrence of B necessarily results in that of $A(B \subset A)$, then $AB = B$ and $P(AB) = P(B)$, which means $P(A|B) = 1$. If A is a union of mutually exclusive events $A_1, A_2, \ldots, A_n$, then the product AB represents a union of mutually exclusive events $A_1 B, A_2 B, \ldots, A_n B$, and according to Axiom 3

$$P(AB) = \sum_{k=1}^{n} P(A_k B)$$

and $P(A|B)$ equals

$$P(A|B) = \frac{P(AB)}{P(B)} = \frac{1}{P(B)} \sum_{k=1}^{n} P(A_k B) = \sum_{k=1}^{n} \frac{P(A_k B)}{P(B)} = \sum_{k=1}^{n} P(A_k|B)$$

Example 2.4

A pair of ordinary dice are thrown. What is the probability of the sum of spots on the upward-landing faces being 7 (event A), given that this sum is odd (event B)?

The sample space is composed of 36 outcomes:

$$\Omega = \{(1, 1), (1, 2), (1, 3), (1, 4), (1, 5), (1, 6),$$

$$(2, 1), (2, 2), (2, 3), (2, 4), (2, 5), (2, 6),$$

$$(3, 1), (3, 2), (3, 3), (3, 4), (3, 5), (3, 6),$$

$$(4, 1), (4, 2), (4, 3), (4, 4), (4, 5), (4, 6),$$

$$(5, 1), (5, 2), (5, 3), (5, 4), (5, 5), (5, 6),$$

$$(6, 1), (6, 2), (6, 3), (6, 4), (6, 5), (6, 6)\}$$

The number of outcomes favorable to A is 6, and hence the unconditional probability is

$$P(A) = \tfrac{6}{36} = \tfrac{1}{6}$$

If B has taken place, then one of 18 events took place (a "new" sample space with 18 points), and the conditional probability is

$$P(A|B) = \tfrac{6}{18} = \tfrac{1}{3}$$

The probability of event B is

$$P(B) = \tfrac{18}{36} = \tfrac{1}{2}$$

and $P(A|B)$ is also obtained from the general formula (2.10):

$$P(A|B) = \frac{P(AB)}{P(B)} = \frac{\tfrac{6}{36}}{\tfrac{1}{2}} = \tfrac{1}{3}$$

Note that the definition of conditional probability enables us to find the probability of a product. From Eq. (2.10), it follows immediately that

$$P(AB) = P(A|B)P(B) \tag{2.12}$$

That is, the probability of product AB is calculated by constructing the product of the conditional probability, under event A, of event B occurring and the (unconditional) probability of event B.
    On the other hand,

$$P(AB) = P(B|A)P(A)$$

and

$$P(AB) = P(A|B)P(B) = P(B|A)P(A) \qquad (2.13)$$

That is, the probability of the product of two events is calculated by constructing the product of the conditional probability of one of these events under the condition of another event occurring and the (unconditional) probability of the latter event. Formula (2.12) is readily extended by induction to n events $A_1, A_2, \ldots, A_n$:

$$P(A_1 A_2 A_3 \cdots A_{n-1} A_n) = P(A_1|A_2 A_3 \cdots A_n)P(A_2 A_3 \cdots A_n)$$

$$= P(A_1|A_2 A_3 \cdots A_n)P(A_2|A_3 \cdots A_n)$$

$$\times P(A_3 \cdots A_n)$$

$$= \cdots$$

$$= P(A_1|A_2 A_3 \cdots A_n)P(A_2|A_3 \cdots A_n)$$

$$\times P(A_3|A_4 \cdots A_n) \cdots P(A_{n-1}|A_n)P(A_n) \quad (2.14)$$

Equation (2.14) is known as the *multiplication rule*.

2.7 INDEPENDENT EVENTS

Events A and B are called *independent* if

$$P(A|B) = P(A) \qquad (2.15)$$

that is, if occurrence of B does not affect the probability of A. Note that mutually exclusive events are dependent. In fact, if $AB = \varnothing$, then $P(A|B) = P(\varnothing) = 0 \neq P(A)$ unless $A = \varnothing$.

If event A is independent of B, then, according to Eqs. (2.13) and (2.15) we have

$$P(A)P(B|A) = P(B)P(A|B)$$

$$= P(B)P(A)$$

Thus, $P(B|A) = P(B)$, implying that event B is equally independent of A or, in other words, that the property of independence is mutual.

The probability of a product of independent events is readily calculated:

$$P(AB) = P(A|B)P(B) = P(A)P(B)$$

This is often used as a definition of independence.

n events, $A_1, A_2, \ldots, A_n$ are *individually* independent if and only if

$$P(A_i A_j) = P(A_i)P(A_j), \quad i \neq j$$

$$P(A_i A_j A_k) = P(A_i)P(A_j)P(A_k), \quad i \neq j, j \neq k, i \neq k$$

$$\vdots$$

$$P\left(\prod_{i=1}^{n} A_i \right) = \prod_{i=1}^{n} P(A_i) \tag{2.16}$$

Thus, pairwise independence is *not* sufficient for the component events to be *individually* independent. This is illustrated by the following example, due to N. S. Bernstein.

Example 2.5

Given events A, B, C such that

$$P(A) = P(B) = P(C) = \tfrac{1}{2}$$

and

$$P(AB) = P(AC) = P(BC) = P(ABC) = \tfrac{1}{4}$$

These events are pairwise independent:

$$P(AB) = P(A)P(B)$$

$$P(AC) = P(A)P(C)$$

$$P(BC) = P(B)P(C)$$

but not *individually* independent, since

$$P(ABC) \neq P(A)P(B)P(C).$$

Note: A, B, and C have a "physical meaning"—as in the case of an "honest" tetrahedron, with one face red (event A), another green (event B), the third blue (event C), and the fourth in all three colors.

Example 2.6. Series System

A device consists of n components (Fig. 2.8a) connected in *series*, that is, all components are so interrelated that failure of any one of them implies failure of the entire system. These component failures are taken as independent

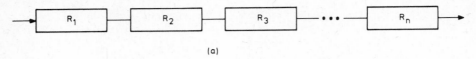

(a)

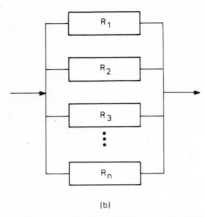

(b)

Fig. 2.8.   (a) Series system. (b) Parallel system.

random events A_i. Denoting the reliability, the probability of nonfailure performance, of element E_i by R_i, the reliability of the entire system is then

$$R = P\left(\overline{A}_1\overline{A}_2 \cdots \overline{A}_n\right) = P\left(\overline{A}_1\right)\overline{P}\left(\overline{A}_2\right) \cdots P\left(\overline{A}_n\right)$$

$$= R_1 R_2 \cdots R_n = \prod_{k=1}^{n} R_k \qquad (2.17)$$

As reliability in principle does not exceed unity, multiplication of the component reliabilities makes for a decrease of the overall system reliability R as the number of components increases. In fact, R cannot exceed the reliability of the weakest component:

$$R \leqslant \min(R_1, R_2, \ldots, R_n)$$

Example 2.7.   Parallel System

Consider now the same elements as above, this time connected in *parallel* (see Fig. 2.8*b*). In this case the system fails when *all* components fail; the probability of failure is given by

$$P(A_1 A_2 \cdots A_n) = P(A_1)P(A_2) \cdots P(A_n)$$

$$= (1 - R_1)(1 - R_2) \cdots (1 - R_n)$$

and the reliability of the entire system is

$$R = 1 - (1 - R_1)(1 - R_2) \cdots (1 - R_n) = 1 - \prod_{k=1}^{n} (1 - R_k) \quad (2.18)$$

Note that series or parallel systems do not refer to physical series or parallel connection between the components. For example, let the components ($n = 2$) in Fig. 2.8 be valves, whose probability of proper performance is 0.9. If proper performance of the entire system is defined as allowing flow through it, then the reliability of the physical series system is $0.9^2 = 0.81$, and of a physical parallel system is $1 - (1 - 0.9)^2 = 0.99$; that is, in this case the *physical parallel* system is the more reliable. However, if proper performance is defined as preventing flow through the entire system, then the reliability of the physical series system is 0.99, whereas that of the physical parallel system is 0.81. In this case the *physical series* system is the more reliable.

2.8 RELIABILITY OF STATICALLY DETERMINATE TRUSS

Consider a statically determinate truss consisting of n bars under given specified deterministic loading (Fig. 2.9a, b), with failure (overloading beyond the yield-stress level) of a single bar implying failure of the entire truss. Assuming that the failures of component bars are random events, the reliability of the truss is given by

$$R = 1 - \text{Prob (any bar failing)}$$

$$= 1 - \text{Prob} \left(A_1 + A_2 + \cdots + A_n \right)$$

$$= 1 - \sum_{i=1}^{n} P(A_i) + \sum_{1 \leqslant i < j}^{n} \sum^{n} P(A_i A_j) - \sum_{1 \leqslant i < j < k}^{n} \sum^{n} \sum^{n} P(A_i A_j A_k) + \cdots$$

$$(2.19)$$

where

$$P(A_i A_j) = P(A_i | A_j) P(A_j)$$

$$P(A_i A_j A_k) = P(A_i | A_j A_k) P(A_j | A_k) P(A_k), \text{ etc.}$$

For example, for the case $n = 2$ (Fig. 2.9c):

$$R = 1 - \left[P(A_1) + P(A_2) - P(A_1 A_2) \right]$$

$$= 1 - \left[P(A_1) + P(A_2) - P(A_1 | A_2) P(A_2) \right] \quad (2.20)$$

In order to find the reliability, we must know the conditional probability.

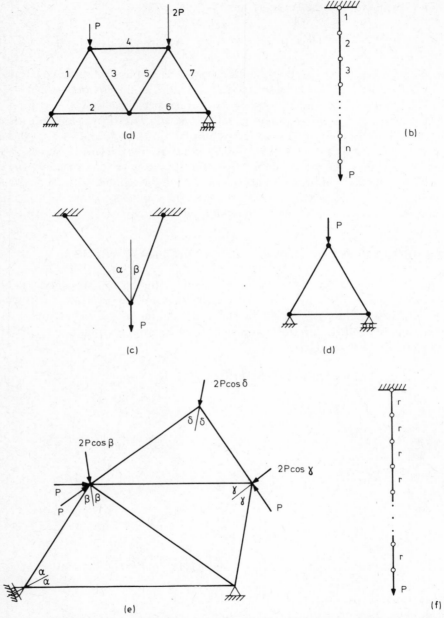

Fig. 2.9. Truss systems: (a, b) With different unconditional and conditional probabilities of bar failure. (c) $n = 2$. (d) $n = 3$. (e, f) Constituent bars have the same probabilities of failure.

For $n = 3$ (Fig. 2.9d), we have

$$R = 1 - \left[P(A_1) + P(A_2) + P(A_3) - P(A_1 A_2) - P(A_1 A_3) \right.$$

$$\left. - P(A_2 A_3) + P(A_1 A_2 A_3) \right]$$

$$= 1 - \left[P(A_1) + P(A_2) + P(A_3) - P(A_1|A_2) P(A_2) - P(A_1|A_3) P(A_3) \right.$$

$$\left. - P(A_2|A_3) P(A_3) + P(A_1|A_2 A_3) P(A_2|A_3) P(A_3) \right] \qquad (2.21)$$

In this case we must know the conditional probabilities of failure $P(A_1|A_2)$, $P(A_1|A_3)$, $P(A_2|A_3)$, and $P(A_1|A_2 A_3)$. Two extreme cases are possible:
(1) *Bar failures represent individually independent random variables.* (The bars can be visualized as manufactured at different plants, using different processes.) In this case it suffices to know the reliabilities of individual bars:

$$P(A_i A_j) = P(A_i) P(A_j) = (1 - R_i)(1 - R_j)$$

$$P(A_i A_j A_k) = (1 - R_i)(1 - R_j)(1 - R_k)$$

and Eq. (2.19) becomes

$$R = 1 - \sum_{i=1}^{n} (1 - R_i) + \sum_{1 \leqslant i < j}^{n \ \ n} (1 - R_i)(1 - R_j)$$

$$- \sum_{1 \leqslant i < j < k}^{n \ \ n \ \ n} (1 - R_i)(1 - R_j)(1 - R_k) + \cdots$$

Another way of finding the reliability is

$$R = P \text{ (no bars fail)} = P(\overline{A}_1 \overline{A}_2 \cdots \overline{A}_n)$$

$$= P(\overline{A}_1) P(\overline{A}_2) \cdots P(\overline{A}_n) = R_1 R_2 \cdots R_n$$

so that in this case the truss behaves like a "series" system (see also Fig. 2.9b). Indeed, for $n = 2$ we have from Eq. (2.19)

$$R = 1 - \left[P(A_1) + P(A_2) - P(A_1) P(A_2) \right]$$

$$= \left[1 - P(A_1) \right]\left[1 - P(A_2) \right] = R_1 R_2$$

For $n = 3$,

$$R = 1 - [P(A_1) + P(A_2) + P(A_3) - P(A_1)P(A_2) - P(A_1)P(A_3)$$

$$- P(A_2)P(A_3) + P(A_1)P(A_2)P(A_3)]$$

$$= [1 - P(A_1)][1 - P(A_2)][1 - P(A_3)] = R_1 R_2 R_3$$

In the particular case where $R_1 = R_2 = \cdots = R_n = r$, we have

$$R = r^n \qquad (2.22)$$

(2) *Each realization of the truss, and therefore of its constituent bars, has the same strength, which changes from realization to realization.* (All bars are manufactured at the same plant from the same material by the same process.)

Reliability then depends generally on the level of the stresses in the bars. Consider the particular case where the stresses are equal in all bars, and equal reliabilities $R_1 = R_2 = \cdots = R_n = r$ are assigned to them (Figs. 2.9e, 2.9f). Failure of any bar indicates failure of all remaining bars, and all conditional probabilities arising in Eqs. (2.19)–(2.21) are unity. A truss behaves as a single bar and its reliability equals that of a single bar:

$$R = r \qquad (2.23)$$

Indeed, for $n = 2$, we have $P(A_1|A_2) = 1$, and Eq. (2.20) yields

$$R = 1 - [P(A_1) + P(A_2) - P(A_2)] = 1 - P(A_1) = R_1 = r \quad (2.24)$$

Consequently, the reliability of a statically determinate truss with equal reliabilities of constituent bars varies between r^n and r, corresponding to the two extreme cases considered above.

It is worth noting that the parallel system model discussed in Example 2.7 occurs in collapse analysis of statically *indeterminate* structures in which failure takes place only after several bars have reached their strength capacity (see, e.g., papers by Moses, and Ang and Ma).

2.9 OVERALL PROBABILITY AND BAYES' FORMULA

Let B be an event in a sample space partitioned by events $A_1, A_2, \ldots, A_n$. Then $B = B\Omega = B(A_1 + A_2 + \cdots + A_n)$, and

$$P(B) = \sum_{i=1}^{n} P(BA_i)$$

Using the multiplication rule for each of $P(BA_i)$, we arrive at

$$P(B) = \sum_{i=1}^{n} P(B|A_i)P(A_i) \tag{2.25}$$

which is known as the *formula of overall probability*. Now, provided $P(B) \neq 0$, $P(A_i|B)$ can be expressed as

$$P(A_i|B) = \frac{P(A_iB)}{P(B)} = \frac{P(B|A_i)P(A_i)}{\sum_{i=1}^{n} P(B|A_i)P(A_i)} \tag{2.26}$$

This formula is due to Bayes. The unconditional probabilities $P(A_i)$ are called *a priori probabilities*, and the conditional probabilities $P(A_i|B)$ are *a posteriori probabilities*.

Example 2.8

Given two boxes, the first containing a white and b black balls and the second c white and d black balls. One ball is removed at random from the first box and placed in the second, after which one ball is removed from the latter. What is the probability of this ball being white?

Denote the following events: A = white ball removed from second box, H_1 = white ball placed in second box, H_2 = black ball placed in second box:

$$P(H_1) = \frac{a}{a+b} \qquad\qquad P(H_2) = \frac{b}{a+b}$$

$$P(A|H_1) = \frac{c+1}{c+d+1} \qquad P(A|H_2) = \frac{c}{c+d+1}$$

According to the formula of overall probability, we have

$$P(A) = P(A|H_1)P(H_1) + P(A|H_2)P(H_2)$$

$$= \left(\frac{a}{a+b}\right)\left(\frac{c+1}{c+d+1}\right) + \left(\frac{b}{a+b}\right)\left(\frac{c}{c+d+1}\right)$$

In the particular case of both boxes containing equal numbers of white and black balls ($c = a$, $d = b$), we have

$$P(A) = \left(\frac{a}{a+b}\right)\left(\frac{a+1}{a+b+1}\right) + \left(\frac{b}{a+b}\right)\left(\frac{a}{a+b+1}\right) = \frac{a}{a+b}$$

indicating that the probability of a white ball being removed from the second box is unaffected by adding the ball from the first box.

Example 2.9

The structure under a time-dependent load, $P = P(t)$, consists of two components with reliability [defined as nonfailure performance in the time interval

$(0, T)]$ R_1 and R_2, respectively; nonfailure of both is required for nonfailure of the structure. The structure was inspected at the end of time interval T and found to have failed. Find the probability of only the first component having failed, the second not.

Before the experiment, the four following hypotheses were possible:

H_0: both components do not fail
H_1: first component fails, second does not
H_2: first component does not fail, second does
H_3: both components fail

The respective probabilities of these hypotheses are

$$P(H_0) = R_1 R_2 \qquad\qquad P(H_1) = (1 - R_1) R_2$$

$$P(H_2) = R_1(1 - R_2) \qquad P(H_3) = (1 - R_1)(1 - R_2)$$

Event A has taken place, the structure failed, hence

$$P(A|H_0) = 0,$$

$$P(A|H_1) = P(A|H_2) = P(A|H_3) = 1$$

and Bayes' formula yields

$$P(H_1|A) = \frac{(1 - R_1) R_2}{(1 - R_1) R_2 + (1 - R_2) R_1 + (1 - R_1)(1 - R_2)} = \frac{(1 - R_1) R_2}{1 - R_1 R_2}$$

PROBLEMS

2.1. Present a Venn diagram for $\overline{C}$, where $C = B \setminus A$.

2.2. Verify by means of a Venn diagram that a union and an intersection of random events are distributive, that is,

$$(A \cup B) \cap C = (A \cap C) \cup (B \cap C)$$

$$(A \cap B) \cup C = (A \cup C) \cap (B \cup C)$$

2.3. A telephone relay satellite is known to have five malfunctioning channels out of 500 available. If a customer gets one of the malfunctioning channels on first dialing, what is the probability of his hitting on another malfunctioning channel on dialing again?

2.4. Let m items be chosen at random from a lot containing $n > m$ items of which $p(m < p < n)$ are defective. Find the probability of all m items being nondefective. Consider also the particular case $m = 3$, $p = 4$, $n = 8$.

2.5. A single playing card is picked at random from a well-shuffled ordinary deck of 52. Consider the events A, king picked; B, ace picked; C, heart picked. Check whether (1) A and B, (2) A and C, (3) B and C, are dependent or independent.

2.6. In a family of two children the older child is a boy. What is the probability of both of them being boys?

2.7. In a family of two children, at least one of them is a boy. What is the probability of both of them being boys?

2.8. A statically determinate truss consists of n bars, whose failures represent independent random variables with identical probabilities p. Find the permissible value of p such that the probability of failure Q of the entire truss does not exceed some prescribed value q.

2.9. A space shuttle is assigned to visit a TV satellite in a GSO (geosynchronous orbit) with radius 36,000 km (event A_1), and carry out maintenance (event B_1). Alternatively, it may, due to failure of one of the engines, go into an LEO (low Earth orbit) with radius 200 km (event A_2) and carry out a crop survey (event B_2). B_1 and B_2 are regarded as successful performance. The third possibility is failure to take off (event A_3). What is the reliability (probability of successful performance) of the space shuttle if $P(B_1|A_1) = 0.75$, $P(B_2|A_2) = 0.85$, $P(A_1) = 0.80$, and $P(A_2) = 0.15$?

2.10. (Birger) During inspection of a gas turbine, two symptoms are checked: increase of the engine gas temperature at the turbine outlet by more than 50°C (symptom k_1) and increase of the acceleration time of the rpm from minimum to maximum, by more than 5 s (symptom k_2). Assume that for the given type of engine these symptoms are associated either with failure of the fuel flow regulator (state D_1) or with reduction of the radial clearance of the turbine (state D_2). The normal state is denoted by D_3. The following probabilities are given:

$$P(D_1) = 0.05, \quad P(k_1|D_1) = 0.2, \quad P(k_2|D_1) = 0.3$$
$$P(D_2) = 0.15, \quad P(k_1|D_2) = 0.4, \quad P(k_2|D_2) = 0.5$$
$$P(D_3) = 0.80, \quad P(k_1|D_3) = 0.0, \quad P(k_2|D_3) = 0.05$$

Show that

$$P(D_1|k_1k_2) = 0.09, \quad P(D_2|k_1k_2) = 0.91, \quad P(D_3|k_1k_2) = 0.0$$
$$P(D_1|\bar{k}_1k_2) = 0.12, \quad P(D_2|\bar{k}_1k_2) = 0.46, \quad P(D_3|\bar{k}_1k_2) = 0.41$$
$$P(D_1|\bar{k}_1\bar{k}_2) = 0.03, \quad P(D_2|\bar{k}_1\bar{k}_2) = 0.05, \quad P(D_3|\bar{k}_1\bar{k}_2) = 0.92$$

indicating, for example, that in the presence of symptoms k_1 and k_2 the probability of reduction of the radial clearance is about ten times that of failure of the fuel flow regulator.

CITED REFERENCES

Ang, A. H. -S., and Ma, H. -F., "On the Reliability Analysis of Framed Structures," in A. H. -S. Ang and M. Shinozuka, Eds., *Probabilistic Mechanics and Structural Reliability*, ASCE, 1979, pp. 106–111.

Arbocz, J., "The Effect of General Imperfections on the Buckling of Cylindrical Shells," PhD thesis, Cal. Inst. Technology, Pasadena, CA, 1968. Also: J. Arbocz and C. D. Babcock, Jr., *ASME J. Appl. Mech.*, **36**, 28–38 (1969).

Bernstein, S. N., *The Theory of Probability*, 4th ed., Gostekhizdat, Moscow, 1946.

Birger, I. A., "Application of Bayes' Formula in the Technical Diagnostics", *Mashinostroenie*, 1964, pp. 24–25.

Koiter, W. T., "On the Stability of Elastic Equilibrium" (in Dutch), PhD thesis, Delft Univ. Technology, H. J. Paris, Amsterdam; English translations (a) NASA TT-F-10, 1967, (b) AFFDL-TR-70-20, 1970 (translated by E. Riks).

Kolmogorov, A. N., "Ueber die Analytischen Methoden in der Wahrscheinlichkeitsrechnung," *Math. Ann.*, **104**, 415–458 (1931).

Kolmogorov, A. N., "Foundations of the Theory of Probability," *Ergeb. Math. Ihrer Grensgeb.*, **2** (3), 1933 (English translation: Chelsea Publishing Co., New York, 1956).

Moses F., "Reliability of Structural Systems," *J. Structural Division*, Proc. ASCE, **ST9**, 1213–1220 (1974).

Von Mises, R., *Wahrscheinlichkeit, Statistik und Wahrheit*, Springer-Verlag, Vienna, 1936.

RECOMMENDED FURTHER READING

Babillis, R. A., and Smith, A. M., "Application of Bayesian Statistics in Reliability Measurement," *Ann. Reliability Maintainability*, **4**, Practical Techniques and Application (J. de S. Coutinho, Ed.), Fourth Annual Reliability and Maintainability Conference, Spartan Books, 1965, pp. 357–366.

Feller, W., *An Introduction to Probability Theory and its Applications*, 3rd ed., Vol. 1, John Wiley & Sons, New York, 1968. Chap. 1: The Sample Space, pp. 7–25; Chap. 2: Elements of Combinatorial Analysis, pp. 26–66; Chap. 4: Combination of Events, pp. 98–113; Chap. 5: Conditional Probability, Stochastic Independence, pp. 114–145.

Parzen, E., *Modern Probability Theory and its Applications*, John Wiley & Sons, New York, 1960. Chap. 1: Probability Theory as the Study of Mathematical Models of Random Phenomena, pp. 1–31; Chap. 2: Basic Probability Theory, pp. 32–86; Chap. 3: Independence and Dependence, pp. 87–147.

Papoulis, A., *Probability, Random Variables and Stochastic Processes*, Intern. Student Ed., McGraw-Hill Kogakusha, Tokyo, 1965. Chap. 1: The Meaning of Probability, pp. 3–17; Chap. 2: The Axioms of Probability, pp. 18–46.

chapter **3**

Single Random Variable

3.1 RANDOM VARIABLE

A real random variable $X(\omega)$, $\omega \in \Omega$, is a real function mapping a sample space Ω into the real line R, that is, making a real number $X(\omega)$ correspond to every outcome ω of an experiment such that (1) the set $\{X(\omega) \leqslant x\}$ is an event for any real number x, and (2) the probabilities $P\{X(\omega) = +\infty\} = P\{X(\omega) = -\infty\}$ are zero.

Random variables are conventionally denoted by capital letters. The notations $X(\omega)$ and X will hereinafter be used interchangeably, although the former is preferable as an explicit indication of the functional character of the variable.

Example 3.1

Consider a bar of given geometry and strength. The tensile force N may take on two values: n_1 (with probability p) at which the bar fails, and n_2 (with probability $1 - p$) at which it survives. Accordingly, $\Omega = \{\text{fail, survive}\}$. Let $X(\omega) = 1$ if $\omega = $ fail, and $X(\omega) = 0$ if $\omega = $ survive; we thus have a real number $X(\omega)$ corresponding to either outcome of the experiment. We shall show that $\{\omega\colon X(\omega) \leqslant x\}$ is an event.

If $x < 0$, $\{\omega\colon X(\omega) \leqslant x\} = \varnothing$
If $0 \leqslant x < 1$, $\{\omega\colon X(\omega) \leqslant x\} = \{\text{survive}\}$
If $x \geqslant 1$, $\{\omega\colon X(\omega) \leqslant x\} = \Omega = \{\text{fail, survive}\}$

Consequently, for each x the set $\{\omega\colon X(\omega) \leqslant x\}$ is an event, and $X(\omega)$ is a random variable.

Example 3.2

Consider a three-dice experiment. The sample space Ω contains $6^3 = 216$ points $\{(i, j, k);\ i,\ j,\ k = 1, 2, \ldots, 6\}$. Let X denote the sum of spots on the upward-landing faces; then $X(\omega) = i + j + k$ if $\omega = (i, j, k)$. X is a random variable; the counterdomain of $X(\omega)$ is an ensemble of positive numbers between 3 and 18.

3.2   DISTRIBUTION FUNCTION

The (cumulative) distribution function $F_X(x)$ of the random variable X is defined as

$$F_X(x) = P(X \leqslant x) = P\big[\{\omega \in \Omega\colon X(\omega) \leqslant x\}\big] \tag{3.1}$$

for every real number x.

Example 3.3

Consider the random variable described in Example 3.1. The probability distribution is as follows:

If $x < 0$, $F_X(x) = 0$,
since $(X \leqslant x)$ is an impossible event.

If $x \geqslant 1$, $F_X(x) = 1$,
since $\{X(\omega) \leqslant x\}$ is a certain event, because $X(\text{fail}) = 1$, $X(\text{survive}) = 0 < x$.

If $0 \leqslant x < 1$, $F_X(x) = P(X \leqslant x) = P(\text{survive}) = 1 - p$
since $X(\text{fail}) = 1 > x$, $X(\text{survive}) = 0 \leqslant x$, and therefore $\{X(\omega) \leqslant x\} = \{\text{survive}\}$.

The distribution function is shown in Fig. 3.1.

Example 3.4

Consider the experiment of tossing a single "honest" coin:

$$\Omega = \{\text{heads, tails}\}$$

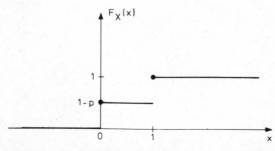

Fig. 3.1.   Probability distribution function of random variable X in Example 3.1.

This example can be reduced to the preceding one by the following formal substitution:

$$\text{fail} \rightarrow \text{heads}, \quad \text{survive} \rightarrow \text{tails}, \quad p \rightarrow \tfrac{1}{2}$$

Example 3.5

Consider the experiment of throwing a single "honest" die. Let X denote the number of spots on the upward-landing face. $X(\omega_i) = i, i = 1, 2, \ldots, 6$. $P(\omega_i) = \tfrac{1}{6}$; $F_X(x) = 0$ for $x < 1$, since the event $\{X < 1\} = \varnothing$, and thus $F_X(x) = P(\varnothing) = 0$. For $1 \leqslant x < 2$ we have

$$F_X(x) = P(X \leqslant x) = P(\omega_1) = \tfrac{1}{6}$$

For $2 \leqslant x < 3$, the event $(X \leqslant x)$ equals (ω_1, ω_2), and due to their mutual exclusiveness, we have

$$F_X(x) = P(\omega_1, \omega_2) = P(\omega_1) + P(\omega_2) = \tfrac{1}{6} + \tfrac{1}{6} = \tfrac{1}{3}$$

Thus,

$$F_X(x) = \begin{cases} 0, & x < 1 \\ \tfrac{1}{6}i, & i \leqslant x < i + 1, i = 1, 2, 3, 4, 5 \\ 1, & 6 \leqslant x \end{cases}$$

The last equality is readily obtainable by noting that for $6 < x$ the event $(X < x)$ equals $(\omega_1, \omega_2, \omega_3, \omega_4, \omega_5, \omega_6)$, that is, it represents a certain event. The distribution function is shown in Fig. 3.2; it may also be formulated as

$$F_X(x) = \tfrac{1}{6} \sum_{i=1}^{6} U(x - i)$$

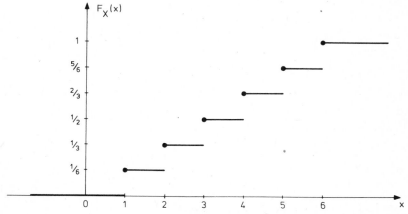

Fig. 3.2. Probability distribution function of number of spots on upward-landing face in single "honest" die-throwing experiment.

where $U(x - i)$ is a unit step function, defined as

$$U(x - i) = \begin{cases} 1, & x \geqslant i \\ 0, & \text{otherwise} \end{cases} \qquad (3.2)$$

This function is shown in Fig. 3.3.

3.3 PROPERTIES OF THE DISTRIBUTION FUNCTION

The distribution function has the following properties

1. $F_X(-\infty) = 0$.
2. $F_X(+\infty) = 1$.
3. $F_X(x)$ is a monotone nondecreasing function, that is, $F_X(a) \leqslant F_X(b)$ if $a < b$.
4. $F_X(x)$ is continuous on the right, meaning that

$$\lim_{0 < \varepsilon \to 0} F_X(x + \varepsilon) = F_X(x^+) = F_X(x)$$

All these properties can be shown to hold, using the definition of the distribution function. In fact, $F_X(-\infty) = P(X \leqslant -\infty) = P(X < -\infty) = 0$, where the last equality arises from the definition of a random variable; and $F_X(+\infty) = \text{Prob}(X \leqslant +\infty) = \text{Prob}(\Omega) = 1$, since the event $\{X \leqslant +\infty\}$ is a certain one. For each outcome, $X(\omega) \leqslant +\infty$.

Let us establish the third property

$$\text{Prob}(X \leqslant b) = \text{Prob}(\{X \leqslant a\} \cup \{a < X \leqslant b\})$$

Since events $\{X \leqslant a\}$ and $\{a < X \leqslant b\}$ are mutually exclusive, Axiom 3 yields

$$F_X(b) = \text{Prob}(X \leqslant a) + \text{Prob}(a < X \leqslant b)$$

$$= F_X(a) + \text{Prob}(a < X \leqslant b)$$

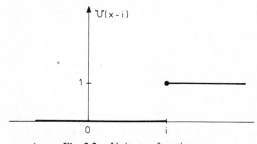

Fig. 3.3. Unit step function.

and

$$F_X(b) - F_X(a) = \text{Prob}(a < X \leqslant b) \tag{3.3}$$

Due to Axiom 1, the last quantity is nonnegative. This yields the desired result

$$F_X(b) - F_X(a) \geqslant 0$$

The last property follows from the definition of $F_X(x)$, as $P(X \leqslant x)$.

It follows from the properties that the distribution function is bounded between zero and unity.

Rewriting Eq. (3.3) for $a - 0$, we have

$$F_X(b) - F_X(a - 0) = \text{Prob}(a - 0 < X \leqslant b)$$

Now, considering the limit when $b \to a + 0$, we have

$$F_X(a + 0) - F_X(a - 0) = \text{Prob}(a - 0 < X \leqslant a + 0) = \text{Prob}(X = a)$$

Since $F_X(a + 0) = F_X(a)$, we obtain

$$F_X(a) = F_X(a - 0) + \text{Prob}(X = a)$$

For $\text{Prob}(X = a) = 0$, the distribution function turns out to be continuous on the left and therefore continuous at a. For $P(X = a) \neq 0$, we have

$$\text{Prob}(X = a) = F_X(a) - F_X(a - 0) \tag{3.4}$$

That is, the probability of the event of the random variable X taking on the value a equals the jump discontinuity of its distribution function at a (see Fig. 3.4). Since the distribution function is bounded between zero and unity, it can

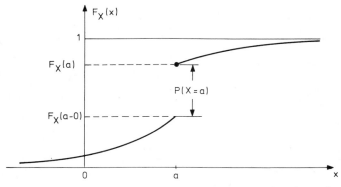

Fig. 3.4. Probability of random variable X taking on value a equal to jump discontinuity of $F_X(x)$ at a.

have at most a countable number of jump discontinuities. In fact, it cannot have more than $2^i - 1$ jumps with values between 2^{-i} and 2^{1-i}.

A random variable X is called *discrete* if the range of X is countable, that is, if there exists a finite or denumerable set of real numbers $x_1, x_2, \ldots$, such that X takes on values only within that set. These values $x_1, x_2, \ldots$, we call *possible* values of the discrete random variable X. In order to characterize the discrete random variable completely, it suffices to know the probabilities $p_i = P(X = x_i)$. The distribution function then becomes

$$F_X(x) = \sum_{i:\, x_i \leqslant x} p_i \tag{3.5}$$

or, using the unit step function defined by Eq. (3.2):

$$F_X(x) = \sum_{\text{for all } x_i} p_i U(x - x_i) \tag{3.6}$$

and summation is taken over all indices of possible values of the random variable X. The distribution function of a discrete random variable is discontinuous, jumping by $P[X = x_i]$ at the discontinuity point x_i (see, e.g., Figs. 3.1 or 3.2).

A random variable X is called *continuous* if there exists a nonnegative function $f_X(x)$ such that

$$F_X(x) = \int_{-\infty}^{x} f_X(x)\, dx \tag{3.7}$$

for every real number x. The function $f_X(x)$ is called the *probability density function of X*. If $F_X(x)$ is absolutely continuous and differentiable at every x, then its derivative equals the probability density function of X:

$$f_X(x) = \frac{dF_X(x)}{dx} = \lim_{\Delta x \to 0} \frac{F_X(x + \Delta x) - F_X(x)}{\Delta x} \tag{3.8}$$

The properties of the probability density function are

1. $f_X(x) \geqslant 0$.
2. For every x_1 and x_2,

$$P\{X \in [x_1, x_2]\} = \int_{x_1}^{x_2} f_X(x)\, dx \tag{3.9}$$

That is, the probability of a random variable taking on a value in any interval in its range equals the integral of its probability density over that interval. In particular, if $f_X(x)$ is continuous in x, then $P(x \leqslant X \leqslant x + \Delta x) \simeq f_X(x)\Delta x$, that is, the probability that X takes on values in a small interval containing the value x, is approximately equal to the width of that interval times $f_X(x)$. Taking the limit as Δx approaches zero, we have

$$\lim_{\Delta x \to 0} P(x \leqslant X < x + \Delta x) = P(X = x) = 0 \qquad (3.10)$$

for any real x. That is, for a continuous random variable, the probability of X taking on any specified value x is zero. This conclusion is also obtainable from Eq. (3.4), since the distribution function of a continuous variable is continuous also on the left, $F_X(a) = F_X(a - 0)$, and $P(X = a) = 0$. This implies that over a large number of trials the random variable X would very seldom take on this value so that the relative frequency of the latter will tend to zero.

3. Since $F_X(+\infty) = 1$, we get from Eq. (3.7)

$$\int_{-\infty}^{\infty} f_X(x)\, dx = 1 \qquad (3.11)$$

We formally define the probability density function of a discrete random variable with a distribution function as per (3.6):

$$f_X(x) = \sum p_i \delta(x - x_i) \qquad (3.12)$$

Summation is again taken over all indices of possible values of the random variable X, and $\delta(x - x_i)$ is Dirac's delta function, so that

$$U(x - x_i) = \int_{-\infty}^{x} \delta(x - x_i)\, dx \qquad (3.13)$$

The basic properties of the Dirac delta function are

$$\delta(x - a) = \begin{cases} 0, & x \neq a \\ \infty, & x = a \end{cases}$$

$$\int_{-\infty}^{\infty} \varphi(x)\delta(x - a)\, dx = \varphi(a) \qquad (3.14)$$

for any continuous function $\varphi(x)$ at $x = a$. The integral in (3.14) "screens" out, as it were, the value of function φ at the argument a; δ is a *functional* or a *generalized function*. We are already familiar with this function from courses in the mechanics of solids (see, e.g., Crandall et al. or Popov).

$$\delta(x - a) = \langle x - a \rangle_*^{-1}, \qquad U(x - a) = \langle x - a \rangle^0 \qquad (3.15)$$

As a "refresher" on its use, consider the beam in Fig. 3.5 with n concentrated forces acting on it. The shear force will then be

$$V_y(x) = R_A \langle x - 0 \rangle^0 - P_1 \langle x - x_1 \rangle^0 - P_2 \langle x - x_2 \rangle^0 - \cdots - P_n \langle x - x_n \rangle^0.$$

and the distributed force is

$$p_y(x) = -\frac{dV_y}{dx} = -R_A \langle x - 0 \rangle_*^{-1} + P_1 \langle x - x_1 \rangle_*^{-1} + P_2 \langle x - x_2 \rangle_*^{-1}$$
$$+ \cdots + P_n \langle x - x_n \rangle_*^{-1}$$

where jumps in the graph of $V_y(x)$ equal the values of the concentrated forces.

Analogously, the jumps in the distribution function of a discrete random variable equal the probability of its taking on the specific value of x_i where the jump takes place. The analogy is obviously incomplete. Jumps in forces can be either positive or negative ($p_i \lessgtr 0$), whereas jumps in the probability distribution function are exclusively positive ($p_i > 0$) and $F_X(x)$ is a nondecreasing function. Figure 3.6 shows a probability density function of Example 3.5. The functions $\langle x - x_i \rangle_*^{-1}$ and $\delta(x - x_i)$ are indicated in Figs. 3.5 and 3.6 by a spike with an open arrowhead.

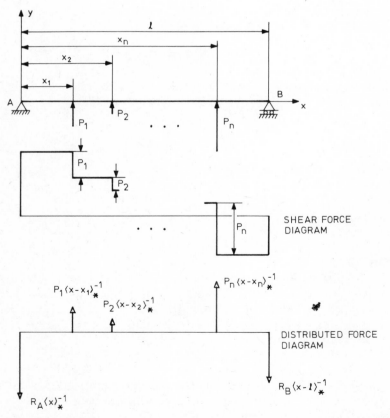

Fig. 3.5. Beam under concentrated forces: Distributed force diagram representing analogue of probability density of discrete random variable.

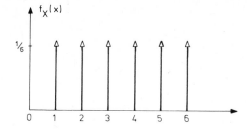

Fig. 3.6. Probability density function of number of spots on upward-landing face in single "honest" die-throwing experiment.

A random variable X is called *mixed* if it is neither purely discrete nor purely continuous, in other words, if it possesses the properties of both discreteness and continuity: It contains jumps, equal $p_1, p_2, \ldots, p_n$, respectively, at points $x_1, x_2, \ldots, x_n$, but is continuous between them. Subtracting, then, the sum

$$\sum_{i=1}^{n} p_i U(x - x_i)$$

from the distribution function, we obtain a continuous function. We can thus express the probability distribution function of a mixed random variable as

$$F_X(x) = F^*(x) + \sum_{i=1}^{n} p_i U(x - x_i) \tag{3.16}$$

where $F^*(x)$ is a continuous function, and p_i $(i = 1, 2, \ldots, n)$ are the jumps of $F_X(x)$ at points x_i $(i = 1, 2, \ldots, n)$. If $F^*(x)$ is absolutely continuous and differentiable at every x, then the probability density function of X may be expressed as

$$f_X(x) = f^*(x) + \sum_{i=1}^{n} p_i \delta(x - x_i) \tag{3.17}$$

where

$$f^*(x) = \frac{dF^*(x)}{dx}$$

Unlike a continuous random variable, a mixed random variable has a countable number of possible values, which it takes on with nonzero probability.

An "analogue" of the probability density of a mixed random variable is represented by the following example. The beam, simply supported at its ends, is subjected to a concentrated moment m (see Fig. 3.7). The bending moment will then be written as

$$M_z(x) = -\frac{m}{L}x + m\langle x - a\rangle^0$$

and the shear force as

$$V_y(x) = \frac{m}{L} - m\langle x - a\rangle_*^{-1}$$

This expression comprises a continuous part m/L and a singular, discontinuous part $-m\langle x - a\rangle_*^{-1}$, whose probability counterparts are obviously positive. For clearer understanding of the notions of "discrete," "continuous," and "mixed" distributions, the following analogy is useful. Let us visualize a string with a mass distribution such that the entire mass equals unity, and its density is given by the probability density function. The discrete case corresponds to the entire mass being lumped at certain points $x_1, x_2, \ldots, x_n$; the continuous case, to a distributed (uniformly, or otherwise) mass without concentrations; and the mixed case, to a combination of continuities and concentrations. This

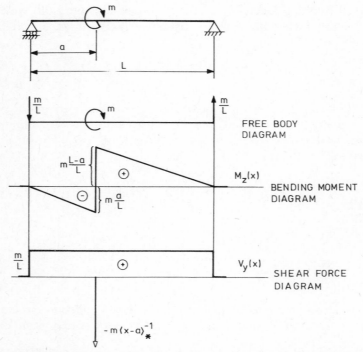

Fig. 3.7. Beam under concentrated moment: Shear force diagram representing analogue of probability density of mixed random variable.

analogy will also be useful in further analysis of random variables. We actually assumed that the distribution function of a continuous random variable has only a countable number of points for which the derivative does not exist. At such a point we assign any positive value to $f_X(x)$ so that it becomes defined for all points x.

3.4 MATHEMATICAL EXPECTATION

While the distribution function provides a complete characterization of the random variable, it is sometimes possible to make do with a simpler, albeit incomplete, characterization based on few numbers. Suppose that we are concerned with a series of trials whose possible outcomes are x_i^*, $i = 1, 2, \ldots, n$; the simplest characteristic of the discrete random variable X in the given series would be the arithmetical mean

$$\bar{x} = \frac{1}{n} \sum_{i=1}^{n} x_i^* \tag{3.18}$$

If some of the n_i values of $x_1^*, x_2^*, \ldots, x_n^*$ taken on by the random variable X in n experiments coincide, the coefficient of the common x_i^* value will be n_i. Denoting the possible values by $x_1, x_2, \ldots, x_m$, and their relative frequencies by n_i/n, Eq. (3.18) becomes

$$\bar{x} = \sum_{i=1}^{m} \frac{n_i}{n} x_i \tag{3.19}$$

We regard a discrete random variable X as having a *mathematical expectation*, if

$$\sum_{i=-\infty}^{\infty} |x_i| P(X = x_i) < \infty \tag{3.20}$$

the expectation being given by the expression

$$E(X) = \sum_{i=-\infty}^{\infty} x_i P(X = x_i) \tag{3.21}$$

with summation over a finite or countable number of possible values x_i, $-\infty < x_i < +\infty$.

If $\varphi(X)$ is a deterministic function of a random variable, then

$$E[\varphi(X)] = \sum_{i=-\infty}^{\infty} \varphi(x_i) P(X = x_i) \tag{3.22}$$

Since the random variable $Y = \varphi(X)$ takes on only values $y_i = \varphi(x_i)$, where x_i are the possible values of X, we have

$$P(Y = y) = \sum_{x_i: \, \varphi(x_i) = y} P(X = x_i)$$

so that

$$E(Y) = \sum_{j = -\infty}^{\infty} y_j P(Y = y_j)$$

$$= \sum_{j = -\infty}^{\infty} y_j \sum_{x_i: \, \varphi(x_i) = y_j} P(X = x_i) = \sum_{i = -\infty}^{\infty} \varphi(x_i) P(x = x_i)$$

Example 3.6

A random variable X takes on values i $(i = 2, 3, \dots)$ with probabilities $1/i^2$. Since

$$\sum_{i=2}^{\infty} i \frac{1}{i^2} = \sum_{i=2}^{\infty} \frac{1}{i} \to \infty$$

this random variable has no mathematical expectation.

Example 3.7

Consider the experiment of throwing an "honest" die, X denoting the number of spots on the upward-landing face. We have

$$E[X] = \sum_{i=1}^{6} iP(X = i) = 1 \cdot \tfrac{1}{6} + 2 \cdot \tfrac{1}{6} + 3 \cdot \tfrac{1}{6} + 4 \cdot \tfrac{1}{6} + 5 \cdot \tfrac{1}{6} + 6 \cdot \tfrac{1}{6} = 3\tfrac{1}{2}$$

Since the result does not coincide with one of the possible values of X, it is seen that the mathematical expectation is not a value to be "expected," but rather an average or a *mean*, as it is actually referred to sometimes.

Note that in the example of a string of unity mass with concentrations, $E(X)$ coincides numerically with the coordinate of the center of gravity (or centroid).

Let X be a continuous random variable with probability density function $f_X(x)$. We will say that X has a mathematical expectation

$$E(X) = \int_{-\infty}^{\infty} x f_X(x) \, dx \tag{3.23}$$

if

$$\int_{-\infty}^{\infty} |x| f_X(x) \, dx < \infty \tag{3.24}$$

For constant c, the following properties of the mathematical expectation are readily established from definition (3.23):

1. $E(c) = c$ (3.25)
2. $E(cX) = cE(X)$ (3.26)

Example 3.8

If the probability density function of a continuous random variable X is given by

$$f_X(x) = \frac{a}{\pi} \frac{1}{a^2 + x^2}, \quad -\infty < x < \infty$$

we say that X has a Cauchy distribution.
 The distribution function $F_X(x)$ is

$$F_X(x) = \int_{-\infty}^{x} \frac{a}{\pi} \frac{dx}{a^2 + x^2} = \frac{1}{2} + \frac{1}{\pi} \tan^{-1}\left(\frac{x}{a}\right)$$

X has no mathematical expectation; in fact, the integral

$$\int_{-\infty}^{\infty} \frac{a}{\pi} \frac{|x|}{a^2 + x^2} dx$$

does not exist.

Example 3.9

If the probability density function of a continuous random variable X is given by

$$f_X(x) = \frac{a}{2} e^{-a|x|}$$

we say that X has a Laplace distribution

$$F_X(x) = \begin{cases} \frac{1}{2} e^{ax}, & -\infty < x < 0 \\ 1 - \frac{1}{2} e^{-ax}, & 0 \leqslant x < \infty \end{cases}$$

and its mathematical expectation is

$$E(X) = \int_{-\infty}^{\infty} \frac{a}{2} x e^{-a|x|} dx = 0$$

In complete analogy with (3.22), the mathematical expectation of $Y = \varphi(X)$ is calculated as follows:

$$E[\varphi(X)] = \int_{-\infty}^{\infty} \varphi(x) f_X(x) dx \qquad (3.27)$$

if that integral is absolutely convergent, that is, if

$$\int_{-\infty}^{\infty} |\varphi(x)| f_X(x)\, dx \tag{3.28}$$

is finite.

3.5 MOMENTS OF RANDOM VARIABLE; VARIANCE

A particular case of Eq. (3.27) is $\varphi(X) = X^k$, where k is zero or a positive integer. For $k = 0$ we have unity, for $k = 1$ the mathematical expectation, and for $k > 1$, the *kth moment*:

$$m_k = E(X^k) = \int_{-\infty}^{\infty} x^k f_X(x)\, dx \tag{3.29}$$

The mathematical expectation is thus a first moment of a random variable.

The mathematical expectation of $[X - E(X)]^k$ is defined as the kth *central moment*:

$$\mu_k = E\{[X - E(X)]^k\} = \int_{-\infty}^{\infty} [x - E(X)]^k f_X(x)\, dx \tag{3.30}$$

Obviously, the zeroth central moment equals unity, and the first central moment, zero. The second central moment of a random variable is called the *variance* (provided the integral in (3.30) is absolutely convergent) and denoted by Var(X):

$$\mathrm{Var}(X) = E\{[X - E(X)]^2\} = \int_{-\infty}^{\infty} [x - E(X)]^2 f_X(x)\, dx \tag{3.31}$$

The mathematical expectation is a measure of the "average" of the values taken on by the random variable, whereas the variance is one of *spread*.

The mean is the center of gravity of density, and variance is the moment of inertia of the same density about an axis through the center of gravity.

For a discrete random variable, variance is defined by

$$\mathrm{Var}(X) = \sum_{i=-\infty}^{\infty} [x_i - E(X)]^2 P(X = x_i) \tag{3.32}$$

if the sum on the right is finite. It follows from the definitions (3.31) and (3.32) that

$$\mathrm{Var}(X) = \mu_2 = E\{[X - E(X)]^2\}$$

$$= m_2 - 2[E(X)]^2 + [E(X)]^2 = m_2 - [E(X)]^2$$

$$= m_2 - m_1^2$$

so that

$$\text{Var}(X) = E(X^2) - [E(X)]^2 \tag{3.33}$$

This is equivalent to the *parallel-axis theorem* or *Steiner's* theorem in statics, which relates the respective moments of inertia of a body about an arbitrary axis and about the central axis parallel to it.

The *standard deviation* of X is defined as $\sqrt{\text{Var}(X)}$, and denoted by σ_X; that is,

$$\sigma_X = \sqrt{\text{Var}(X)} \tag{3.34}$$

For a constant c, the following properties of variance are readily established from the definition (3.31):

1. $\text{Var}(c) = 0$ (3.35)
2. $\text{Var}(cX) = c^2\text{Var}(X)$ (3.36)

Example 3.10

Consider the experiment of throwing an "honest" die, letting X denote the number of spots on the upward-landing face:

$$\text{Var}(X) = \sum_{i=1}^{6} [x_i - E(X)]^2 P(X = x_i)$$

$$= (1 - 3.5)^2 \cdot \tfrac{1}{6} + (2 - 3.5)^2 \cdot \tfrac{1}{6} + (3 - 3.5)^2 \cdot \tfrac{1}{6} + (4 - 3.5)^2 \cdot \tfrac{1}{6}$$

$$+ (5 - 3.5)^2 \cdot \tfrac{1}{6} + (6 - 3.5)^2 \cdot \tfrac{1}{6} = 2\tfrac{11}{12}$$

Example 3.11

It is readily shown that for a random variable with Cauchy distribution, no variance is defined (see also Example 3.8).

Example 3.12

The variance of a random variable X with Laplace distribution is (see Example 3.9)

$$\text{Var}(X) = \int_{-\infty}^{\infty} \frac{a}{2} x^2 e^{-a|x|} \, dx = \frac{2}{a^2}$$

Example 3.13

If the probability density function of a continuous random variable X is given by

$$f_X(x) = \begin{cases} \dfrac{1}{b-a}, & a \leqslant x \leqslant b \\ 0, & \text{otherwise} \end{cases}$$

that is, if it represents a rectangular "pulse," we say that X is uniformly distributed in the interval (a, b).

Using the definition of a distribution function, Eq. (3.7), we obtain

$$F_X(x) = \begin{cases} 0, & x < a \\ \dfrac{x-a}{b-a}, & a \leqslant x < b \\ 1, & b \leqslant x \end{cases}$$

(see Fig. 3.8), and the kth moment is

$$m_k = \frac{1}{b-a} \int_a^b x^k \, dx = \frac{b^{k+1} - a^{k+1}}{(k+1)(b-a)}$$

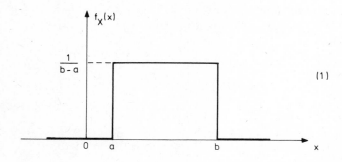

(1)

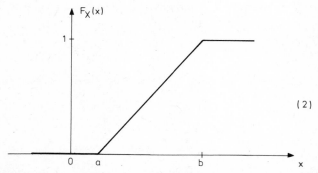

(2)

Fig. 3.8. Probability density function (1) and distribution function (2) of random variable uniformly distributed in interval (a, b).

In particular, the mean is given by

$$E(X) = m_1 = \frac{a+b}{2}$$

the kth central moment is

$$\mu_k = \frac{1}{b-a} \int_a^b \left(x - \frac{a+b}{2} \right)^k dx = \begin{cases} 0, & k \text{ odd} \\ \dfrac{(b-a)^k}{2^k(k+1)}, & k \text{ even} \end{cases}$$

and the variance is

$$\text{Var}(X) = \mu_2 = \frac{(b-a)^2}{12}$$

This value is numerically equal to the central moment of inertia of a rectangular cross section with width $1/(b-a)$ and depth $b-a$.

The *absolute kth moment* is defined by

$$\tilde{m}_k = E(|X|^k) = \int_{-\infty}^{\infty} |x|^k f_X(x)\, dx \tag{3.37}$$

and the *absolute central kth moment* by

$$\tilde{\mu}_k = E[|X - E(X)|^k] = \int_{-\infty}^{\infty} |x - E(X)|^k f_X(x)\, dx \tag{3.38}$$

The generalized kth moments are defined, respectively, by

$$\bar{\mu}_k = E[(X-a)^k], \qquad \bar{\tilde{\mu}}_k = E[|X-a|^k] \tag{3.39}$$

The *median* of a continuous random variable X, denoted by med(X), is defined as the smallest root of the equation

$$F_X[\text{med}(X)] = 0.5$$

Note that the curves $y = F_X(x)$ and $y = 0.5$ may have more than one common point, the minimal of them being the median. The median has the following property:

If the absolute generalized moment $E(|X-a|)$ of a continuous random variable X is treated as a function of a, it has a minimum at $a = \text{med}(X)$.

This property follows immediately from:

$$E(|X - a|) = \begin{cases} E(|X - \text{med}(X)|) + 2\int_{\text{med}(X)}^{a} (a - x)f_X(x)\, dx, \\ \quad a > \text{med}(X) \\ E(|X - \text{med}(X)|) + 2\int_{a}^{\text{med}(X)} (x - a)f_X(x)\, dx, \\ \quad a < \text{med}(X) \end{cases}$$

The *qth quantile* of a continuous random variable X is defined as a smallest root of the equation

$$F_X(\xi_q) = q, \quad \text{for } 0 < q < 1$$

Obviously, the median is the 0.5th quantile.

The point at which $f_X(x)$ attains its maximum is called its *mode*.

Consider now the third central moment

$$\mu_3 = \int_{-\infty}^{\infty} [x - E(X)]^3 f_X(x)\, dx$$

$$= \int_{-\infty}^{\infty} y^3 f_X[y + E(X)]\, dy$$

$$= \int_{0}^{\infty} y^3 \{ f_X[E(X) + y] - f_X[E(X) - y] \}\, dy$$

where $y = x - E(X)$.

If the probability density $f_X(x)$ is *symmetric* about $E(X)$, then

$$f_X[E(X) + y] = f_X[E(X) - y]$$

and consequently $\mu_3 = 0$. Therefore, if the probability density function is symmetric about the mean, the third central moment vanishes. The opposite is not always valid: For the probability distribution to be symmetric about the mean, all odd central moments must vanish. Despite this, the third moment μ_3 is sometimes regarded as a measure of asymmetry or *skewness*, the *coefficient of skewness* being defined as the ratio $\gamma_{1X} = \mu_3/\mu_2^{3/2} = \mu_3/\sigma_X^3$. Figure 3.9 shows examples of probability densities with positive, zero, and negative coefficients of skewness, respectively.

The fourth central moment μ_4 is used to estimate the steepness of the peak of the probability density near its center. The value

$$\gamma_{2X} = \frac{\mu_4}{\sigma_X^4} - 3$$

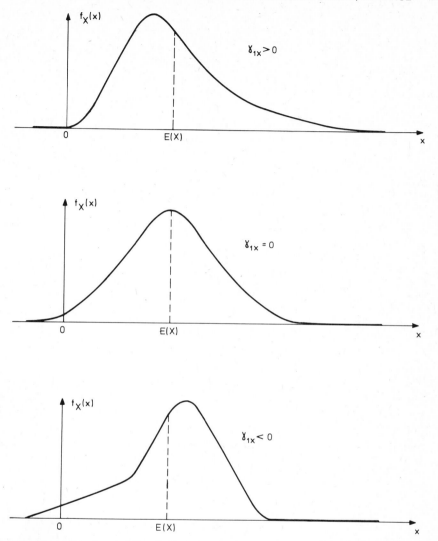

Fig. 3.9. Examples of probability density functions with positive, zero, and negative coefficients of skewness.

is called the *coefficient of excess* or *kurtosis*. γ_{2x} is constructed so as to equal zero for the very important normal probability density curve (see Sec. 4.10). Probability density curves that are flatter at the peak than the normal have *negative* kurtosis, while those that are steeper than the latter have *positive* kurtosis. Figure 3.10 shows an example of a normal distribution curve ($\gamma_{2x} = 0$) and probability densities with positive and negative kurtosis, respectively.

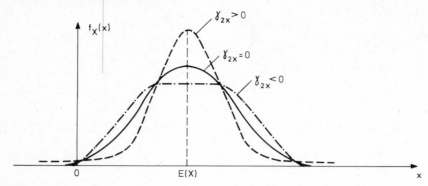

Fig. 3.10. Examples of probability density functions with positive, zero, and negative coefficients of kurtosis.

3.6   CHARACTERISTIC FUNCTION*

The *characteristic function* of a random variable X, denoted by $M_X(\theta)$, is defined as the mathematical expectation of the complex variable, $\exp(i\theta X)$, treated as a function of θ:

$$M_X(\theta) = E\left[e^{i\theta X}\right] = \int_{-\infty}^{\infty} e^{i\theta x} f_X(x)\, dx \tag{3.40}$$

in other words, the characteristic function is the Fourier transform of the probability density function. Since $|e^{i\theta x}| = 1$ for every real θ and x, due to a property of the probability density function (3.11), the characteristic function does not exceed unity in its absolute value and equals unity at $\theta = 0$. Since $f_X(x)$ is nonnegative, the probability density function can be found as the inverse Fourier transform of the characteristic function

$$f_X(x) = \frac{1}{2\pi} \int_{-\infty}^{\infty} M_X(\theta) e^{-i\theta x}\, d\theta \tag{3.41}$$

Example 3.14

The characteristic function of a random variable, distributed uniformly in the interval (a, b), equals

$$M_X(\theta) = \frac{1}{b-a} \int_a^b e^{i\theta x}\, dx = \frac{1}{i\theta(b-a)} \left(e^{i\theta b} - e^{i\theta a} \right)$$

Example 3.15

The characteristic function of a random variable with Laplace distribution equals

$$M_X(\theta) = \frac{a}{2} \int_{-\infty}^{\infty} e^{i\theta x - a|x|}\, dx = \frac{a^2}{a^2 + \theta^2}$$

*This section may be omitted at first reading.

Example 3.16

The characteristic function $M_Y(\theta)$ of a random variable $Y = aX + b$, where a and b are real numbers and X is a continuous random variable, is found as

$$M_Y(\theta) = M_X(a\theta)e^{ib\theta} \qquad (3.42)$$

since

$$M_Y(\theta) = E(e^{i\theta Y}) = E[e^{i\theta(aX+b)}] = E[e^{ia\theta X}e^{i\theta b}] = M_X(a\theta)e^{i\theta b}$$

Here property (3.26) of the mathematical expectation was used.

If a random variable X has an absolute moment of kth order, its characteristic function is differentiable k times. Conversely, the value of the kth derivative of the characteristic function $M_X(\theta)$ at $\theta = 0$ determines the kth moments m_k of a random variable X. Indeed, after k-fold differentiation of $M_X(\theta)$ we obtain

$$\frac{d^k M_X(\theta)}{d\theta^k} = i^k \int_{-\infty}^{\infty} x^k e^{i\theta x} f_X(x)\, dx$$

This derivative can be estimated by its absolute value

$$\left| \frac{d^k M_X(\theta)}{d\theta^k} \right| = \left| \int_{-\infty}^{\infty} x^k e^{i\theta x} f_X(x)\, dx \right| \leqslant \int_{-\infty}^{\infty} \left| x^k e^{i\theta x} f_X(x) \right| dx = \int_{-\infty}^{\infty} |x|^k f_X(x)\, dx$$

Since the latter integral is finite by assumption, the derivative $d^k M_X(\theta)/d\theta^k$ exists.

Equation (3.42) yields

$$\left[\frac{d^k M_X(\theta)}{d\theta^k} \right]_{\theta=0} = i^k \int_{-\infty}^{\infty} x^k f_X(x)\, dx = i^k m_k$$

Hence

$$m_k = \frac{1}{i^k} \left[\frac{d^k M_X(\theta)}{d\theta^k} \right]_{\theta=0} \qquad (3.43)$$

The mean of a random variable equals, then,

$$m_1 = E(X) = \frac{1}{i} \left[\frac{dM_X(\theta)}{d\theta} \right]_{\theta=0} \qquad (3.44)$$

If all derivatives of the characteristic function $M_X(\theta)$ exist at $\theta = 0$, the function can be represented by a Maclaurin series expansion

$$M_X(\theta) = \sum_{k=0}^{\infty} \frac{1}{k!} \left[\frac{d^k M_X(\theta)}{d\theta^k} \right]_{\theta=0} \theta^k$$

or, with Eq. (3.43) taken into account, we obtain

$$M_X(\theta) = 1 + \sum_{k=1}^{\infty} \frac{(i\theta)^k}{k!} m_k \tag{3.45}$$

The principal value of the logarithm of a characteristic function is called a *log-characteristic function*:

$$\psi_X(\theta) = \ln M_X(\theta) \tag{3.46}$$

Let us differentiate $\psi_X(\theta)$ and put $\theta = 0$. We have

$$\left[\frac{d\psi_x(\theta)}{d\theta} \right]_{\theta=0} = \frac{1}{M_X(0)} \left[\frac{dM_X(\theta)}{d\theta} \right]_{\theta=0} = im_1 = iE(X)$$

$$\left[\frac{d^2\psi_X(\theta)}{d\theta^2} \right]_{\theta=0} = \frac{1}{M_X^2(0)} \left\{ \frac{d^2 M_X(\theta)}{d\theta^2} M_X(\theta) - \left[\frac{dM_X(\theta)}{d\theta} \right]^2 \right\}_{\theta=0}$$

$$= -m_2 + m_1^2 = -\text{Var}(X)$$

We used here the fact that $M_X(0) = 1$ and Eq. (3.43). As a result we obtain

$$E(X) = -i\left[\frac{d\psi_X(\theta)}{d\theta} \right]_{\theta=0} \tag{3.47}$$

$$\text{Var}(X) = -\left[\frac{d^2\psi_X(\theta)}{d\theta^2} \right]_{\theta=0} = -\left[\frac{d^2 \ln M_X(\theta)}{d\theta^2} \right]_{\theta=0} \tag{3.48}$$

Further differentiation of $\psi_X(\theta)$ yields

$$\left[\frac{d^3\psi_X(\theta)}{d\theta^3} \right]_{\theta=0} = -\frac{m_3 - 3m_1 m_2 + 2m_1^3}{i^3}$$

$$\left[\frac{d^4\psi_X(\theta)}{d\theta^4} \right]_{\theta=0} = m_4 - 4m_1 m_3 - 3m_2^2 + 12m_1^2 m_2 - 6m_1^4$$

For the coefficients of skewness and kurtosis, we obtain, respectively,

$$\gamma_{1X} = \frac{\psi_X^{III}(0)}{\left[\psi_X^{II}(0)\right]^{3/2}} \tag{3.49}$$

$$\gamma_{2X} = \frac{\psi_X^{(IV)}(0)}{\left[\psi_X^{II}(0)\right]^2} \tag{3.50}$$

The value $i^k d^k \psi_X(\theta)/d\theta^k$ is called the kth *cumulant* or *semiinvariant* of X. With semiinvariants known, the various moments of a random variable are readily obtainable. It should be noted that moments (if they exist) are uniquely determined through the probability density function (or the characteristic function, or the log-characteristic function). Accordingly, the question arises whether the set of moments determines uniquely the probability density function of a random variable, and the answer is generally no; this is known as the *problem of moments* and will not be discussed here. (The reader may consult, for example, Kendall and Stuart.)

3.7 CONDITIONAL PROBABILITY DISTRIBUTION AND DENSITY FUNCTIONS

First we recall Eq. (2.10), which states that if B is an event with nonzero probability, then the conditional probability of an event A, knowing that B has taken place, is

$$P(A|B) = \frac{P(AB)}{P(B)}$$

The *conditional probability distribution function* of the random variable X under condition B is defined as the probability of the event $\{X \leqslant x\}$:

$$F_X(x|B) = P(X \leqslant x|B) = \frac{P(X \leqslant x, B)}{P(B)} \tag{3.51}$$

where $\{X \leqslant x, B\}$ is the product of events $\{X \leqslant x\}$ and B.

All properties of the unconditional probability distribution functions are preserved in the conditional distribution function:

$$F_X(+\infty|B) = 1, \qquad F_X(-\infty|B) = 0$$

and

$$P(x_1 < X \leqslant x_2|B) = F_X(x_2|B) - F_X(x_1|B) \tag{3.52}$$

If X is a continuous random variable, we will define the *conditional probability density function* $f_X(x|B)$ as the derivative of $F_X(x|B)$:

$$f_X(x|B) = \frac{dF_X(x|B)}{dx} = \lim_{\Delta x \to 0} \frac{P\{x \leqslant X \leqslant x + \Delta x|B\}}{\Delta x}$$

If events $B_1, B_2, \ldots, B_n$ partition the sample space Ω, we have, by (2.25),

$$P(X \leqslant x) = P(X \leqslant x|B_1)P(B_1) + P(X \leqslant x|B_2)P(B_2)$$
$$+ \cdots + P(X \leqslant x|B_n)P(B_n)$$

In view of the definition (3.51), we arrive at

$$F_X(x) = \sum_{i=1}^{n} F_X(x|B_i)P(B_i) \tag{3.53}$$

Example 3.17

Assume that the event B is the following:

$$\{B\} = \{X \leqslant a\}, \qquad P(X \leqslant a) = F_X(a) \neq 0$$

We seek the conditional probability distribution function $F_X(x|X \leqslant a)$:

$$F_X(x|X \leqslant a) = P(X \leqslant x|X \leqslant a)$$

Note that if $x \geqslant a$, then the event $(X \leqslant x|X \leqslant a)$ is a certain one and

$$F_X(x|X \leqslant a) = 1$$

If $x < a$, then

$$F_X(x|X \leqslant a) = P(X \leqslant x|X \leqslant a)$$

$$= \frac{P\{(X \leqslant x),(X \leqslant a)\}}{P(X \leqslant a)} = \frac{P(X \leqslant x)}{P(X \leqslant a)} = \frac{F_X(x)}{F_X(a)}$$

The conditional probability density function is found by differentiating the latter equation:

$$f_X(x|X \leqslant a) = \begin{cases} \dfrac{f_X(x)}{F_X(a)}, & x < a \\ 0, & x \geqslant a \end{cases}$$

As an example, consider the random variable distributed uniformly in the interval $(0, 2a)$. Then

$$F_X(x) = \begin{cases} 0, & x < 0 \\ \dfrac{x}{2a}, & 0 \leqslant x < 2a \\ 1, & 2a \leqslant x \end{cases}$$

The conditional probability distribution is (see Fig. 3.11)

$$F_X(x|X \le a) = \begin{cases} 0, & x < 0 \\ \dfrac{x}{a}, & 0 \le x < a \\ 1, & a \le x \end{cases}$$

since $F_X(a) = \frac{1}{2}$.

The *conditional mathematical expectation* of a random variable X, under condition B, is defined as

$$E(X|B) = \int_{-\infty}^{\infty} x f_X(x|B)\, dx \qquad (3.54)$$

where $f_X(x|B)$ is the conditional probability density function. Differentiating Eq. (3.53) and taking (3.54) into account, we obtain the relation between the unconditional mathematical expectation $E(X)$ and the conditional expecta-

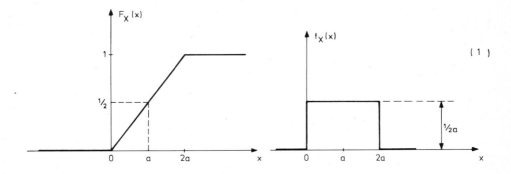

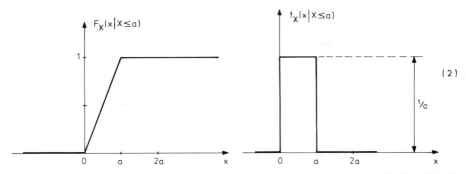

Fig. 3.11. (1) Unconditional probability distribution and density functions. (2) Conditional probability distribution and density functions.

tions $E(X|B_i)$ under conditions B_i, which are respectively random variables partitioning Ω:

$$E(X) = \sum_{i=1}^{n} E(X|B_i)P(B_i) \tag{3.55}$$

3.8 INEQUALITIES OF BIENAYMÉ AND TCHEBYCHEFF

Let Y be a random variable taking on only nonnegative values. Then the following inequality is valid, provided α is a positive constant:

$$P(Y \geqslant \alpha) \leqslant \frac{E(Y)}{\alpha} \tag{3.56}$$

Indeed, if $f_Y(y)$ is a probability density function of a continuous random variable Y

$$E(Y) = \int_{-\infty}^{\infty} yf_Y(y)\, dy = \int_{0}^{\infty} yf_Y(y)\, dy = \int_{0}^{\alpha} yf_Y(y)\, dy + \int_{\alpha}^{\infty} yf_Y(y)\, dy$$

$$\geqslant \int_{\alpha}^{\infty} yf_Y(y)\, dy \geqslant \alpha \int_{\alpha}^{\infty} f_Y(y)\, dy = \alpha P(Y \geqslant \alpha)$$

which was to be proved. Here the fact was used that $f_Y(y)$ vanishes for negative values of y. Consider now the random variable $Y = |X - a|^n$, where n is a positive integer. This random variable takes on only nonnegative values irrespective of the sign of X. Therefore for $\alpha = \varepsilon^n$ ($\varepsilon > 0$),

$$P(|X - a|^n \geqslant \varepsilon^n) \leqslant \frac{E[|X - a|^n]}{\varepsilon^n}$$

that is,

$$P(|X - a| \geqslant \varepsilon) \leqslant \frac{E[|X - a|^n]}{\varepsilon^n} \tag{3.57}$$

This inequality, named after Bienaymé, signifies that

$$P(|X - a| \geqslant \varepsilon) \leqslant \min_{n} \frac{E[|X - a|^n]}{\varepsilon^n}$$

If a random variable has a finite variance σ_X^2, then we may put $a = E(X)$, $n = 2$, and $\varepsilon = k\sigma_X$ in the inequality to get

$$P(|X - E(X)| \geqslant k\sigma_X) \leqslant \frac{1}{k^2} \tag{3.58}$$

This inequality, named after Tchebycheff, signifies also that

$$P(|X - E(X)| < k\sigma_X) > 1 - \frac{1}{k^2}$$

For example, for $k = 2$ we obtain

$$P\{E(X) - 2\sigma_X < X < E(X) + 2\sigma_X\} > \tfrac{3}{4}$$

for any random variable X with finite variance. For $k = 3$,

$$P[E(X) - 3\sigma_X < X < E(X) + 3\sigma_X] > \tfrac{8}{9}$$

For any random variable X with finite variance, the latter inequality signifies that the probability of X falling within three standard deviations of its mean is at least $\tfrac{8}{9}$. This bound is independent of the distribution of X, provided it has a finite variance.

PROBLEMS

3.1. A spring-mass system is subjected to harmonic sinusoidal excitation with specified frequency ω. Suppose the spring coefficient is a random variable with given probability distribution $F_K(k)$, where the mass m is a specified quantity. What is the probability of no resonance occurring in a system picked up at random?

3.2. A random variable X is said to have a *triangular distribution* if its density is

$$f_X(x) = \begin{cases} A\left(1 - \dfrac{|x|}{a}\right), & -a < x < a \\ 0, & \text{otherwise} \end{cases}$$

Find:

(a) the value of A,
(b) the mean value, $E(X)$
(c) the median, med(X)
(d) the *coefficient of variation*, $\gamma_X = \sigma_X/E(X)$
(e) the 0.9th quantile, $\xi_{0.9}$.

3.3. A random variable X is said to have a *beta distribution* if its density is

$$f_X(x) = \begin{cases} Ax^{\alpha-1}(1 - x)^{\beta-1}, & 0 < x < 1 \\ 0, & \text{otherwise} \end{cases}$$

Find

(a) the value of A
(b) the distribution function associated with this probability density
(c) the probability of X taking on values less than 0.1
(d) the values of α and β for which a uniformly distributed random variable is obtained from X.

3.4. The *hazard function* $h(t)$ is defined as the conditional instantaneous failure rate. That is, $h(t)\,dt$ is the probability of the system failing in the time interval $(t, t + dt)$, given that the system has not failed prior to time t. It may be interpreted as the rate at which the system population still under test at time t is failing. Let T denote the random time of failure with density $f_T(t)$, such that $f_T(t)\,dt$ represents the probability of the system failing in the interval $(t, t + dt)$.

(a) Show that $h(t) = f_T(t|T \geqslant t)$.
(b) Show that the hazard function at time t equals the probability density of the time of failure divided by the reliability $R(t)$, the probability of the system surviving up to t:

$$h(t) = \frac{f_T(t)}{R(t)}, \quad f_T(t) = F_T'(t), \quad R(t) = 1 - F_T(t)$$

(c) Verify that

$$R(t) = \left[1 - F_T(0)\right]\exp\left[-\int_0^t h(t)\,dt\right]$$

$$f_T(t) = \left[1 - F_T(0)\right]h(t)\exp\left[-\int_0^t h(t)\,dt\right]$$

where $F_T(0)$ is the probability of failure at $t = 0$.

Remark Since $0 \leqslant R \leqslant 1$, $h(t) \geqslant f_T(t)$. By analogy, the probability of a specimen subjected to a fatigue test with sufficiently high amplitude of the repeated load fracturing between 10^9 cycles and $10^9 + 10$ cycles [corresponding to $f_T(t)\,dt$] is very small. The probability of fracture in the same interval, provided the specimen survived up to 10^9 cycles [corresponding to $h(t)\,dt$] is much higher.

3.5. Find the probability density of the time of failure $f_T(t)$, if the hazard function is constant $h(t) = a$, and the probability of initial failure is zero, $F_T(0) = 0$. Show that a is the reciprocal of the mean time of failure, $E(T)$.

3.6. Find the conditional probability of a system failing in the time interval (t_1, t_2), assuming that it did not fail prior to time t_1; $h(t) = a$.

3.7. Find the conditional probability of a system surviving in the time interval (t_1, t_2) assuming that it survived up to time t_1 [probability of prolongation by an additional time interval $\Delta t = t_2 - t_1$]; $h(t) = a$.

3.8. Verify that if X is uniformly distributed in the interval (a, b), the probability of $X \leqslant a + p(b - a)$, where $0 < p < 1$, equals p.

3.9. Following the steps used in the text, prove the inequalities of Bienaymé and Tchebycheff for a discrete random variable.

3.10. X is uniformly distributed in the interval [8, 12]. Calculate the probability $P\{E(X) - \sigma_X < X < E(X) + \sigma_X\}$ and compare it with the upper bound furnished by Tchebycheff's inequality.

3.11. Show, using Tchebycheff's inequality, that if $E[(X - a)^2] = 0$, where a is a deterministic constant, then $X = a$ with unity probability.

CITED REFERENCES

Crandall, S. H., Dahl, N. C., and Lardner, T. J., Eds., *An Introduction to the Mechanics of Solids*, 2nd ed., McGraw-Hill, New York, 1969, pp. 164–169.

Kendall, M. G., and Stuart, A., *The Advanced Theory of Statistics*, Vol. 1, *Distribution Theory*, Charles Griffin, London, 1958, pp. 109–115.

Popov, E., *Introduction to Mechanics of Solids*, SI ed., McGraw-Hill, New York, 1975, pp. 47–53.

RECOMMENDED FURTHER READING

Gelfand, I. M., and Shilow, G. E., *Generalized Functions*, Vol. 1, Academic Press, New York, 1964. Chap. 1: Definition and Simplest Properties of Generalized Functions, pp. 1–18; Chap. 2: Differentiation and Integration of Generalized Functions, pp. 18–44.

Gnedenko, B. N., *Theory of Probability*, Chelsea Publ. Co., New York, 1962. Chap. 4: Random Variables and Distribution Functions, pp. 155–200; Chap. 5: Numerical Characteristics of Random Variables, pp. 201–235; Chap. 6: The Law of Large Numbers, pp. 236–265; Chap. 7: Characteristic Functions, pp. 266–301.

Leighthill, M. J., *Introduction to Fourier Analysis and Generalised Functions*, Students' ed., Cambridge at the University Press, 1962. Chap. 1: Introduction, pp. 1–14; Chap. 2: The Theory of Generalised Functions and Their Fourier Transforms, pp. 15–29.

Papoulis, A., *Probability, Random Variables, and Stochastic Processes*, International Student ed., McGraw-Hill, Kogakusha, Tokyo, 1965, Chap. 4: The Concept of a Random Variable, pp. 83–115; Chap. 5: Function of One Random Variable, Sec. 5. 5-4, 5-5, pp. 138–164.

chapter 4

Examples of Probability Distribution and Density Functions. Functions of a Single Random Variable

In this chapter we present some widely used discrete and continuous probability distribution and density functions.

4.1 CAUSAL DISTRIBUTION

In order to represent the constant c probabilistically, we use a *causally distributed* random variable with a probability density function (Fig. 4.1)

$$f_X(x) = \delta(x - c) \tag{4.1}$$

that is, the random variable X takes on the value c with probability unity. The distribution function is readily obtained by integration of (4.1):

$$F_X(x) = U(x - c)$$

The mathematical expectation is obviously

$$E(X) = c$$

and the variance

$$\text{Var}(X) = 0$$

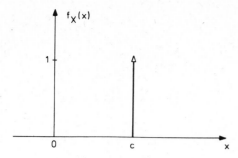

Fig. 4.1. Probability density of causally distributed random variable represented by Dirac's delta function.

The characteristic function is

$$M_X(\theta) = E(e^{i\theta X}) = e^{ic\theta}$$

4.2   DISCRETE UNIFORM DISTRIBUTION

A random variable X has a discrete uniform distribution, if its probability density function reads (Fig. 4.2)

$$f_X(x) = \frac{1}{n} \sum_{i=1}^{n} \delta(x - c_i) \tag{4.2}$$

that is, X can take on only values $c_1, c_2, \ldots, c_n$ each with probability $1/n$. The distribution function is

$$F_X(x) = \frac{1}{n} \sum_{i=1}^{n} U(x - c_i)$$

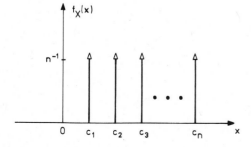

Fig. 4.2. Probability density of random variable with discrete uniform distribution represented by combination of Dirac's delta functions.

The mathematical expectation is

$$E(X) = \sum_{i=1}^{n} c_i P(X = c_i)$$

For the particular case $c_i \equiv i$, we get for the mathematical expectation

$$E(X) = \sum_{i=1}^{n} i \frac{1}{n} = \frac{n+1}{2}$$

and for the variance

$$\text{Var}(X) = E(X^2) - [E(X)]^2 = \sum_{i=1}^{n} i^2 \frac{1}{n} - \left(\frac{n+1}{2}\right)^2$$

$$= \frac{n(n+1)(2n+1)}{6n} - \frac{(n+1)^2}{4} = \frac{n^2-1}{12}$$

4.3 BINOMIAL OR BERNOULLI DISTRIBUTION

Independent trials, each of which involves an event A, occurring with positive probability $p = P(A)$, are called *Bernoulli trials*. The event itself is referred to as a "success," and the complementary event $\bar{A}$, which occurs in each of the trials with probability $q = 1 - p$, as a "failure." In other words, $p = P(\text{success})$, $q = P(\text{failure})$.

If n trials are considered, then each elementary outcome ω can be described by a mixed sequence of "successes" and "failures," for example, $(s, s, f, f, f, s, \ldots, f, s)$. The probability $P(\omega)$ of each elementary outcome ω, at which there are exactly m successes and $n - m$ failures, is given, in view of the mutual independence of the outcomes, by

$$P(\omega) = p^m q^{n-m}$$

As can be seen, the elementary outcomes are equiprobable if $p = q = \frac{1}{2}$.

Consider now the random variable X, which denotes the number of successes in n Bernoulli trials. $X(\omega) = m$ if an elementary outcome indicates exactly m successes. The number of different outcomes ω resulting in m successes in *any sequence* equals that of combinations of m s's and $n - m$ f's, which in turn equals that of combinations of m objects drawn from an ensemble of n objects:

$$\binom{n}{m} = \frac{n!}{m!(n-m)!}$$

All these outcomes have the same probability $P(\omega)$, hence the event $X = m$ has the probability

$$P(X = m) = \binom{n}{m} p^m q^{n-m}, \quad m = 0, 1, 2, \ldots, n \tag{4.3}$$

The probability density function of X is

$$f_X(x) = \sum_{m=0}^{n} \binom{n}{m} p^m q^{n-m} \delta(x - m) \tag{4.4}$$

Note that since all possible mutually exclusive outcomes of n trials consist in a success occurring 0 times, once, twice, ..., n times, it is obvious that

$$\sum_{m=0}^{n} P(X = m) = 1$$

which also follows from the equality

$$\sum_{m=0}^{n} \binom{n}{m} p^m q^{n-m} = (p + q)^n = 1$$

The mathematical expectation of X is

$$E(X) = \sum_{m=0}^{n} m \binom{n}{m} p^m q^{n-m} = \sum_{m=1}^{n} m \binom{n}{m} p^m q^{n-m}$$

Since, however,

$$\binom{n}{m} = \binom{n-1}{m-1} \frac{n}{m}$$

we have

$$E(X) = np \sum_{m=1}^{n} \binom{n-1}{m-1} p^{m-1} q^{n-1-(m-1)}$$

$$= np \sum_{k=0}^{n-1} \binom{n-1}{k} p^k q^{n-1-k} = np$$

The variance is $\mathrm{Var}(X) = E(X^2) - [E(X)]^2$, so that

$$E(X^2) = \sum_{m=1}^{n} m^2 \binom{n}{m} p^m q^{n-m} = np \sum_{m=1}^{n} m \binom{n-1}{m-1} p^{m-1} q^{n-1-(m-1)}$$

$$= np \sum_{k=0}^{n-1} (k+1) \binom{n-1}{k} p^k q^{n-k-1}$$

$$= np \sum_{k=0}^{n-1} k \binom{n-1}{k} p^k q^{n-k-1} + np \sum_{k=0}^{n-1} \binom{n-1}{k} p^k q^{n-k-1}$$

$$= np[(n-1)p] + np \tag{4.5}$$

It was taken into account that the first sum in Eq. (4.5) represents the mathematical expectation of X in $n-1$ trials, and the second sum equals unity as the probability of a certain event. As a result,

$$\mathrm{Var}(X) = np(n-1)p + np - (np)^2 = np(1-p) = npq$$

Example 4.1

A device consists of five components, the reliability of which equals p and the failures of which are mutually independent. Then the probability of none of the components failing equals $P_0 = p^5$, that of at least one of the components failing equals $1 - p^5$. The probability of exactly one component failing is

$$P_1 = \binom{5}{4} p^4 q = 5p^4 q$$

that of two components failing is

$$P_2 = \binom{5}{3} p^3 q^2 = 10p^3 q^2$$

That of at least two components failing is

$$1 - P_0 - P_1 = 1 - p^5 - 5p^4 q$$

4.4   POISSON DISTRIBUTION

A random variable X is said to have a *Poisson distribution* if it takes on values $0, 1, 2, \ldots$, with probabilities

$$P(X = m) = \frac{a^m}{m!} e^{-a}, \quad m = 0, 1, \ldots \tag{4.6}$$

where a is a positive constant.

The probability density function is

$$f_X(x) = e^{-a} \sum_{m=0}^{\infty} \frac{a^m}{m!} \delta(x - m)$$

The characteristic and the log-characteristic functions are, respectively,

$$M_X(\theta) = E(e^{i\theta X}) = e^{-a} \sum_{m=0}^{\infty} e^{i\theta m} \frac{a^m}{m!} = e^{-a} \sum_{m=0}^{\infty} \frac{(ae^{i\theta})^m}{m!}$$

$$= e^{-a} e^{ae^{i\theta}} = e^{ae^{i\theta} - a},$$

and

$$\psi_X(\theta) = ae^{i\theta} - a$$

The mathematical expectation and variance are obtainable, for example, from Eqs. (3.47) and (3.48), respectively,

$$E(X) = -i\psi'_X(0) = a$$

$$\mathrm{Var}(X) = -\psi''_X(0) = a$$

The Poisson distribution may be regarded as a limiting case of the binomial distribution, where the number of trials is large and the probability of success very small but the mean number of successes $a = np$ not too small, in which case it can be shown that

$$\binom{n}{m} p^m q^{n-m} \sim \frac{a^m}{m!} e^{-a}$$

Indeed, as is known

$$\lim_{n \to \infty} \left(1 - \frac{a}{n}\right)^n = e^{-a}$$

and since $p = a/n$, we obtain from Eq. (4.3)

$$P(X = 0) = q^n = (1 - p)^n = \left(1 - \frac{a}{n}\right)^n \sim e^{-a}$$

and moreover,

$$\frac{P(X = m)}{P(X = m - 1)} = \frac{np - (m - 1)p}{mq} \sim \frac{a}{m}, \quad m = 1, 2, \ldots$$

when $n \to \infty$. Therefore,

$$P(X = 1) \sim \frac{a}{1} P(X = 0) \sim \frac{a}{1} e^{-a}$$

$$P(X = 2) \sim \frac{a}{2} P(X = 1) \sim \frac{a^2}{1 \cdot 2} e^{-a}$$

$$\vdots$$

$$P(X = m) \sim \frac{a}{m} P(X = m - 1) \sim \frac{a^m}{m!} e^{-a}$$

which is the Poisson distribution.

4.5   RAYLEIGH DISTRIBUTION

A continuous random variable X is said to have a *Rayleigh distribution* if its probability density function is given by (Fig. 4.3)

$$f_X(x) = \begin{cases} 0, & x < 0 \\ \dfrac{x}{a^2} e^{-x^2/2a^2}, & x \geqslant 0 \end{cases} \tag{4.7}$$

with parameter a^2.

The distribution function is

$$F_X(x) = \left[1 - e^{-x^2/2a^2} \right] U(x)$$

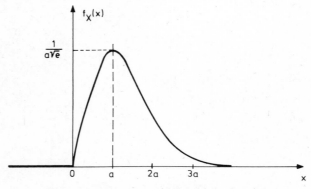

Fig. 4.3.   Rayleigh probability density function.

The mathematical expectation and the variance are, respectively,

$$E(X) = \frac{a\sqrt{\pi}}{\sqrt{2}} \simeq 1.25a \qquad \text{Var}(X) = \frac{4-\pi}{2}a^2 \simeq 0.43a^2$$

The Rayleigh distribution is an example of a nonsymmetric distribution. The third moment m_3 is

$$m_3 = \frac{1}{a^2}\int_0^\infty x^4 e^{-x^2/2a^2}\,dx = 3a^3\left(\frac{\pi}{2}\right)^{1/2}$$

and the third central moment is

$$\mu_3 = 3a^3\left(\frac{\pi}{2}\right)^{1/2} - 3a\left(\frac{\pi}{2}\right)^{1/2}2a^2 + 2a^3\left(\frac{\pi}{2}\right)^{3/2}$$

$$= (\pi - 3)\left(\frac{\pi}{2}\right)^{1/2}a^3$$

The coefficient of skewness is

$$\gamma_{1X} = 2\left(\frac{\pi-3}{4-\pi}\right)\left(\frac{\pi}{4-\pi}\right)^{1/2} \simeq 0.63$$

4.6   EXPONENTIAL DISTRIBUTION

A random variable X is said to have an *exponential distribution* if its density $f_X(x)$ is given by

$$f_X(x) = \begin{cases} 0, & x < 0 \\ ae^{-ax}, & x \geqslant 0,\ a > 0 \end{cases} \qquad (4.8)$$

the distribution function being

$$F_X(x) = [1 - e^{-ax}]U(x)$$

and the characteristic function

$$M_X(\theta) = \frac{a}{a - i\theta}$$

The mathematical expectation and the variance are, respectively,

$$E(X) = \frac{1}{a} \quad \text{and} \quad \text{Var}(X) = \frac{1}{a^2}$$

4.7 χ^2 (CHI-SQUARE) DISTRIBUTION WITH m DEGREES OF FREEDOM

A random variable is said to have a chi-square distribution if its density is given by

$$f_X(x) = \begin{cases} 0, & x < 0 \\ \dfrac{e^{-x/2}x^{\lambda-1}}{2^\lambda\Gamma(\lambda)}, & x \geqslant 0, \lambda = m/2 \end{cases} \tag{4.9}$$

where m is the number of degrees of freedom, and $\Gamma(x)$ is a gamma function defined as

$$\Gamma(x) = \int_0^\infty e^{-t}t^{x-1}\,dt, \quad x > 0$$

For $x > 1$ we obtain, after integration by parts,

$$\Gamma(x) = (x-1)\Gamma(x-1)$$

Moreover, $\Gamma(1) = 1$; hence, for x a positive integer, we have

$$\Gamma(x) = (x-1)!$$

The mathematical expectation and the variance are, respectively,

$$E(X) = m \quad \text{and} \quad \text{Var}(X) = 2m$$

4.8   GAMMA DISTRIBUTION

A random variable X is said to have a *gamma distribution* if its density is given by (Fig. 4.4)

$$f_X(x) = \begin{cases} 0, & x < 0 \\ \dfrac{1}{\beta^{\alpha+1}\Gamma(\alpha+1)}x^\alpha e^{-x/\beta}, & x > 0 \end{cases} \tag{4.10}$$

where $\alpha > -1$, $\beta > 0$ are constants. For positive integer α and $x \geqslant 0$, Eq. (4.10) takes the form

$$f_X(x) = \frac{x^\alpha e^{-x/\beta}}{\alpha!\beta^{\alpha+1}}U(x)$$

The characteristic function reads

$$M_X(\theta) = \frac{1}{(1 - i\theta\beta)^{\alpha+1}}$$

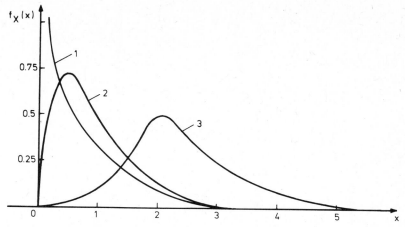

Fig. 4.4. Gamma probability density function 1) $\alpha = -1/2, \beta = 1, 2)\ \alpha = 1, \beta = 1/2, 3)\ \alpha = 10, \beta = 1/5$.

The mathematical expectation and the variance are, respectively,

$$E(X) = (\alpha + 1)\beta \quad \text{and} \quad \text{Var}(X) = (\alpha + 1)\beta^2$$

Note that for $\alpha = 0$, a gamma distribution reduces to an exponential one with $a = 1/\beta$; for $\beta = 2$, $\alpha = m/2 - 1$, it reduces to a χ^2 distribution with m degrees of freedom.

4.9 WEIBULL DISTRIBUTION

A random variable X is said to have a *Weibull distribution* if its density is given by

$$f_X(x) = \begin{cases} 0, & x < 0 \\ \alpha\beta x^{\beta-1}e^{-\alpha x^\beta}, & x \geq 0 \end{cases} \tag{4.11}$$

where α and β are positive constants. The mathematical expectation and the variance are, respectively,

$$E(X) = \frac{1}{\alpha^{1/\beta}}\Gamma\left(1 + \frac{1}{\beta}\right)$$

$$\text{Var}(X) = \frac{1}{\alpha^{2/\beta}}\left[\Gamma\left(1 + \frac{2}{\beta}\right) - \Gamma^2\left(1 + \frac{1}{\beta}\right)\right]$$

Note that for $\beta = 1$, the Weibull distribution reduces to an exponential distribution, and for $\beta = 2$, $\alpha = 1/2a^2$, to a Rayleigh distribution.

4.10 NORMAL OR GAUSSIAN DISTRIBUTION

A random variable X is said to have a *normal or Gaussian distribution* if its density is given by

$$f_X(x) = \frac{1}{\sigma_X\sqrt{2\pi}} \exp\left[-\tfrac{1}{2}\left(\frac{x-a}{\sigma_X}\right)^2\right], \quad -\infty < x < \infty, \sigma_X > 0 \quad (4.12)$$

The relevant distribution function (inexpressible in terms of elementary functions—polynomial, trigonometric, or exponential) is

$$F_X(x) = \frac{1}{\sigma_X\sqrt{2\pi}} \int_{-\infty}^{x} \exp\left[-\tfrac{1}{2}\left(\frac{x-a}{\sigma_X}\right)^2\right] dx$$

$$= \tfrac{1}{2} + \text{erf}\left(\frac{x-a}{\sigma_X}\right)$$

$$F_X(a) = \tfrac{1}{2} \quad (4.13)$$

where $\text{erf}(x)$ is the *error function*, defined as

$$\text{erf}(x) = \frac{1}{\sqrt{2\pi}} \int_0^x e^{-y^2/2}\, dy \quad (4.14)$$

Note that this function is odd:

$$\text{erf}(-x) = \frac{1}{\sqrt{2\pi}} \int_0^{-x} e^{-y^2/2}\, dy = -\frac{1}{\sqrt{2\pi}} \int_0^x e^{-z^2/2}\, dz = -\text{erf}(x)$$

In view of the familiar integral (see Appendix A)

$$\int_{-\infty}^{\infty} e^{-\alpha y^2}\, dy = \sqrt{\frac{\pi}{\alpha}} \quad (4.15)$$

it can be shown that

$$\text{erf}(\infty) = \frac{1}{\sqrt{2\pi}} \int_0^{\infty} e^{-y^2/2}\, dy = \tfrac{1}{2}$$

Comprehensive tables of $\text{erf}(x)$ are readily available; some values are listed in Appendix B. Note that the error function is often (e.g., in computer subroutines) defined in a different way, namely,

$$\text{erf}^*(x) = \frac{2}{\sqrt{\pi}} \int_0^x e^{-y^2}\, dy \quad (4.16)$$

The connection between erf(x) and erf*(x) is as follows:

$$\text{erf}(x) = \tfrac{1}{2}\,\text{erf*}\!\left(\frac{x}{\sqrt{2}}\right) \tag{4.17}$$

A normal distribution is often denoted by $N(a, \sigma_X^2)$, and the parameters a and σ_X^2 can be shown to represent the mathematical expectation and variance of X. Indeed,

$$E(X) = \int_{-\infty}^{\infty} x f_X(x)\, dx = \frac{1}{\sigma_X \sqrt{2\pi}} \int_{-\infty}^{\infty} x \exp\!\left[-\tfrac{1}{2}\!\left(\frac{x-a}{\sigma_X}\right)^{\!2}\right] dx$$

or, introducing a new variable $\xi = (x - a)/\sigma_X$,

$$E(X) = \frac{1}{\sqrt{2\pi}} \int_{-\infty}^{\infty} (\sigma_X \xi + a) \exp(-\tfrac{1}{2}\xi^2)\, d\xi$$

$$= \frac{\sigma_X}{\sqrt{2\pi}} \int_{-\infty}^{\infty} \xi \exp\!\left[-\tfrac{1}{2}\xi^2\right] d\xi + \frac{a}{\sqrt{2\pi}} \int_{-\infty}^{\infty} \exp\!\left[-\tfrac{1}{2}\xi^2\right] d\xi$$

the first integral vanishes (the integrand being odd), and the second equals $\sqrt{2\pi}$ by Eq. (4.15). Thus,

$$E(X) = a \tag{4.18}$$

This also follows immediately from the fact that a normal density (4.12) is symmetric about $x = a$. For the variance we obtain

$$\text{Var}(X) = \int_{-\infty}^{\infty} (x-a)^2 \frac{1}{\sqrt{2\pi}\,\sigma_X} \exp\!\left[-\frac{(x-a)^2}{2\sigma_X^2}\right] dx$$

$$= \frac{\sigma_X^2}{\sqrt{2\pi}} \int_{-\infty}^{\infty} \xi^2 \exp\!\left[-\tfrac{1}{2}\xi^2\right] d\xi = \sigma_X^2 \tag{4.19}$$

thus justifying use of the symbol σ_X^2 for this parameter. The graph of a normal density function is shown in Fig. 4.5. The curve is symmetrical about a and descends steeply as $x - a$ increases. At $x = a$ the curve has a maximum equal to $1/\sigma_X\sqrt{2\pi}$; this ordinate increases as σ_X decreases.

A normal random variable with zero mean and unity variance $N(0, 1)$ is called *standard* or *normalized*, and its density and distribution functions are,

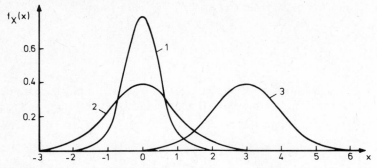

Fig. 4.5. Probability density function of normal distribution with (1) $a = 0$, $\sigma_X = \frac{1}{2}$, (2) $a = 0$, $\sigma_X = 1$, (3) $a = 3$, $\sigma_X = 1$.

respectively,

$$\varphi(x) = \frac{1}{\sqrt{2\pi}} \exp\left(-\tfrac{1}{2}x^2\right) \qquad (4.20a)$$

$$P(x) = \int_{-\infty}^{x} \varphi(x)\, dx = \tfrac{1}{2} + \mathrm{erf}(x) \qquad (4.20b)$$

For $P(x)$, the following approximate formula holds (see Abramowitz and Stegun):

$$P(x) = 1 - \varphi(x) \sum_{i=1}^{5} b_i t^i$$

where

$$t = \frac{1}{1 + px}$$

and $p = 0.2316419$, $b_1 = 0.319381530$, $b_2 = -0.356563782$, $b_3 = 1.781477937$, $b_4 = -1.821255978$, and $b_5 = 1.330274429$.

The central kth moment of a normally distributed random variable is given by

$$\mu_k = \int_{-\infty}^{\infty} (x - a)^k f_X(x)\, dx = \sigma_X^k \int_{-\infty}^{\infty} \frac{\xi^k}{\sqrt{2\pi}} \exp\left[-\tfrac{1}{2}\xi^2\right] d\xi$$

Integration by parts yields

$$\mu_k = (k-1)\sigma_X^2 \left\{ \sigma_X^{k-2} \int_{-\infty}^{\infty} \frac{\xi^{k-2}}{\sqrt{2\pi}} \exp\left[-\tfrac{1}{2}\xi^2\right] d\xi \right\}$$

or

$$\mu_k = (k-1)\sigma_X^2 \mu_{k-2}$$

Since $\mu_1 = 0$ and $\mu_2 = \sigma_X^2$, all odd central moments vanish, while the even ones are

$$\mu_4 = 3\sigma_X^4 = 3\mu_2^2$$
$$\mu_6 = 15\sigma_X^6$$
$$\vdots$$
$$\mu_{2k} = (2k-1)!!\sigma_X^{2k}, \quad k = 1,2,\ldots$$

(4.21)

where

$$(2k-1)!! = 1 \cdot 3 \cdot 5 \cdots (2k-1)$$

The absolute kth moments $\tilde{m}_k$ as per Eq. (3.37) are

$$\tilde{m}_k = E(|X|^k) = \begin{cases} (2r-1)!!\sigma_X^{2r}, & k = 2r \\ \left(\dfrac{2}{\pi}\right)^{1/2} 2^r r! \sigma_X^{2r+1}, & k = 2r+1 \end{cases}$$

Note that for a normal random variable μ_k and $\tilde{m}_k$ are functions of the variance.

The characteristic function of a normally distributed random variable is

$$M_X(\theta) = \frac{1}{\sigma_X\sqrt{2\pi}} \int_{-\infty}^{\infty} \exp\left[i\theta x - \frac{(x-a)^2}{2\sigma_X^2}\right] dx$$

$$= \frac{e^{i\theta a}}{\sigma_X\sqrt{2\pi}} \int_{-\infty}^{\infty} \exp\left[i\theta t - \frac{t^2}{2\sigma_X^2}\right] dt$$

Recalling the equality (see Appendix A)

$$\int_{-\infty}^{\infty} e^{\alpha t - \beta^2 t^2} dt = \frac{\sqrt{\pi}}{\beta} \exp\left(\frac{\alpha^2}{4\beta^2}\right)$$

(4.22)

we have

$$M_X(\theta) = \exp\left(i\theta a - \tfrac{1}{2}\sigma_X^2\theta^2\right)$$

$$\psi_X(\theta) = i\theta a - \tfrac{1}{2}\sigma_X^2\theta^2 \tag{4.23}$$

These formulas can also be used for the moments of X.
From (4.13) it follows that

$$P(x_1 \leqslant X \leqslant x_2) = F_X(x_2) - F_X(x_1) = \mathrm{erf}\left(\frac{x_2 - a}{\sigma_X}\right) - \mathrm{erf}\left(\frac{x_1 - a}{\sigma_X}\right)$$

$$\tag{4.24}$$

In the particular case where the interval (x_1, x_2) is symmetric about the mathematical expectation, x_1 and x_2 may be represented as

$$x_1 = a - \varepsilon \qquad x_2 = a + \varepsilon$$

where $\varepsilon > 0$, and (4.24) takes the form

$$P(|X - a| < \varepsilon) = \mathrm{erf}\left(\frac{\varepsilon}{\sigma_X}\right) - \mathrm{erf}\left(-\frac{\varepsilon}{\sigma_X}\right)$$

or, since $\mathrm{erf}(-x) = -\mathrm{erf}(x)$,

$$P(|X - a| < \varepsilon) = 2\,\mathrm{erf}\left(\frac{\varepsilon}{\sigma_X}\right)$$

In particular,

$$P(|X - a| < k\sigma_X) = 2\,\mathrm{erf}(k)$$

and (see Appendix B)

$$P(|X - a| < \sigma_X) = 2\,\mathrm{erf}(1) = 2(0.34134) = 0.68268$$

$$P(|X - a| < 2\sigma_X) = 2\,\mathrm{erf}(2) = 2(0.47725) = 0.95450$$

$$P(|X - a| < 3\sigma_X) = 2\,\mathrm{erf}(3) = 2(0.49865) = 0.99730$$

$$P(|X - a| < 4\sigma_X) = 2\,\mathrm{erf}(4) = 2(0.499968) = 0.999936 \tag{4.25}$$

Accordingly, the probability of X falling outside the interval $(a - \sigma_X, a + \sigma_X)$ is 0.31732; outside $(a - 2\sigma_X, a + 2\sigma_X)$, 0.0455; outside $(a - 3\sigma_X, a + 3\sigma_X)$, 0.0027; and outside $(a - 4\sigma_X, a + 4\sigma_X)$, 0.000064. It is thus seen that the probability of large deviations from a decreases steeply as k increases and that for "practical" purposes, "possible" values are confined to the range $(a - 3\sigma_X, a + 3\sigma_X)$ (Fig. 4.6).

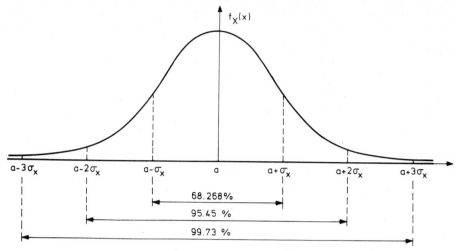

Fig. 4.6.   Area under normal probability density curve.

4.11   TRUNCATED NORMAL DISTRIBUTION

A random variable is said to have a truncated normal distribution if its density is given by

$$
f_X(x) = \begin{cases} \dfrac{A}{\sigma\sqrt{2\pi}} \exp\left[-\dfrac{(x - x_0)^2}{2\sigma^2} \right], & x_1 \leqslant x \leqslant x_2 \\ 0, & \text{otherwise} \end{cases} \tag{4.26}
$$

It is completely similar to its normal counterpart except for being confirmed to the interval $[x_1, x_2]$ (Fig. 4.7). A is found from the requirement (3.11)

$$
A = \left[\operatorname{erf}\left(\frac{x_2 - x_0}{\sigma} \right) - \operatorname{erf}\left(\frac{x_1 - x_0}{\sigma} \right) \right]^{-1} \tag{4.27}
$$

Hence, the distribution function is given by

$$
F_X(x) = \begin{cases} 0, & -\infty < x < x_1 \\[2mm] \dfrac{\operatorname{erf}\left(\dfrac{x - x_0}{\sigma} \right) - \operatorname{erf}\left(\dfrac{x_1 - x_0}{\sigma} \right)}{\operatorname{erf}\left(\dfrac{x_2 - x_0}{\sigma} \right) - \operatorname{erf}\left(\dfrac{x_1 - x_0}{\sigma} \right)}, & x_1 \leqslant x < x_2 \\[2mm] 1, & x_2 \leqslant x < \infty \end{cases} \tag{4.28}
$$

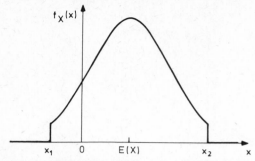

Fig. 4.7.   Probability density function of truncated normal distribution.

Also,

$$E(X) = x_0 + B\sigma \tag{4.29a}$$

$$B = \frac{1}{A}\left\{ \exp\left[-\frac{(x_1 - x_0)^2}{2\sigma^2} \right] - \exp\left[-\frac{(x_2 - x_0)^2}{2\sigma^2} \right] \right\} \tag{4.29b}$$

$$\frac{\text{Var}(X)}{\sigma^2} = 1 - B^2 - A\left\{ \frac{(x_2 - x_0)}{\sigma\sqrt{2\pi}} \exp\left[-\frac{(x_2 - x_0)^2}{2\sigma^2} \right] \right.$$

$$\left. - \frac{(x_1 - x_0)}{\sigma\sqrt{2\pi}} \exp\left[-\frac{(x_1 - x_0)^2}{2\sigma^2} \right] \right\} \tag{4.29c}$$

If $x_1 \to -\infty$ and $x_2 \to +\infty$, then $A \to 1$ and X becomes a normally distributed random variable with $E(X) = x_0$ and $\text{Var}(X) = \sigma^2$.

A normal variable is said to have a *symmetrical truncated distribution* if in (4.26) we have $x_1 = x_0 - k\sigma$ and $x_2 = x_0 + k\sigma$. A then becomes $A = [2\,\text{erf}(k)]^{-1}$. From Appendix B we have for $k = 2$, $\text{erf}(2) = 0.47725$, and A differs only slightly from unity.

4.12   FUNCTION OF A RANDOM VARIABLE

We now consider the mapping of a random variable into another random variable by means of some deterministic relationship.

Let X be a random variable. This signifies that to each experimental outcome ω we assign a real number $X(\omega)$. The set $(X \leqslant x)$ is an event for any real number x, and the probabilities are

$$P(X = +\infty) = P(X = -\infty) = 0$$

Suppose that a real function $\varphi(x)$ of a real variable x is given. We construct the function

$$Y = \varphi(X) \qquad (4.31)$$

Suppose that the set $(Y \leqslant y)$ is an event for any real number y, and the probabilities are

$$P(Y = +\infty) = P(Y = -\infty) = 0$$

To every outcome ω of the experiment we assign the real number $\varphi[X(\omega)]$. Then $Y = \varphi(X)$ is a random variable.

4.13 MOMENTS OF A FUNCTION OF A RANDOM VARIABLE

According to the definitions given by (3.29) and (3.30),

$$m_k(Y) = E(Y^k) \qquad (4.32)$$

$$\mu_k(Y) = E\{[Y - E(Y)]^k\} \qquad (4.33)$$

Substituting (4.31) in (4.32) and (4.33), we obtain

$$m_k(Y) = E[\varphi(X)]^k \qquad (4.34)$$

$$\mu_k(Y) = E\big(\{\varphi(X) - E[\varphi(X)]\}^k\big) \qquad (4.35)$$

and therefore, for a continuously distributed X,

$$m_k(Y) = \int_{-\infty}^{\infty} [\varphi(x)]^k f_X(x)\, dx \qquad (4.36)$$

$$\mu_k(Y) = \int_{-\infty}^{\infty} \{\varphi(x) - E[\varphi(X)]\}^k f_X(x)\, dx \qquad (4.37)$$

As is seen from Eqs. (4.32)–(4.37), the moments of a function of a random variable can be determined directly through $f_X(x)$ without recourse to $F_Y(y)$. For the mathematical expectation and variance of Y, we obtain, respectively,

$$E(Y) = \int_{-\infty}^{\infty} \varphi(x) f_X(x)\, dx \qquad (4.38)$$

$$\text{Var}(Y) = \int_{-\infty}^{\infty} \varphi^2(x) f_X(x)\, dx - \left[\int_{-\infty}^{\infty} \varphi(x) f_X(x)\, dx\right]^2 \qquad (4.39)$$

4.14 DISTRIBUTION AND DENSITY FUNCTIONS OF A FUNCTION OF A RANDOM VARIABLE (SPECIAL CASE)

We seek the distribution function $F_Y(y)$ of Y. By definition,

$$F_Y(y) = P(Y \leqslant y) = P[\varphi(X) \leqslant y] \qquad (4.40)$$

Consider first the case where $y = \varphi(x)$ is a strictly monotone increasing function (Fig. 4.8). A straight line parallel to the abscissa intersects the graph of $\varphi(x)$ at a point with the abscissa depending on y, and the interval where the inequality $Y = \varphi(x) \leqslant y$ holds is marked off. $x = \psi(y)$ is then a unique inverse function

$$F_Y(y) = \int_{-\infty}^{\psi(y)} f_X(x)\, dx = F_X[\psi(y)] \qquad (4.41)$$

The probability density function $f_Y(y)$ of Y is obtained as

$$f_Y(y) = \frac{dF_Y(y)}{dy} = f_X[\psi(y)] \frac{d\psi(y)}{dy} \qquad (4.42)$$

Next, consider the case where $\varphi(x)$ is a strictly monotone decreasing function (Fig. 4.9). Here

$$F_Y(y) = \int_{\psi(y)}^{\infty} f_X(x)\, dx = 1 - F_X[\psi(y)] \qquad (4.43)$$

and the probability density function is

$$f_Y(y) = \frac{dF_Y(y)}{dy} = -f_X[\psi(y)] \frac{d\psi(y)}{dy} \qquad (4.44)$$

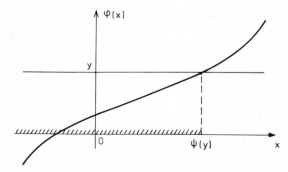

Fig. 4.8. $\varphi(x)$-strictly monotone increasing function.

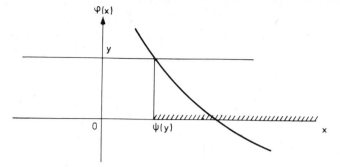

Fig. 4.9. $\varphi(x)$-strictly monotone decreasing function.

Since in the latter case $\psi'(y) < 0$, the two cases can be combined:

$$f_Y(y) = f_X[\psi(y)]\left|\frac{d\psi(y)}{dy}\right| \qquad (4.45)$$

4.15 LINEAR FUNCTION OF A RANDOM VARIABLE

If a random variable Y is a linear function of X, $Y = \alpha X + \beta$, and the probability density function $f_X(x)$ of X is known, then

$$y = \varphi(x) = \alpha x + \beta$$

$$x = \psi(y) = \frac{y - \beta}{\alpha} \qquad \psi'(y) = \frac{1}{\alpha}$$

Substituting these in (4.45), we obtain

$$f_Y(y) = \frac{1}{|\alpha|}f_X\left(\frac{y - \beta}{\alpha}\right) \qquad (4.46)$$

Now let $\alpha = 1/\sigma_X$ and $\beta = -E(X)/\sigma_X$, where $E(X)$ is the mathematical expectation of X and σ_X its standard deviation. Then,

$$Y = \frac{X - E(X)}{\sigma_X}$$

is the *normalized random variable*, and

$$f_Y(y) = \sigma_X f_X[\sigma_X y + E(X)] \qquad (4.47)$$

Consider two special cases.

Example 4.2

Let X be uniformly distributed in the interval (x_1, x_2), that is

$$f_X(x) = \begin{cases} \dfrac{1}{x_2 - x_1}, & x_1 < x < x_2 \\ 0, & \text{otherwise} \end{cases}$$

Then we have

$$f_Y(y) = \begin{cases} \dfrac{1}{|\alpha|} \dfrac{1}{x_2 - x_1}, & x_1 < \dfrac{y - \beta}{\alpha} < x_2 \\ 0, & \text{otherwise} \end{cases} \tag{4.48}$$

In other words, Y is likewise uniformly distributed in the interval $(\alpha x_1 + \beta, \alpha x_2 + \beta)$ for $\alpha > 0$, and in the interval $(\alpha x_2 + \beta, \alpha x_1 + \beta)$ for $\alpha < 0$.

Example 4.3

If X is $N(a, \sigma_X^2)$, then

$$f_X(x) = \frac{1}{\sigma_X\sqrt{2\pi}} \exp\left[-\frac{(x - a)^2}{2\sigma_X^2}\right]$$

and

$$f_Y(y) = \frac{1}{\sigma_X|\alpha|\sqrt{2\pi}} \exp\left[-\frac{[(y - \beta)/\alpha - a]^2}{2\sigma_X^2}\right]$$

or

$$f_Y(y) = \frac{1}{\sigma_X|\alpha|\sqrt{2\pi}} \exp\left[-\frac{[y - (\alpha a + \beta)]^2}{2\alpha^2\sigma_X^2}\right] \tag{4.49}$$

demonstrating that a linear function of a normally distributed random variable is likewise normal, $N(\alpha a + \beta, \sigma_X^2\alpha^2)$.

4.16 EXPONENTS AND LOGARITHMS OF A RANDOM VARIABLE

Now let

$$Y = e^X \tag{4.50}$$

The function $y = e^x$ is strictly monotone increasing, and we have

$$F_Y(y) = F_X[\psi(y)] = F_X(\ln y)U(y)$$

$$f_Y(y) = \frac{1}{y}f_X[\ln y]U(y) + F_X(\ln y)\delta(y) \tag{4.51}$$

In particular, when X is $N(a, \sigma_X^2)$, we obtain

$$f_Y(y) = \begin{cases} 0, & y \leqslant 0 \\ \dfrac{1}{y\sigma_X\sqrt{2\pi}} \exp\left[-\dfrac{(\ln y - a)^2}{2\sigma_X^2}\right], & y > 0 \end{cases} \tag{4.52}$$

The random variable Y is said to have a *logarithmic-normal* or for brevity, *log-normal distribution*. We can show that the probability of Y falling within an interval $[y_1, y_2]$ is $(y_1, y_2 \geqslant 0)$

$$P(y_1 \leqslant Y \leqslant y_2) = \text{erf}\left(\frac{\ln y_2 - a}{\sigma_X}\right) - \text{erf}\left(\frac{\ln y_1 - a}{\sigma_X}\right) \tag{4.53}$$

In fact,

$$P(y_1 \leqslant Y \leqslant y_2) = \int_{y_1}^{y_2} \frac{1}{y\sigma_X\sqrt{2\pi}} \exp\left[-\frac{(\ln y - a)^2}{2\sigma_X^2}\right] dy$$

Substituting in the integral

$$\frac{\ln y - a}{\sigma_X} = z$$

we obtain

$$P(y_1 \leqslant Y \leqslant y_2) = \frac{1}{\sqrt{2\pi}} \int_{(\ln y_1 - a)/\sigma_X}^{(\ln y_2 - a)/\sigma_X} \exp\left(-\frac{z^2}{2}\right) dz$$

which immediately leads to the desired result (4.53).

We now seek the mean $E(Y)$ and the variance $\text{Var}(Y)$, using Eqs. (4.38) and (4.39):

$$E(Y) = \frac{1}{\sigma_X\sqrt{2\pi}} \int_{-\infty}^{\infty} e^x \exp\left[-\frac{(x - a)^2}{2\sigma_X^2}\right] dx$$

$$= \frac{1}{\sigma_X\sqrt{2\pi}} \int_{-\infty}^{\infty} \exp\left[x - \frac{(x - a)^2}{2\sigma_X^2}\right] dx$$

which, in view of Eq. (4.15), transforms into

$$E(Y) = \exp\{a + \tfrac{1}{2}\sigma_X^2\} \tag{4.54}$$

In order to find the variance, we first calculate in Eq. (4.39)

$$\int_{-\infty}^{\infty} \varphi^2(x) f_X(x)\, dx = \int_{-\infty}^{\infty} \frac{e^{2x}}{\sigma_X \sqrt{2\pi}} \exp\left[-\frac{(x-a)^2}{2\sigma_X^2} \right] = \exp[2(a + \sigma_X^2)]$$

whence

$$\mathrm{Var}(Y) = \exp(2a + \sigma_X^2)\left[\exp(\sigma_X^2) - 1\right] \tag{4.55}$$

If instead of (4.50) we have $Y = 10^x$ and Eq. (4.52) becomes

$$f_Y(y) = \frac{\log_{10} e}{y} f_X[\log_{10} y] U(y) \tag{4.56}$$

where $\log_{10} e \simeq 0.4343$.

Consider now the *Napierian logarithm* of a random variable

$$U = \ln V$$

where V is a positive-valued random variable with probability density $f_V(v)$. Then, since the function $u = \ln v$ is strictly monotone increasing, we have

$$f_U(u) = e^u f_V[e^u]$$

In particular, if V has a log-normal distribution as per Eq. (4.52) then U has a normal distribution $N(a, \sigma_X^2)$, as anticipated.

4.17 DISTRIBUTION AND DENSITY FUNCTIONS OF A FUNCTION OF A RANDOM VARIABLE (GENERAL CASE)

Consider now the case when $y = \varphi(x)$ is not a monotone function (see Fig. 4.10). On the abscissa axis, the interval

$$Y = \varphi(X) \leqslant y \tag{4.57}$$

is marked off. The number of intervals where (4.57) is satisfied depends on y. Let this number be n; the condition (4.57) is then equivalent to the event of X

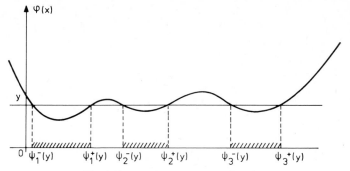

Fig. 4.10. $\varphi(x)$-nonmonotone function.

taking on values in one of the intervals $[\psi_i^-(y), \psi_i^+(y)]$. Thus,

$$F_Y(y) = P[\varphi(X) \leqslant y]$$
$$= P\{[\psi_1^-(y) \leqslant X \leqslant \psi_1^+(y)]$$
$$\cup [\psi_2^-(y) \leqslant X \leqslant \psi_2^+(y)]$$
$$\cup \cdots \cup [\psi_n^-(y) \leqslant X \leqslant \psi_n^+(y)]\} \tag{4.58}$$

Since the random events in (4.58) are mutually exclusive, we may write

$$F_Y(y) = \sum_{i=1}^{n} P[\psi_i^-(y) \leqslant X \leqslant \psi_i^+(y)] = \sum_{i=1}^{n} \int_{\psi_i^-(y)}^{\psi_i^+(y)} f_X(x)\, dx$$

$$F_Y(y) = \sum_{i=1}^{n} \{F_X[\psi_i^+(y)] - F_X[\psi_i^-(y)]\} \tag{4.59}$$

the probability density function is

$$f_Y(y) = \frac{dF_Y(y)}{dy} = \sum_{i=1}^{n} \left\{ f_X[\psi_i^+(y)] \frac{d\psi_i^+(y)}{dy} - f_X[\psi_i^-(y)] \frac{d\psi_i^-(y)}{dy} \right\}$$

$$\tag{4.60}$$

Equations (4.59) and (4.60), respectively, represent in general terms, the distribution and probability density function of a function of a random variable.

For a strictly monotone *increasing* function (Fig. 4.8), $n = 1$, the interval being $\psi_i^-(y) = -\infty$, $\psi_i^+(y) = \psi(y) = \varphi^{-1}(y)$, the inverse of φ, and the equation reduces to (4.42). For a strictly monotone *decreasing* function (Fig.

4.9), again $n = 1$, the interval this time being $\psi_i^+(y) = +\infty$, $\psi_i^-(y) = \varphi^{-1}(y)$, and the equation reduces to (4.44).

Consider now the case where $\varphi(x)$ is a constant in some interval (x_1, x_2) (Fig. 4.11). Comparison of Figs. 4.11b and 4.11c shows that

$$F_Y(y)|_{y=c+0} - F_Y(y)|_{y=c-0} = \text{Prob}(x_1 < X < x_2) = \int_{x_1}^{x_2} f_X(x)\, dx$$

and therefore the probability distribution function $F_Y(y)$ has a jump discontinuity at c.

If the range of $y[= \varphi(x)]$ is not the entire real line but only part of it, say the interval (x_1, x_2), we put $f_Y(y) = 0$ outside that interval.

Example 4.4

Let $Y = X^2$, where X has a Rayleigh distribution, with

$$f_X(x) = \frac{x}{a^2} \exp\left[-\frac{x^2}{2a^2}\right] U(x)$$

The function $y = x^2$ $(0 < x < \infty)$ is strictly monotone increasing, and (4.42) may be applied. The inverse function is $\psi(y) = +\sqrt{y}$, and we have

$$f_Y(y) = \frac{1}{2a^2} \exp\left[-\frac{y}{2a^2}\right] U(x)$$

that is, an exponentially distributed random variable.

Example 4.5

Consider once again the transformation (Fig. 4.12) $Y = X^2$, where X can now take on both positive and negative values. In this case Y cannot take on negative values, so

$$F_Y(y) = 0, \quad f_Y(y) = 0, \quad y < 0$$

For $y \geq 0$ there is a single interval where $x^2 \leq y$, with

$$\psi_1^-(y) = -\sqrt{y} \qquad \psi_1^+(y) = \sqrt{y}$$

and in accordance with (4.59) and (4.60) we write for $y > 0$

$$F_Y(y) = F_X(\sqrt{y}) - F_X(-\sqrt{y}) \tag{4.61}$$

$$f_Y(y) = \frac{1}{2\sqrt{y}}\left[f_X(\sqrt{y}) + f_X(-\sqrt{y})\right] \tag{4.62}$$

and $F_Y(y) = 0$, $f_Y(y) = 0$ for $y \leq 0$.

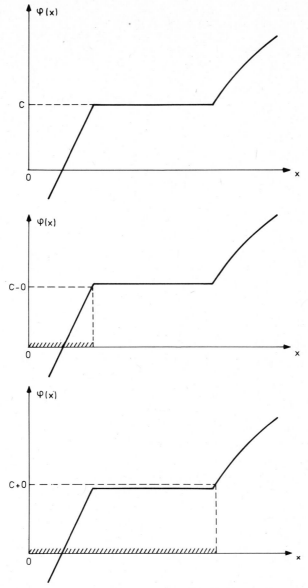

Fig. 4.11. $\varphi(x)$ constant in some interval (x_1, x_2). Probability distribution function $F_Y(y)$ of random variable $Y = \varphi(X)$ undergoing jump discontinuity equal to integral of $f_X(x)$ over interval (x_1, x_2).

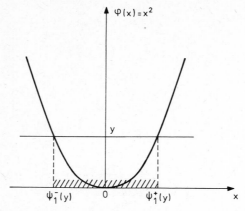

Fig. 4.12.   Determination of the probability density function of the square of a random variable.

If, for example, X is a standard normal variable $N(0, 1)$, Eq. (4.20.1), we find

$$f_Y(y) = \frac{1}{\sqrt{2\pi}} y^{-1/2} e^{-y/2}, \quad y \geqslant 0$$

which represents a one-degree-of-freedom chi-square probability density (4.9), with $\lambda = \frac{1}{2}$, $[\Gamma(\frac{1}{2}) = \sqrt{\pi}]$.

Example 4.6

Consider now the function $Y = |X|$ (Fig. 4.13). In this case we again have a single interval where $|x| \leqslant y$ is satisfied with

$$\psi_1^-(y) = -y \qquad \psi_1^+(y) = y$$

As in the preceding example,

$$F_Y(y) = 0, \quad f_Y(y) = 0, \quad y < 0$$

Combining (4.59) and (4.60), we obtain

$$F_Y(y) = F_X(y) - F_X(-y) \tag{4.63}$$

$$f_Y(y) = f_X(y) + f_X(-y) \tag{4.64}$$

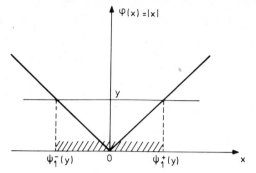

Fig. 4.13. Determination of the probability density function of the absolute value of a random variable.

If X is $N(a, \sigma_X^2)$ we obtain for $F_Y(y)$ and $f_Y(y)$, respectively,

$$f_Y(y) = \frac{1}{\sigma_X\sqrt{2\pi}} \left\{ \exp\left[-\frac{(y-a)^2}{2\sigma_X^2} \right] + \exp\left[-\frac{(y+a)^2}{2\sigma_X^2} \right] \right\} \quad (4.65)$$

$$F_Y(y) = \left[\operatorname{erf}\left(\frac{y-a}{\sigma_X} \right) + \operatorname{erf}\left(\frac{y+a}{\sigma_X} \right) \right] U(y) \quad (4.66)$$

for $y > 0$ and $f_Y(y) = 0$ for $y < 0$.

For $a = 0$ we obtain

$$f_Y(y) = \begin{cases} 0, & y \leq 0 \\ \left(\dfrac{2}{\pi\sigma_X^2} \right)^{1/2} \exp\left(-\dfrac{y^2}{2\sigma_X^2} \right), & y > 0 \end{cases} \quad (4.67)$$

$$F_Y(y) = 2\operatorname{erf}\left(\frac{y}{\sigma_X} \right) U(y) \quad (4.68)$$

These functions are shown in Fig. 4.14.

Sometimes a random variable Y with density (4.67) is said to have a *one-sided normal distribution* with parameter σ_X. Note that σ_X does not exceed the standard deviation of Y.

Example 4.7

A random variable X is uniformly distributed in the interval $[0, 1]$. The random variable Y is a strictly monotone increasing function of X: $Y = \varphi(X)$. We are interested in $F_Y(y)$ and $f_Y(y)$. We have

$$f_X(x) = \begin{cases} 1, & 0 \leq x \leq 1 \\ 0, & \text{otherwise} \end{cases} \quad (4.69)$$

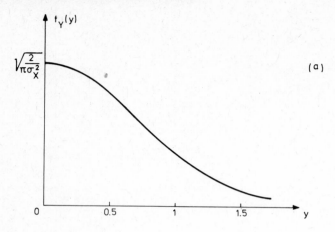

(a)

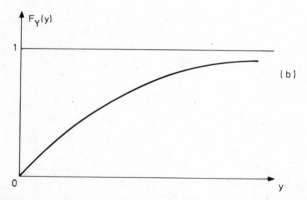

(b)

Fig. 4.14. Probability density (a), and distribution function (b) of the square of a normally distributed random variable.

Let $\psi(y)$ be the inverse function. Then

$$f_Y(y) = \begin{cases} f_X[\psi(y)]\psi'(y) = \psi'(y), & 0 < \psi(y) < 1 \\ 0, & \text{otherwise} \end{cases}$$

which yields

$$F_Y(y) = \begin{cases} 0, & \psi(y) < 0 \\ \psi(y), & 0 \leqslant \psi(y) < 1 \\ 1, & \psi(y) \geqslant 1 \end{cases} \qquad (4.70)$$

that is, the desired distribution function of Y equals the inverse of φ. By this means, different random variables may be derived from a uniformly distributed one. Consider, as an illustration, determination of an exponentially distributed random variable as a transformation of a uniformly distributed one:

$$f_Y(y) = \begin{cases} 0, & y < 0 \\ ae^{-ay}, & y \geqslant 0 \end{cases}$$

We have then to put $Y = \varphi(X)$, where φ is an inverse $F_Y(y) = 1 - e^{-ay}$,

$$x = 1 - e^{-ay}$$

and therefore

$$Y = -\frac{1}{a}\ln(1 - X), \quad 0 \leqslant X \leqslant 1 \tag{4.71}$$

In complete analogy, consider the following problem. Let $Y = \varphi(X)$, φ being strictly monotone increasing with $f_X(x)$ and $f_Y(y)$ given; we seek φ.

We write in accordance with (4.41),

$$F_X(x) = \int_{-\infty}^{\varphi(x)} f_Y(y)\, dy = F_Y[\varphi(x)]$$

$$\varphi(x) = F_Y^{-1}[F_X(x)]$$

where F_Y^{-1} is the inverse of F_Y. Thus,

$$Y = F_Y^{-1}[F_X(x)] \tag{4.72}$$

We now consider a concrete example of how probabilistic considerations can be incorporated directly in engineering design.

4.18 EXAMPLE OF APPLICATION OF THE PROBABILISTIC APPROACH IN AN ENGINEERING DECISION PROBLEM

The Netherlands is a low-lying country bordering on the North Sea with nearly half its land below the mean sea level (Fig. 4.15) and only partially shielded by natural sand dunes. In these circumstances, permanent habitation became possible only when the population learned how to set up protective dikes against the incursions of the sea and control the water level in the areas so protected (known as "polders", see Fig. 4.16). The protective system of dikes and dunes has to sustain extreme loads especially during winter storms; and as perfect safety is unattainable, the country was repeatedly devastated by floods

Fig. 4.15. Hatched area of the map of the Netherlands is below mean sea level (lowest point minus 6.7 meters, north of the city of Rotterdam).

in the course of its history. The most recent disaster (Fig. 4.17) occurred on February 1, 1953, mainly hitting the southwestern region (parts of England and Belgium were also affected but to a smaller extent). It took 1800 human lives. In addition, over 150,000 hectares of land were inundated, more than 9000 buildings demolished and 38,000 damaged. The dike system was breached at 67 points and damaged over stretches hundreds of kilometers long. The total economic loss was estimated at 1.5 to 2 billion Dutch guilders.

In the wake of this disaster, a governmental committee (known as the "Delta Committee") was appointed with a view to re-answering the age-old question: How high must the dikes be? Until a few decades ago, dike height had been set so as to exceed the peak tide mark hitherto observed at the site in

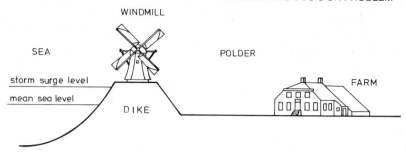

Fig. 4.16. Schematic vertical profile of Netherlands terrain.

question. Since then, however, probabilistic considerations of the frequency of floods of different levels have been introduced. We use $F_H(h)$ to denote the (cumulative) distribution function of high tides. The *exceedance probability* (i.e., the probability of the height h of high tide being exceeded in any given year) is then

$$P(H_m > h) = 1 - F_{H_m}(h) \tag{4.73}$$

Fig. 4.17. Village of Nieuwerkerk on Schouwen-Duiveland island in Zeeland, one of the areas hardest hit during the disastrous flood on the night of 31 January–1 February 1953. Two months later the water level was still high, the village was accessible only by rowing boats and the few inhabitants who had remained behind camped in their attics. Today the damage has been repaired, the dead have been buried, and Nieuwerkerk is once more a perfectly ordinary Dutch village; but the memory of the disaster lives on as strong as ever. (Reproduced with permission of B. V. Uitgeversmaatschappij Elsevier, from Scherer K., and Werkman, E., "Holland in Close-Up", 1979, Elsevier-Amsterdam, p. 117).

where H_m is the maximum storm surge above the mean sea level. In 1939 Wemelsfelder found that the annual exceedance frequencies during high tide at the Hook of Holland in the period 1888–1937 followed the exponential distribution

$$P(H_m > h) = \exp\left(-\frac{h - \alpha}{\beta}\right) \tag{4.74}$$

where α and β are some constants.

We confine ourselves to the question of optimal dike height from the economical point of view, presenting the solution by Van Danzig. The economic decision problem can then be formulated as follows: *Taking into account the cost of constructing the dike, the material losses when a breach occurs, and the distribution of different sea levels, determine the optimal height of the dike.*

Assume that future dikes at the site in question will all have the same height H above a given standard level, so that the amount x by which the dikes must be elevated is

$$x = H - H_0 \tag{4.75}$$

The cost C of elevating the dikes from H_0 to H is a function of x.

The simplest assumption about the possibility of losses is as follows. Let h denote the sea level along the dikes at any moment; then, no loss is incurred so long as $h \leqslant H$; if $h > H$, one may disregard the possibility of partial losses and reckon with total loss only, that is, with the circumstance that all houses, farms, livestock, industries, etc., in the polder are lost. Let V be the total value of the property in the polder and assume that the consequent loses (migration costs of the population and livestock, slump in production, etc.) are included in it. We then have

$$S = \begin{cases} 0, & \text{if } h \leqslant H \\ V, & \text{if } h > H \end{cases} \tag{4.76}$$

Van Danzig treated the problem as one of insurance; he assumed that a sum L would be reserved to cover all future losses. If L is invested at a certain rate of interest i (in percent), it must cover the expected values of all future losses, $P(H_m > H)V$ per annum, and we have

$$L = P(H_m > H)V \sum_{t=0}^{\infty} (1 + 0.01i)^{-t}$$

$$\simeq P(H_m > H)V \int_0^{\infty} e^{-(it)/100} \, dt = \frac{100}{i} P(H_m > H)V \tag{4.77}$$

Cost of elevating the dikes is

$$I = I_0 + kx \tag{4.78}$$

where I_0 is the initial cost and k the subsequent cost of elevation per meter. Adding (4.77) and (4.78), we obtain

$$I + L = I_0 + kx + 100\frac{V}{i}e^{-(H-\alpha)/\beta} \qquad (4.79)$$

Now we determine x so that $I + L$ is minimal, that is,

$$\frac{dI}{dx} + \frac{dL}{dx} = 0 \qquad (4.80)$$

or

$$k - \frac{100\ Ve^{-(H-\alpha)/\beta}}{\beta i} = 0 \qquad (4.81)$$

yielding the optimal dike elevation increment

$$x = \alpha + \beta \ln\left(\frac{100V}{k\beta i}\right) - H_0 \qquad (4.82)$$

With constants $\alpha = 1.96$ m, $\beta = 0.33$ m, $V = 24 \cdot 10^9$ Guilders, $k = 40 \cdot 10^6$ Guilder/m, $i = 1.5\%$, $H_0 = 4.25$ m, van Danzig's analysis yielded the optimal increment of 1.57 m.

PROBLEMS

4.1. A dike is designed with 5 m freeboard above the mean sea level. The probability of its being topped by waves in 1 yr is 0.005. What is the probability of waves exceeding 5 m within 200 yrs?

4.2. Using Eq. (4.23) for the characteristic function of a normally distributed random variable, show that

$$E(X) = a \qquad \text{Var}(X) = \sigma_X^2$$

4.3. Suppose that the duration of successful performance (lifetime, in years) of a piece of equipment is normally distributed with a mean of 8 yrs. What is the largest value its standard deviation may have if the operator requires at least 95% of the population to have lifetimes exceeding 6 yrs? What is the probability of a piece of equipment turning out faulty at delivery?

4.4. A system is activated at $t = 0$. Its time of failure is the random variable T with distribution function $F_T(t)$ and density $f_T(t)$. Denote the hazard function by $h(t)$ (see Prob. 3.4).

(a) Does T in Prob. 3.4 have an exponential distribution?

(b) Show that if $h(t) = \alpha t$, T has a Rayleigh distribution.

(c) Show that if $h(t) = \alpha\beta t^{\beta-1}$, T has a Weibull distribution.

(d) Find $f_T(t)$ if $h(t) = \alpha\gamma e^{\gamma t}$.

Remark The resulting $f_T(t)$ is called an *extreme-value* density function. For interesting failure models leading to the extreme-value distribution, see Epstein (1958).

4.5. A system consisting of n elements in parallel with equal reliabilities $R(t)$ fails only when all elements fail simultaneously.

(a) Show that the reliability $R_n(t)$ of such a system is

$$R_n(t) = [1 - R(t)]^n$$

(b) Find $R_n(t)$ when the conditional failure rate of each element is constant, $h(t) = a$.

(c) Find the *allowable operation time* t_r of the system, such that $R(t_r) = r$, where r is a required reliability, for $n = 1, 2, 3$.

(d) Show that the mean lifetime of the system is

$$E(T) = \frac{1}{a}\left(1 + \frac{1}{2} + \cdots + \frac{1}{n}\right)$$

and interpret the result for $n \to \infty$.

4.6. Repeat problem 4.5, but with the elements arranged in series instead of in parallel.

4.7. Suppose that the reliabilities of n individual elements are identical and are given by

$$R(t) = t/\tau \qquad \text{for } 0 \leqslant t \leqslant \tau$$

Show that the mean lifetime of the system is

$$E(T) = \begin{cases} r/(n+1) & \text{for elements in series} \\ rn/(n+1) & \text{for elements in parallel} \end{cases}$$

Interpret the result for $n \to \infty$.

4.8. The random variables X and Y are linked the following functional relationship

$$Y = \begin{cases} -a^3, & x < -a \\ X^3, & -a \leqslant x \leqslant a \\ a^3, & x > a \end{cases}$$

X has a uniform distribution in the interval $(-a, a)$. Find $F_Y(y)$ and $f_Y(y)$, and present them graphically.

4.9. Given the functions of Problem 4.8 but with X having a uniform distribution in the interval $(-2a, 2a)$, show that Y is a mixed random variable.

4.10. X has a Cauchy distribution. Show that $Y = 1/X$ also has a Cauchy distribution.

CITED REFERENCES

Abramowitz, M., and Stegun, I. A., Eds., *Handbook of Mathematical Functions*, Dover Publ., New York, 1970, p. 932, formula 26.2.17.

Epstein, B., "The Exponential Distribution and Its Role in Life Testing," *Ind. Quality Control*, **15** (6), 4–9 (1958).

National Bureau of Standards, *Tables of Normal Probability Functions*, *Appl. Math. Series*, **23**, 1953.

Van Danzig, D., "Economic Decision Problems for Flood Prevention," *Econometrica*, **24**, 276–287 (1956).

Wemelsfelder, P. J., "Wetmatigheden in het Optreden van Stormvloeden," *De Ingenieur*, No. 9, (1939). (In Dutch.)

RECOMMENDED FURTHER READING

Melsa, J. L., and Sage, A. P., *An Introduction to Probability and Stochastic Processes*, Prentice-Hall, Englewood Cliffs, NJ, 1973; Chap. 3: Random Variables, pp. 46–109; Chap. 4: Functions of Random Variables, pp. 110–188.

Papoulis, A., *Probability*, *Random Variables*, *and Stochastic Processes*, Intern. Student Ed., Mc-Graw-Hill Kogakusha, Tokyo, 1965. Chap. 3: Repeated Trials, pp. 47–82; Chap. 4: The Concept of a Random Variable, pp. 83–115; Chap. 5: Functions of One Random Variable, pp. 116–164.

chapter 5

Reliability of Structures Described by a Single Random Variable

The scope of the knowledge acquired by us so far suffices for reliability analysis (recalling that reliability is the probability of nonfailure performance) of the simplest structures, described by a single random variable. We are concerned with generally differing structures called upon to realize different functional assignments and will not consider nonfunctional ones. In other words, reliability is associated with a purpose—exploitation of the structure in accordance with defined goals. The *acceptability* criterion consists in the reliability exceeding some specified level.

We now proceed to concrete examples of structural reliability.

5.1 A BAR UNDER RANDOM FORCE

Consider a bar of constant cross-sectional area a, under a tensile force N, which is a random variable with the probability distribution function $F_N(n)$, $n \geq 0$ (Fig. 5.1). The conventional strength requirement is for the normal stress Σ, with values σ, to be less than or equal to the allowable stress σ_{allow}:

$$\Sigma = \frac{N}{a} \leq \sigma_{\text{allow}} \tag{5.1}$$

We simplify the analysis by assuming that both a and σ_{allow} are deterministic quantities. The reliability R will then be defined as the probability of event (5.1):

$$R = P(\Sigma \leq \sigma_{\text{allow}}) = P\left(\frac{N}{a} \leq \sigma_{\text{allow}}\right) \tag{5.2}$$

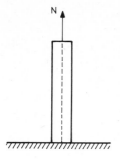

Fig. 5.1. Bar under tensile force.

or

$$R = F_N(\sigma_{\text{allow}}a) \tag{5.3}$$

In other words, the reliability of a bar equals the probability distribution function of the tensile force N at the $\sigma_{\text{allow}}a$ level.

If, for example, N has a uniform distribution

$$f_N(n) = \begin{cases} \dfrac{1}{n_2 - n_1}, & n_1 \leqslant n \leqslant n_2 \\ 0, & \text{otherwise} \end{cases} \tag{5.4}$$

then R is

$$R = \begin{cases} 0, & \sigma_{\text{allow}}a < n_1 \\ \dfrac{\sigma_{\text{allow}}a - n_1}{n_2 - n_1}, & n_1 \leqslant \sigma_{\text{allow}}a < n_2 \\ 1, & n_2 \leqslant \sigma_{\text{allow}}a \end{cases} \tag{5.5}$$

These results are readily understood: If the minimal possible value n_1 of the random tensile force exceeds its maximal deterministically allowable counterpart $\sigma_{\text{allow}}a$, failure is a certain event. In that case the *unreliability* of the structure

$$Q = 1 - R \tag{5.6}$$

defined as the probability of failure, is unity and its reliability zero. If, however, the maximal possible value n_2 of the force does *not* exceed its maximal deterministically allowable counterpart, the structure never fails, its unreliability is zero, and its reliability unity. These situations are illustrated in Fig. 5.2, where the shaded areas represent the reliability of the tensile member.

With R known we can solve the following two problems, noting that the structure functions acceptably if the reliability exceeds (or equals) a specified

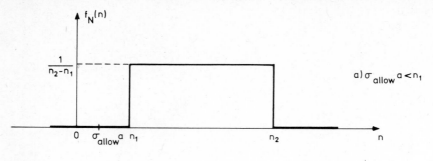

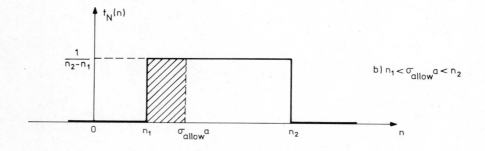

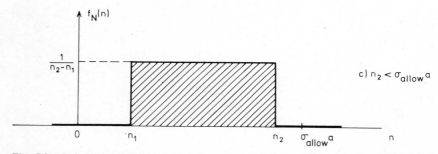

Fig. 5.2. Reliability of bar under random tension having uniform distribution; shaded area represents reliability.

probability r:

$$R \geqslant r, \quad 0 < r \leqslant 1 \tag{5.7}$$

The first problem is concerned with checking whether this inequality is satisfied; the second, with the design of the bar. Suppose we have to choose a deterministic quantity a so that Eq. (5.7) is satisfied. The value of a that satisfies the equality

$$R = r \tag{5.8}$$

is called the *minimal permissible value*.

Comparison of Eqs. (5.3) and (5.8) immediately indicates that $\sigma_{\text{allow}} a$ is the rth quantile of N, and that the required area is obtained simply by dividing this rth quantile by σ_{allow}.

Once the reliability is known, the problem of checking the strength becomes trivial. We now proceed to the design problem. Note that by choosing

$$a \geqslant \frac{n_2}{\sigma_{\text{allow}}} \equiv a_{\text{worst}} \qquad (5.9)$$

we have $R = 1$, and inequality (5.7) is satisfied for any r. This design is, according to the maximal possible value n_2 of the tensile force N, the "worst" case consideration. Concentrate now on the "short-of-the-worst" case

$$n_1 \leqslant \sigma_{\text{allow}} a \leqslant n_2$$

Equation (5.8) then reads

$$\frac{\sigma_{\text{allow}} a - n_1}{n_2 - n_1} = r$$

yielding the required area:

$$a_{\text{req}} = \frac{r(n_2 - n_1) + n_1}{\sigma_{\text{allow}}} = \frac{n_2 - (n_2 - n_1)(1 - r)}{\sigma_{\text{allow}}}$$

$$= a_{\text{worst}} - \frac{n_2 - n_1}{\sigma_{\text{allow}}}(1 - r) \qquad (5.10)$$

Comparison of Eqs. (5.9) and (5.10) shows that the required area is less than that corresponding to the "worst" case. The gain in area, however, may be very small for high values of r:

$$a_{\text{worst}} - a_{\text{req}} \leqslant a_{\text{worst}}(1 - r)$$

equality is achieved only if $n_1 = 0$. If, for example, $r = 0.99$, then the gain in area is only 1% of a_{worst}.

We may express a_{req} in terms of the mean $E(N)$ and the standard deviation $\sigma_N = \sqrt{\text{Var}(N)}$:

$$E(N) = \frac{1}{2}(n_1 + n_2) \qquad \sigma_N = \frac{1}{\sqrt{12}}(n_2 - n_1)$$

to yield

$$a_{\text{req}} = \frac{E(N) + (r - 0.5)\sqrt{12}\,\sigma_N}{\sigma_{\text{allow}}} \qquad (5.11)$$

If $n_2 \to n_1 = n$, $\sigma_N \to 0$, and N tends to become a random variable with *causal* distribution, it takes on the value n with probability 1. We return to the deterministic design load:

$$a = \frac{E(N)}{\sigma_{\text{allow}}} = \frac{n}{\sigma_{\text{allow}}} \tag{5.12}$$

otherwise Eq. (5.11) has to be used for the sought area. Equation (5.11) indicates that the "design" according to mean load is unconservative if $r > 0.5$. It may only be used if $\sigma_N \ll E(N)$.

Thus the deterministic design is a particular case of the probabilistic design.

Note that Eq. (5.11) tells us that when a reliability as low as $r = 0.5$ is required, the mean load-based design may be used, because then the second term in Eq. (5.11) vanishes and $a_{\text{req}} = E(N)/\sigma_{\text{allow}}$. This, however, implies, in accordance with the statistical interpretation of probability, that nearly half the ensemble of structures so designed will fail (which is quite a lot!). Note also that Eq. (5.3) is valid for all positive-valued distributions of N, since then the associated deterministic strength requirement is given by (5.2). If, for example, N has an exponential distribution (Sec. 4.6):

$$F_N(n) = \left\{ 1 - \exp\left[-\frac{n}{E(N)} \right] \right\} U(n)$$

then

$$R = 1 - \exp\left[-\frac{\sigma_{\text{allow}} a}{E(N)} \right] \tag{5.13}$$

and the required area is

$$a_{\text{req}} = \frac{E(N)}{\sigma_{\text{allow}}} \ln \frac{1}{1 - r} \tag{5.14}$$

Thus, if the required reliability is 0.99, then the required area is $\ln 100 = 4.605$ times that calculated according to the mean. Note that now the calculation according to the "worst" case is ruled out: It would yield an infinite a_{req}. The determination of a_{req} is illustrated in Fig. 5.3.

Consider now the case where the force N can take on negative values as well. Assume also that the bar is not slender, so that there is no possibility of buckling failure. Then the strength requirement is written as

$$|\sigma| = \frac{|N|}{a} \leqslant \sigma_{\text{allow}} \tag{5.15}$$

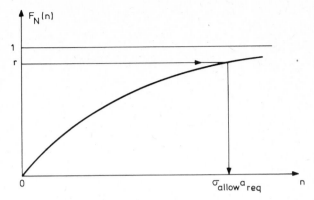

Fig. 5.3.   Calculation of required cross-sectional area.

The reliability is given by

$$P\left(\frac{|N|}{a} \leqslant \sigma_{\text{allow}}\right) = P(|N| \leqslant \sigma_{\text{allow}}a)$$

$$= P(-\sigma_{\text{allow}}a \leqslant N \leqslant \sigma_{\text{allow}}a)$$

$$= F_N(\sigma_{\text{allow}}a) - F_N(-\sigma_{\text{allow}}a) \qquad (5.16)$$

and the strength requirement becomes

$$F_N(\sigma_{\text{allow}}a) - F_N(-\sigma_{\text{allow}}a) \geqslant r \qquad (5.17)$$

where the equality yields the required area a. Note that in this case also the design according to the "worst" case ensures any level of reliability. Indeed, if $n_2 > -n_1$ (Fig. 5.4a), we may choose $a = n_2/\sigma_{\text{allow}}$, yielding

$$F_N(\sigma_{\text{allow}}a) = 1, \qquad F_N(-\sigma_{\text{allow}}a) = 0$$

and $R = 1$. If, however, $n_2 < -n_1$, (Fig. 5.4b), we may choose $a = -n_1/\sigma_{\text{allow}}$ with

$$F_N(-\sigma_{\text{allow}}a) = F_N(n_1) = 0$$

$$F_N(\sigma_{\text{allow}}a) = F_n(-n_1) = 1$$

and still $R = 1$. If $n_2 = -n_1$ (Fig. 5.4c), choosing $a = n_2/\sigma_{\text{allow}}$ again yields $R = 1$.

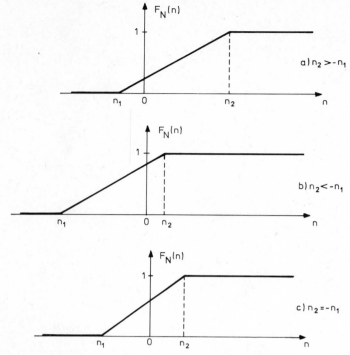

Fig. 5.4. Worst-case design: Worst situation governed by deterministic load equal to (a) n_2, (b) n_1, (c) either n_1 or n_2.

If, however, the required reliability is not too high, the bar will be "overdesigned." For $n_2 = -n_1$, we have from (5.17):

$$F_N(\sigma_{\text{allow}}a) - F_N(-\sigma_{\text{allow}}a) = r$$

but

$$F_N(-\sigma_{\text{allow}}a) = 1 - F_N(\sigma_{\text{allow}}a)$$

yielding

$$F_N(\sigma_{\text{allow}}a) = \tfrac{1}{2}(1 + r)$$

or

$$\frac{\sigma_{\text{allow}}a + n_2}{2n_2} = \tfrac{1}{2}(1 + r) \qquad \sigma_{\text{allow}}a = n_2 r \qquad a_{\text{req}} = \frac{n_2 r}{\sigma_{\text{allow}}}$$

and

$$a_{req} = \frac{n_2 r}{\sigma_{allow}} = a_{worst} r$$

For $r = 0.9$, the required area is only $0.9 a_{worst}$; however, for $r = 0.99$, it is $0.99 a_{worst}$; that is, a_{worst} may be used as the required area.

Consider now the case where N has a symmetrical truncated normal distribution with zero mean, in the interval (n_1, n_2) (Sec. 4.11):

$$n_1 = -k\bar{\sigma} \qquad n_2 = k\bar{\sigma}$$

where $\bar{\sigma}$ is a parameter of the distribution, and

$$F_N(n) = \begin{cases} 0, & -\infty < n < n_1 \\ \dfrac{\text{erf}(n/\bar{\sigma}) + \text{erf}(k)}{2\,\text{erf}(k)}, & n_1 \leqslant n < n_2 \\ 1, & n_2 \leqslant n \end{cases}$$

Equation (5.17) then takes the form for a_{req}:

$$\text{erf}\left(\frac{\sigma_{allow} a_{req}}{\bar{\sigma}}\right) = r\,\text{erf}(k) \tag{5.18}$$

For $k = 3$, we have

$$\text{erf}\left(\frac{\sigma_{allow} a_{req}}{\bar{\sigma}}\right) = 0.49865 r \tag{5.19}$$

for $r = 0.9$, $\text{erf}(\sigma_{allow} a_{req}/\bar{\sigma}) = 0.448785$, and $a_{req} = 1.63\bar{\sigma}/\sigma_{allow}$; for $r = 0.99$, $\text{erf}(\sigma_{allow} a/\bar{\sigma}) = 0.4936635$, and $a_{req} = 2.49\bar{\sigma}/\sigma_{allow}$. Note that calculation according to the "worst" load $n_2 = 3\bar{\sigma}$ yields $a_{req} = 3\bar{\sigma}/\sigma_{allow}$, implying 84.05 and 20.48% overdesign, for $r = 0.9$ and $r = 0.99$, respectively.

Note that if we assume a normal distribution of the load with zero mean and standard deviation σ_N, we obtain, instead of (5.19),

$$\text{erf}\left(\frac{\sigma_{allow} a_{req}}{\sigma_N}\right) = 0.5 r \tag{5.20}$$

which is coincident with the result obtained from that associated with a truncated normal distribution with $k \to \infty$.

Design according to "mean plus three standard deviations," that is, choice of

$$a_{req} = \frac{3\sigma_N}{\sigma_{allow}}$$

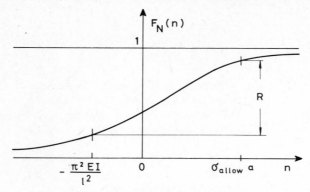

Fig. 5.5. Calculation of reliability of a slender bar.

corresponds to a reliability $R = 2\,\mathrm{erf}(3) = 0.9973$. Design according to "two standard deviations" is equivalent to choice of

$$a_{\mathrm{req}} = \frac{2\sigma_N}{\sigma_{\mathrm{allow}}}$$

yielding $R = 2\,\mathrm{erf}(2) = 0.9545$.

Note that calculation of the required area for a bar under an exponentially distributed tensile force according to the "mean plus α times standard deviation" yields, according to (5.13), $R = 1 - e^{-(1+\alpha)}$ so that in order to achieve the required reliability, say $r = 0.9973$, we now have to use $\alpha = 4.9145$ instead of $\alpha = 3$ for the normally distributed force.

For the case where the bar is slender, and buckling in compression is possible, the strength requirement (5.15) reads

$$-\frac{\pi^2 EI}{4l^2} \leqslant N \leqslant \sigma_{\mathrm{allow}}a$$

where the left-hand term represents the buckling load of a clamped-free bar. Reliability is determined as the probability of the above random event:

$$R = P\left(-\frac{\pi^2 EI}{4l^2} \leqslant N \leqslant \sigma_{\mathrm{allow}}a\right)$$

$$= F_N(\sigma_{\mathrm{allow}}a) - F_N\left(-\frac{\pi^2 EI}{4l^2}\right)$$

which may be used either for determining the reliability or for designing the bar with the desired level of reliability (Fig. 5.5).

5.2   A BAR WITH RANDOM STRENGTH

Consider now a bar under a deterministic load n, and having a random strength with given continuous probability distribution $F_{\Sigma_{\text{allow}}}(\sigma_{\text{allow}})$. We confine ourselves here to the case where Σ_{allow} takes on only positive values, and moreover $n > 0$. Then reliability is

$$R = P\left(\frac{n}{a} \leqslant \Sigma_{\text{allow}}\right) = 1 - F_{\Sigma_{\text{allow}}}\left(\frac{n}{a}\right) \qquad (5.21)$$

Assuming an exponential distribution for Σ_{allow},

$$F_{\Sigma_{\text{allow}}}(\sigma_{\text{allow}}) = 1 - \exp\left[-\frac{\sigma_{\text{allow}}}{E(\Sigma_{\text{allow}})}\right]$$

we obtain immediately

$$R = \exp\left[-\frac{n}{E(\Sigma_{\text{allow}})a}\right]$$

The reliability R is given by the shaded area in Fig. 5.6.

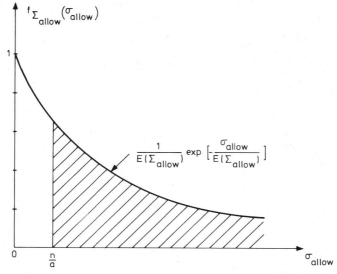

Fig. 5.6.   Shaded area equals the reliability of the structure with random strength subjected to a deterministic load.

The condition $R = r$ yields the required area:

$$a_{\text{req}} = \frac{n}{E(\Sigma_{\text{allow}})} \left(\ln \frac{1}{r} \right)^{-1}$$

So, for the desired reliability $r = 0.99$, we have

$$a_{\text{req}} = 99.5 \frac{n}{E(\Sigma_{\text{allow}})}$$

almost a hundred times the value $n/E(\Sigma_{\text{allow}})$ obtained from the mean value approach.

The results obtained for a single bar can be generalized in a simple way for the truss system of Sec. 2.8. Consider a statically determinate truss, assuming that the stresses in the bars $\sigma_1, \sigma_2, \ldots, \sigma_n$ are deterministic quantities and that the distribution functions of the allowable stresses $F_{\Sigma_{\text{allow}}(j)}(\sigma_{\text{allow}}(j))$ for each bar ($j = 1, 2, \ldots, n$) are given. Then $F_{\Sigma_{\text{allow}}(j)}(\sigma_j)$ is the probability of failure of the jth bar, and its reliability is

$$R_j = 1 - F_{\Sigma_{\text{allow}}(j)}(\sigma_j)$$

If bar failures represent independent random events, the reliability of the entire truss is

$$R = \prod_{j=1}^{n} \left[1 - F_{\Sigma_{\text{allow}}(j)}(\sigma_j) \right]$$

In the particular case where the allowable stresses are distributed identically, this formula becomes

$$R = \prod_{j=1}^{n} \left[1 - F_{\Sigma_{\text{allow}}}(\sigma_j) \right]$$

and in the subcase where the stresses are equal, $\sigma_j = \sigma$ ($j = 1, 2, \ldots, n$) (see Fig. 2.9e), we obtain

$$R = \left[1 - F_{\Sigma_{\text{allow}}}(\sigma) \right]^n$$

instead of Eq. (2.22).

5.3 A BAR WITH A RANDOM CROSS-SECTIONAL AREA

Consider now a circular bar with given σ_{allow} under a given tensile force n, its cross-sectional area A being a random variable with continuous distribution function $F_A(a)$, $a > 0$. The strength requirement reads

$$\Sigma = \frac{n}{A} \leqslant \sigma_{\text{allow}}$$

and the reliability is

$$R = P\left(A \geqslant \frac{n}{\sigma_{\text{allow}}}\right) = 1 - F_A\left(\frac{n}{\sigma_{\text{allow}}}\right) \tag{5.22}$$

The maximal allowable tensile force n_{allow} is then determined from the equality

$$F_A\left(\frac{n_{\text{allow}}}{\sigma_{\text{allow}}}\right) = 1 - r$$

Randomness of the cross-sectional area is due to that of its radius c, which has a continuous distribution function $F_C(c)$; then

$$\Sigma = \frac{n}{\pi c^2} \leqslant \sigma_{\text{allow}} \qquad R = F_\Sigma(\sigma_{\text{allow}})$$

The function $\sigma = n/\pi c^2$ is strictly monotone decreasing so that, according to Eq. (4.43),

$$F_\Sigma(\sigma) = \int_{\psi(\sigma)}^{\infty} f_C(c)\, dc = 1 - F_C[\psi(\sigma)]$$

where $\psi(\sigma) = (n/\pi\sigma)^{1/2}$. The reliability then becomes

$$R = 1 - F_C\left(\sqrt{\frac{n}{\pi\sigma_{\text{allow}}}}\right)$$

and the maximum allowable tensile force is determined as the root of the equation

$$F_C\left(\sqrt{\frac{n}{\pi\sigma_{\text{allow}}}}\right) = 1 - r$$

5.4 A BEAM UNDER A RANDOM DISTRIBUTED FORCE

We now seek the reliability of a beam of span l, with a uniform symmetrical cross section, simply supported at its ends, under a distributed force $p(x) = G\varphi(x)$, where G is a continuous random variable with the probability density function $f_G(g)$, and $\varphi(x)$ is a given (deterministic) function representing the load pattern (Fig. 5.7).

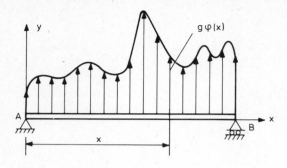

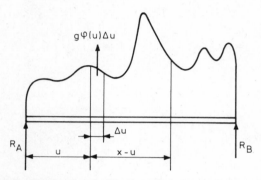

Fig. 5.7. Simply supported beam under random distributed force.

The bending moment is given by

$$M_z(x) = R_A x + \int_0^x G\varphi(u)(x - u)\, du$$

but $M_z(l) = 0$, so that

$$M_z(x) = -\frac{x}{l}\int_0^l G\varphi(u)(l - u)\, du + \int_0^x G\varphi(u)(x - u)\, du$$

This expression can be written as

$$M_z(x) = G\psi(x) \tag{5.23a}$$

$$\psi(x) = -\frac{x}{l}\int_0^l \varphi(u)(l - u)\, du + \int_0^x \varphi(u)(x - u)\, du \tag{5.23b}$$

Denote by Σ' the extremal normal stress within the beam cross section x. The reliability is defined as the probability of the stress Σ', not exceeding in absolute value the σ_{allow} at *any* section of the beam:

$$R = P\{|\Sigma'(x)| \leqslant \sigma_{\text{allow}}, \quad 0 \leqslant x \leqslant l\} \tag{5.24}$$

where

$$\Sigma'(x) = \frac{M_z(x)}{S}$$

with S denoting the section modulus. The requirement (5.24) can be replaced by

$$R = P(|\Sigma| \leqslant \sigma_{allow}) \qquad (5.25)$$

where

$$\Sigma = \frac{M_z(a)}{S}$$

and in section $x = a$, the function $|\psi(a)|$ has a maximum.

Consider now the case where a stiffness requirement is imposed on the beam: Its transverse displacement must not exceed in absolute value some given allowable deflection w_{allow}. We first find the displacement, by integrating the differential equation

$$EI_z \frac{d^2 w}{dx^2} = M_z(x)$$

where E is the modulus of elasticity of the beam material and I_z is the central moment of inertia of the cross section about the z axis. By virtue of Eq. (5.23a), we have:

$$EI_z w = G \int_0^x \int_0^v \psi(u) \, du \, dv + C_1 x + C_2$$

The constants are determined by the boundary conditions:

$$w = 0 \quad \text{at} \quad x = 0 \quad \text{and} \quad x = l$$

which finally yield $EI_z w = G\Psi(x)$, where

$$\Psi(x) = \int_0^x \int_0^v \psi(u) \, du \, dv - \frac{x}{l} \int_0^l \int_0^v \psi(u) \, du \, dv \qquad (5.26)$$

Let $|\Psi(x)|$ reach maximum at $x = b$. Denoting the displacement at this section, which is a random variable, by W, the reliability is then defined as

$$R = P(|W| \leqslant w_{allow}) \qquad (5.27)$$

Sometimes both stress and stiffness requirements are imposed on the beam. In these circumstances the reliability is

$$R = P(B_1 \cap B_2) \qquad (5.28)$$

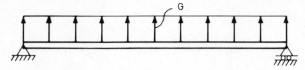

Fig. 5.8.   Beam, simply supported at its ends, under random distributed load.

where random events B_1, B_2 are

$$\{B_1\} = \{|\Sigma| \leqslant \sigma_{\text{allow}}\} \qquad \{B_2\} = \{|W| \leqslant w_{\text{allow}}\} \tag{5.29}$$

and

$$
\begin{aligned}
R &= P(B_1), & B_1 \subset B_2 \\
R &= P(B_2), & B_2 \subset B_1 \\
R &= P(B_1) = P(B_2), & B_1 = B_2
\end{aligned}
\tag{5.30}
$$

Now for some concrete examples.

Example 5.1

Let $\varphi(x) = 1$ (Fig. 5.8). Then the bending moment reaches its extremum at the middle section, $a = l/2$:

$$M_z\left(\frac{l}{2}\right) = \frac{Gl^2}{8}$$

and the reliability is

$$R = P\left(\frac{|G|l^2}{8S} \leqslant \sigma_{\text{allow}}\right)$$

$$= F_G\left(\frac{8S\sigma_{\text{allow}}}{l^2}\right) - F_G\left(-\frac{8S\sigma_{\text{allow}}}{l^2}\right) \tag{5.31}$$

The extremal displacement also occurs at this section, namely,

$$W = \frac{5|G|l^4}{384EI}$$

and the reliability (5.27) according to the stiffness requirement is

$$R = P\left(\frac{5|G|l^4}{384EI_z} \leqslant w_{\text{allow}}\right)$$

$$= F_G\left(\frac{384EI_z w_{\text{allow}}}{5l^4}\right) - F_G\left(-\frac{384EI_z w_{\text{allow}}}{5l^4}\right) \tag{5.32}$$

If both strength and stiffness requirements have to be satisfied,

$$R = P\left[\left(|G| \leqslant \frac{8S\sigma_{\text{allow}}}{l^2}\right) \cap \left(|G| \leqslant \frac{384EI_z w_{\text{allow}}}{5l^4}\right)\right] \qquad (5.33)$$

and if

$$\frac{8S\sigma_{\text{allow}}}{l^2} < \frac{384EI_z w_{\text{allow}}}{5l^4} \qquad (5.34)$$

then

$$\left\{|G| \leqslant \frac{8S\sigma_{\text{allow}}}{l^2}\right\} \subset \left\{|G| \leqslant \frac{384EI_z w_{\text{allow}}}{5l^4}\right\}$$

and the reliability is as per Eq. (5.31). In the case

$$\frac{8S\sigma_{\text{allow}}}{l^2} > \frac{384EI_z w_{\text{allow}}}{5l^4} \qquad (5.35)$$

then,

$$\left\{|G| \leqslant \frac{384EI_z w_{\text{allow}}}{5l^4}\right\} \subset \left\{|G| \leqslant \frac{8S\sigma_{\text{allow}}}{l^2}\right\}$$

and the reliability is as per Eq. (5.32). If, however,

$$\frac{8S\sigma_{\text{allow}}}{l^2} = \frac{384EI_z w_{\text{allow}}}{5l^4} \qquad (5.36)$$

either equation may be used.

We proceed now to the design of the beam. The requirement is as before, $R \geqslant r$, whereas the equality $R = r$ yields the required beam dimensions. Let the cross section be circular with radius c. Then

$$I_z = \frac{\pi c^4}{4} \qquad S = \frac{I_z}{c} = \frac{\pi c^3}{4}$$

Determination of c according to either the strength or the stiffness requirement is straightforward. When both are imposed, we calculate first the reliability as a function of c:

$$R(c) = P\left[\left(|G| \leqslant \frac{96\pi Ew_{\text{allow}}}{5l^2}c^4\right) \cap \left(|G| \leqslant \frac{2\pi\sigma_{\text{allow}}}{l^2}c^3\right)\right]$$

as described above, and then solve the equation $R(c) = r$, which yields c_{req}.

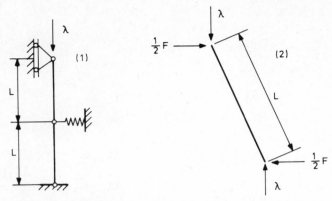

Fig. 5.9. (*1*) Idealized column. (*2*) Equilibrium of single bar.

5.5 STATIC IMPERFECTION SENSITIVITY OF A NONLINEAR MODEL STRUCTURE*

We now proceed to buckling problems, concentrating on the simple model structure proposed by Budiansky and Hutchinson for illustrating Koiter's deterministic imperfection-sensitivity notion. Such a model of an idealized column is shown in Fig. 5.9. The three-hinge, rigid-rod system is constrained laterally by a nonlinear spring with the mass concentrated at the hinge joining the two rods. We suppose that the restoring force F is related to the end shortening (or elongation) x of the spring by

$$F = k_1\xi + k_2\xi^2 + k_3\xi^3 \tag{5.37}$$

where $\xi = x/L$. Refraining temporarily from a restriction as to the sign of k_2 or k_3, we assume $k_1 > 0$. Figure 5.9(1) shows the system in its straight (undisplaced) state, and Fig. 5.9(2), in its displaced state. The horizontal reactions at the hinges equal $F/2$. Furthermore, the moment about the middle hinge must vanish. This requirement leads to the equilibrium equation:

$$\lambda\xi = \tfrac{1}{2}F\sqrt{1 - \xi^2} = \tfrac{1}{2}\big(k_1\xi + k_2\xi^2 + k_3\xi^3\big)\sqrt{1 - \xi^2} \tag{5.38}$$

and for small values of ξ, we obtain the following asymptotic result

$$\lambda\xi = \lambda_c\big(\xi + a\xi^2 + b\xi^3 + \cdots\big) \tag{5.39}$$

$$\lambda_c = \tfrac{1}{2}k_1 \tag{5.40}$$

$$a = \frac{k_2}{k_1}, \qquad b = \frac{k_3}{k_1} - \tfrac{1}{2} \tag{5.41}$$

---

*This section is an extension of the author's paper (1980), from which Figs. 5.9 and 5.12–5.18 are also reproduced.

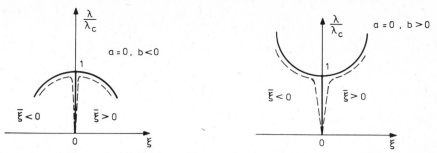

Fig. 5.10. Nondimensional load versus additional displacement relationship for symmetric structure.

Equation (5.39) represents the load-displacement relationship. Now the trivial solution

$$\xi = 0 \tag{5.42}$$

satisfies this equation for all values of the load λ, which merely confirms that the straight, vertical state is one of equilibrium. However, when ξ is not zero, we can still satisfy this equation, i.e. have equilibrium, if

$$\lambda = \lambda_c \left(1 + a\xi + b\xi^2 + \cdots \right) \tag{5.43}$$

The solid curves in Figs. 5.10–5.13 show the equilibrium load λ plotted against the displacement ξ. At point $\xi = 0$, $\lambda = \lambda_c$, there is a *branching (bifurcation)* of the curve, one branch continuing along the straight line $\xi = 0$ and the others moving out along a parabola (5.43). This splitting into branches indicates the onset of instability; therefore the load λ_c at the branching point is called the *classical bifurcation buckling* load.

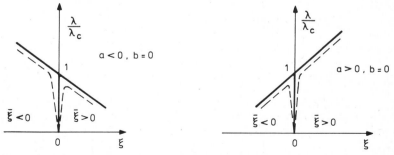

Fig. 5.11. Nondimensional load versus additional displacement relationship for asymmetric structure.

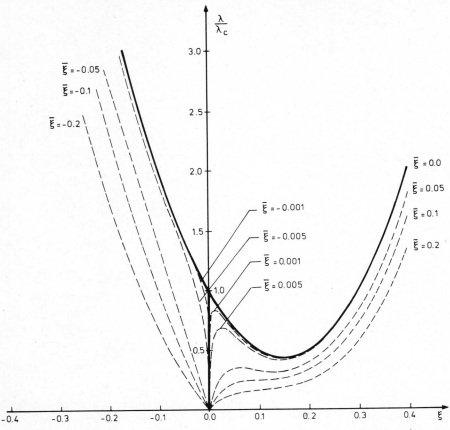

Fig. 5.12. Nondimensional load versus additional displacement curves for nonsymmetric structure ($a = -7.5$, $b = 25$).

The structure is designated as *nonsymmetric* in the general case $a \neq 0$, $b \neq 0$; as *symmetric* if $a = 0$, $b \neq 0$; and as *asymmetric* if $a \neq 0$, $b = 0$. In the latter case the parabola degenerates into a straight line:

$$\frac{\lambda}{\lambda_c} = a\xi + 1 \tag{5.44}$$

We now proceed to the realistic *imperfect* structure. Assuming that an unloaded structure has an initial displacement $\bar{x} = L\bar{\xi}$, then equilibrium dictates, instead of Eq. (5.38), the following relationship:

$$\lambda(\xi + \bar{\xi}) = \tfrac{1}{2}F\left[1 - (\xi + \bar{\xi})^2\right]^{1/2} = \tfrac{1}{2}\left(k_1\xi + k_2\xi^2 + k_3\xi^3\right)\left[1 - (\xi + \bar{\xi})^2\right]^{1/2}$$

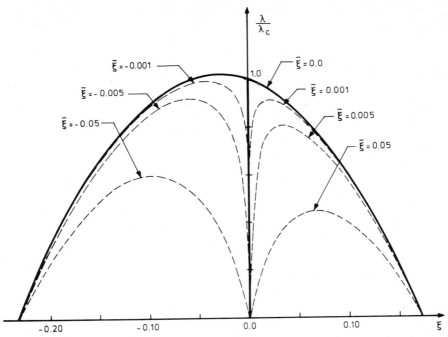

Fig. 5.13. Nondimensional load versus additional displacement curves for nonsymmetric structure ($a = -1.5$, $b = -25$).

where ξ is an additional displacement and $\bar{\xi} + \xi$ a total displacement. For small values of ξ, we arrive at the following asymptotic result:

$$\lambda(\xi + \bar{\xi}) = \lambda_c[\xi + a\xi^2 + b\xi^3 + O(\xi^2\bar{\xi})] \qquad (5.45)$$

Equation (5.45) indicates that ξ and $\bar{\xi}$ have the same sign (i.e., additional displacement ξ of the system is such that the total displacement $\xi + \bar{\xi}$ is increased by its absolute value). Otherwise, the assumption $\xi\bar{\xi} < 0$ would imply $\lambda < 0$ for $0 < |\xi| < \bar{\xi}$, the presence of tension, which is contrary to our formulation of the problem. Note also that the graph λ/λ_c vs. ξ for an imperfect structure issues from the origin of the coordinates. Additional zeroes of λ/λ_c coincide with the zero points $(-a \pm \sqrt{a^2 - 4b})/2b$ of the parabola (5.43) representing the behavior of a perfect structure. The dashed curves in Figures 5.10–5.13 show the equilibrium load λ plotted against the additional displacement ξ, in the imperfect structure.

We now seek the *static buckling load* λ_s, which is defined as the maximum of λ on the branch of the solution $\lambda - \xi$ originating at zero load, for specified $\bar{\xi}$.[†]

[†] The point with coordinates (λ_{max}, ξ_0) where ξ_0 is the additional displacement corresponding to λ_{max}, is sometimes called the *limit point*, and λ_{max} itself the *limit load*.

We refer to a structure as *imperfection-sensitive* if an imperfection results in *reduced* values of the maximum load the structure is able to support; otherwise, we designate the structure *imperfection-insensitive*.

To be able to conclude whether or not a structure is sensitive to initial imperfections, we have to find whether the first derivative of λ with respect to ξ

$$\frac{d\lambda}{d\xi} = \frac{\lambda_c}{(\xi + \bar{\xi})^2} \left[2b\xi^3 + (a + 3b\bar{\xi})\xi^2 + 2a\bar{\xi}\xi + \bar{\xi} \right]$$

has at least one real root. For our purpose, it suffices to examine the numerator

$$\varphi(\xi) = 2b\xi^3 + (a + 3b\bar{\xi})\xi^2 + 2a\bar{\xi}\xi + \bar{\xi}$$

The structure buckles if the equation

$$\varphi(\xi) = 0 \qquad\qquad (5.46)$$

has at least one real positive root for $\bar{\xi} > 0$, or at least one real negative root for $\bar{\xi} < 0$.

In some cases, Descartes' rule of signs provides an answer on buckling of the structure. This rule states that if the coefficients $a_0, a_1, a_2, \ldots, a_n$ of the polynomial

$$\varphi(\xi) = a_0\xi^n + a_1\xi^{n-1} + \cdots + a_{n-1}\xi + a_n$$

have v variations of sign, the number of positive roots of the polynomial equation $\varphi(\xi) = 0$ does not exceed v and is of the same parity. The number of negative roots of this equation equals that of positive roots of equation $\varphi(-\xi) = 0$.

Consider the case $b < 0$, $\bar{\xi} > 0$. We wish to know whether there is at least one positive root between those of $\varphi(\xi) = 0$. In the subcase $a < 0$, we have

$$a_0 = 2b < 0 \qquad a_1 = a + 3b\bar{\xi} < 0 \qquad a_2 = 2a\bar{\xi} < 0 \qquad a_3 = \bar{\xi} > 0$$

so there is a single change in sign, indicating the occurrence of buckling. In the subcase $a > 0$, we have

$$a_0 = 2b < 0 \qquad a_2 = 2a\bar{\xi} > 0 \qquad a_3 > 0$$

so, irrespective of the sign of $a_1 = a + 3b\bar{\xi}$, there is again a single change in sign. Thus the structure has a buckling load if $b < 0$, $\bar{\xi} > 0$.

Consider now the case $b < 0$, $\bar{\xi} < 0$. The question is now whether $\varphi(\xi) = 0$ has at least one negative root. Then

$$\varphi(-\xi) = -2b\xi^3 + (a + 3b\bar{\xi})\xi^2 - 2a\bar{\xi}\xi + \bar{\xi}$$

and for $a > 0$ we have

$$a_0 = -2b > 0, \qquad a_1 = a + 3b\bar{\xi} > 0 \qquad a_2 = -2a\bar{\xi} > 0 \qquad a_3 = \bar{\xi} > 0$$

that is, a single change in sigh. For $a < 0$, we have

$$a_0 = -2b > 0 \qquad a_2 = -2a\bar{\xi} < 0 \qquad a_3 = \bar{\xi} < 0$$

and irrespective of the sign of a_1 we again have a single change in sign, the conclusion being that for $b < 0$ (irrespective of the signs of a or $\bar{\xi}$) the structure carries a finite maximum load. In compete analogy, it can be shown that the structure is imperfection-insensitive for $b > 0$ and $a\bar{\xi} > 0$ (this is left as exercise for the reader).

For $b > 0$ and $a\bar{\xi} < 0$, neither Descartes' rule nor the Routh-Hurwitz criterion for the number of roots with a positive real part (used in conjunction with the fact that $\phi(\xi) = 0$ always has one real root) suffice for a conclusion. This case, however, can be treated by Evans' root-locus method (see, e.g., Ogata), frequently used in control theory.

Let us consider first the particular subcase $b > 0$, $a < 0$, and $\bar{\xi} > 0$. The formal substitution $\xi \to s$, where $s = \mathrm{Re}(s) + i\,\mathrm{Im}(s)$ is a complex variable, in Eq. (5.46)

$$1 + \bar{\xi}\psi(s) = 0 \qquad \psi(s) = \frac{3bs^2 + 2as + 1}{s^2(2bs + a)} \tag{5.47}$$

We now construct the root-locus plot with $\bar{\xi}$ varying from zero to infinity (obviously, only $\bar{\xi} \ll 1$ has physical significance). For $\bar{\xi}$ approaching zero, the roots of Eq. (5.47) are the poles of $\psi(s)$, marked with crosses (X's):

$$s_1 = s_2 = 0 \qquad s_3 = -\frac{a}{2b}$$

The $\bar{\xi} \to \infty$ points of the root loci approach the zeros of $\psi(s)$, marked with circles (O's):

$$s_{1,2} = \frac{1}{3b}\left(-a \pm \sqrt{a^2 - 3b}\right)$$

$\psi(s)$ has three poles: one double at zero, and another at $(-a/2b) > 0$. A root locus issues from each pole as $\bar{\xi}$ increases above zero; a root locus arrives at each zero of $\psi(s)$ or at infinity as $\bar{\xi}$ approaches infinity. For the case $a^2 = 3b$, the two "circles" coincide. As is seen from Fig. 5.14, Eq. (5.47) has two real positive roots, and the structure buckles for any $\bar{\xi} > 0$. For $a^2 < 3b$ both "circles" are complex (Fig. 5.15); for a certain value of $\bar{\xi}$ (called the critical value, $\bar{\xi}_{\mathrm{cr},\,s}$) a pair of loci break away from the real axis. For $\bar{\xi} > \bar{\xi}_{\mathrm{cr},\,s}$, Eq. (5.47) has no real positive root, and consequently the structure is imperfec-

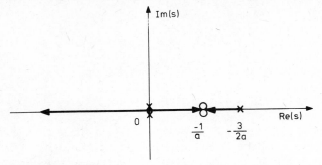

Fig. 5.14.   Root-locus plot $(a < 0, b > 0, \bar{\xi} > 0, a^2 = 3b)$.

tion-insensitive. The breakaway point is found as the root of the equation

$$\frac{dC}{ds} = 0 \qquad C(s) = \frac{1}{\psi(s)} = \frac{s^2(2bs + a)}{3bs^2 + 2as + 1}$$

This equation has only one real root, $s_0 = -a/3b$. The appropriate value of $\bar{\xi}$ equals $-C(s_0)$:

$$\bar{\xi}_{cr,s} = -\frac{a^3}{9b}\frac{1}{3b - a^2} \tag{5.48}$$

The static buckling load associated with $\xi = s_0$ and $\bar{\xi} = \bar{\xi}_{cr,s}$ is

$$\frac{\lambda_s}{\lambda_c} = \frac{a}{3b}\left(\tfrac{2}{9}\frac{a^2}{b} - 1\right)\left(\bar{\xi}_{cr,s} - \frac{a}{3b}\right)^{-1}$$

For example, for $b/a^2 = \tfrac{2}{3}$, we have $\bar{\xi}_{cr,s} = -1/6a$ and $\lambda_s/\lambda_c = \tfrac{1}{2}$, and for $\bar{\xi} > \bar{\xi}_{cr,s}$ static buckling does not occur.

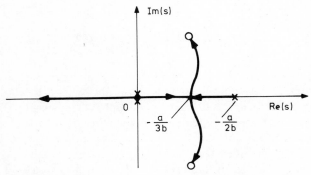

Fig. 5.15.   Root-locus plot $(a < 0, b > 0, \bar{\xi} > 0, a^2 < 3b)$.

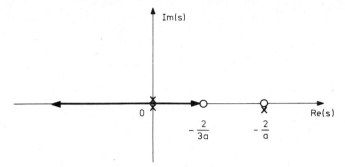

Fig. 5.16. Root-locus plot $\left(a < 0, b > 0, \bar{\xi} > 0, a^2 = 4b \right)$.

For the case $a^2 > 3b$ (see Figs. 5.16–5.18), there are always two real positive roots to $\varphi(\xi) = 0$, and the structure buckles. Consequently, the structure turns out to buckle for $a^2 \geqslant 3b$. In the range $a^2 < 3b$ the structure buckles if $\bar{\xi} \leqslant \bar{\xi}_{cr, s}$.

We next consider the case $a > 0$, $b > 0$, $\bar{\xi} < 0$. It is readily shown that the *inverse root loci* for $-\infty < \bar{\xi} \leqslant 0$ are the mirror images of the original root loci for $0 \leqslant \bar{\xi} < \infty$ with respect to the imaginary axis. The system buckles for $a^2 \geqslant 3b$, and also for $a^2 < 3b$ if $\bar{\xi} \geqslant \bar{\xi}_{cr, s}$ as indicated by Eq. (5.48).

For the particular case $a = 0$ the structure buckles if $b < 0$ (for both $\bar{\xi} > 0$ and $\bar{\xi} < 0$) and is insensitive if $b > 0$. For $b = 0$, the structure buckles if $a\bar{\xi} < 0$ and does not buckle in the opposite case.

As for the structure that buckles, differentiating Eq. (5.45) with respect to ξ and setting (for $b \neq 0$)

$$\frac{d\lambda}{d\xi} = 0 \qquad \lambda = \lambda_s \tag{5.49}$$

we obtain the relation between the buckling load λ_s and initial imperfection

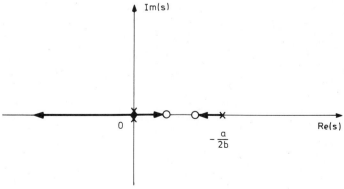

Fig. 5.17. Root-locus plot $\left(a < 0, b > 0, \bar{\xi} > 0, 3b < a^2 < 4b \right)$.

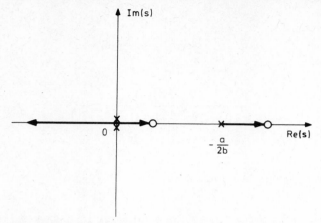

Fig. 5.18.   Root-locus plot $\left(a < 0, b > 0, \bar{\xi} > 0, a^2 > 4b \right)$.

amplitude $\bar{\xi}$:

$$\left(1 - \frac{\lambda_s}{\lambda_c} - \frac{a^2}{3b}\right)^3 = -\tfrac{27}{4}b\left[\frac{a}{3b}\left(1 - \frac{\lambda_s}{\lambda_c}\right) - \tfrac{2}{27}\frac{a^3}{b^2} + \frac{\lambda_s}{\lambda_c}\bar{\xi}\right]^2 \qquad (5.50)$$

For $b < 0$ and $a = 0$, we get from (5.50)

$$\left(1 - \frac{\lambda_s}{\lambda_c}\right)^{3/2} - \frac{3\sqrt{3}}{2}|\bar{\xi}|\sqrt{-b}\,\frac{\lambda_s}{\lambda_c} = 0 \qquad (5.51)$$

Note that the displacement ξ corresponding to λ_s/λ_c is given by

$$\xi_{1,2} = \frac{1}{3b}\left[-a \pm \sqrt{a^2 - 3b\left(1 - \frac{\lambda_s}{\lambda_c}\right)}\,\right] \qquad (5.52)$$

where $\xi_{1,2}$ depend on $\bar{\xi}$ via λ_s/λ_c. The static buckling load can be obtained from Eq. (5.50), given the initial imperfection $\bar{\xi}$. The meaningful root λ_s/λ_c of Eq. (5.50) is the greatest of those that meet the requirement $\xi\bar{\xi} > 0$.

The case $b = 0$ has to be considered separately. Equation (5.46) reduces to

$$\varphi(\xi) = a\xi^2 + 2a\bar{\xi}\xi + \bar{\xi}$$

Descartes' rule then immediately yields the conclusion that the structure is imperfection-sensitive if $a\bar{\xi} < 0$. Equation (5.49) then leaves us with

$$\left(1 - \frac{\lambda_s}{\lambda_c}\right)^2 + 4a\bar{\xi}\frac{\lambda_s}{\lambda_c} = 0 \tag{5.53}$$

Equations (5.50), (5.51), or (5.53) permit us now to find the probabilistic characteristics of the random buckling load Λ_s (with possible values λ_s) provided the initial imperfection $\bar{X}$ (with possible values $\bar{\xi}$) is a random variable with given probability distribution $F_{\bar{X}}(\bar{\xi})$. We seek the reliability of the structure, which is defined in these new circumstances as the probability of the event of the nondimensional buckling load Λ_s/λ_c exceeding the given nondimensional load α:

$$R(\alpha) = P\left(\frac{\Lambda_s}{\lambda_c} > \alpha\right) \tag{5.54}$$

Consider for example, the symmetric structure ($a = 0$), and assume that the initial imperfection $\bar{X}$ is normally distributed $N(m_{\bar{X}}, \sigma_{\bar{X}}^2)$. Figure 5.19a, which shows schematically the solution of Eq. (5.51), indicates that λ_s/λ_c exceeds α when $\bar{X}$ falls within the interval $(-\bar{\xi}', \bar{\xi}')$, where

$$\bar{\xi}' = \frac{2}{3\sqrt{3}} \frac{(1 - \alpha)^{3/2}}{\alpha\sqrt{-b}}$$

Therefore the reliability is

$$R(\alpha) = P(-\bar{\xi}' < \bar{X} < \bar{\xi}') \tag{5.55}$$

In conclusion, we have (see Eq. (4.24))

$$R(\alpha) = \mathrm{erf}\left(\frac{\bar{\xi}'(\alpha) - m_{\bar{X}}}{\sigma_{\bar{X}}}\right) - \mathrm{erf}\left(\frac{-\bar{\xi}'(\alpha) - m_{\bar{X}}}{\sigma_{\bar{X}}}\right)$$

$$= \mathrm{erf}\left(\frac{\bar{\xi}'(\alpha) - m_{\bar{X}}}{\sigma_{\bar{X}}}\right) + \mathrm{erf}\left(\frac{\bar{\xi}'(\alpha) + m_{\bar{X}}}{\sigma_{\bar{X}}}\right) \tag{5.56}$$

Note that this expression is symmetric with respect to the sign of the mean imperfection $m_{\bar{X}}$. Indeed, as is seen from Eq. (5.51) for a symmetric structure, the buckling load depends on $|\bar{\xi}|$. For $m_{\bar{X}} = 0$ we obtain

$$R(\alpha) = 2\,\mathrm{erf}\left(\frac{\bar{\xi}'(\alpha)}{\sigma_{\bar{X}}}\right)$$

In Fig. 5.19b, for example, the shaded area equals the reliability at the load

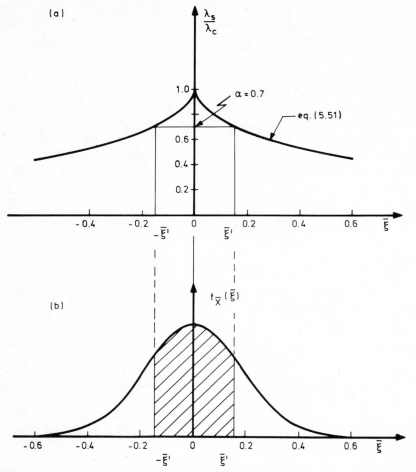

Fig. 5.19. (a) Buckling-load parameter as a function of initial imperfection amplitude. (b) Probability density of initial imperfection amplitude (shaded area equals the reliability of the structure at nondimensional load level α).

level $\alpha = 0.7$ for $m_{\bar{X}} = 0$. Figures 5.20 and 5.21 show the reliability function $R(\alpha)$ versus α for the different standard deviations $\sigma_{\bar{X}}$. Figure 5.20 is associated with $m_{\bar{X}} = 0.05$ and Fig. 5.21 with $m_{\bar{X}} = 0$. Consider specifically the case $m_{\bar{X}} > 0$. For $\sigma_{\bar{X}}$ approaching zero, all structures tend to buckle at some constant level α^*. This α^* satisfies the equation

$$(1 - \alpha^*)^{3/2} - \frac{3\sqrt{3}}{2} m_{\bar{X}} \sqrt{-b} \, \alpha^* = 0$$

which is obtained from Eq. (5.51) by substituting α^* and $m_{\bar{X}}$ for λ_s/λ_c and $\bar{\xi}$,

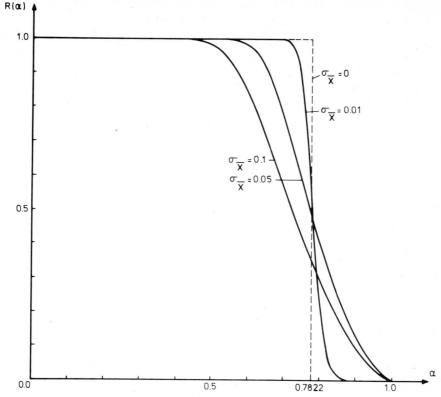

Fig. 5.20.   Influence of standard deviation of initial imperfections on reliability; mean imperfection nonzero ($m_{\bar{X}} = 0.05, b = -1$).

respectively. Indeed, for $\alpha < \alpha^*$, the difference $\bar{\xi}'(\alpha) - m_{\bar{X}}$ is a postive quantity, so that when $\sigma_{\bar{X}} \to 0$

$$\frac{\bar{\xi}'(\alpha) - m_{\bar{X}}}{\sigma_{\bar{X}}} \to +\infty$$

and

$$\mathrm{erf}\left(\frac{\bar{\xi}'(\alpha) - m_{\bar{X}}}{\sigma_{\bar{X}}}\right) \to \tfrac{1}{2}$$

However, since $\bar{\xi}'(\alpha) + m_{\bar{X}} > 0$, irrespective of α, we also have

$$\mathrm{erf}\left(\frac{\bar{\xi}'(\alpha) + m_{\bar{X}}}{\sigma_{\bar{X}}}\right) \to \tfrac{1}{2}$$

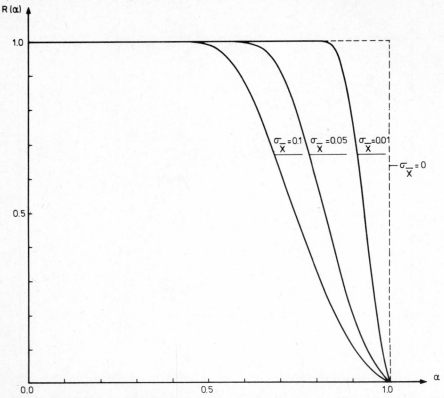

Fig. 5.21.   Influence of standard deviation of initial imperfection on reliability; mean imperfection identically zero ($b = -1$).

and for $\alpha < \alpha^*$, $R(\alpha) \to \frac{1}{2} + \frac{1}{2} = 1$. As for $\alpha > \alpha^*$, the difference $\bar{\xi}'(\alpha) - m_{\bar{X}}$ is a negative quantity. With $\sigma_{\bar{X}} \to 0$,

$$\frac{\bar{\xi}'(\alpha) - m_{\bar{X}}}{\sigma_{\bar{X}}} \to -\infty$$

and

$$\text{erf}\left(\frac{\bar{\xi}'(\alpha) - m_{\bar{X}}}{\sigma_{\bar{X}}}\right) \to -\frac{1}{2}$$

Consequently, $R(\alpha) \to -\frac{1}{2} + \frac{1}{2} = 0$. The reliability function, shown in Fig. 5.20 by the dashed line, has the shape of a rectangular "pulse," α^* being equal to 0.7822. Such a rectangular reliability function is characteristic of a deterministic structure. In fact, for the deterministic imperfection, all realizations of the structure are identical with the same buckling load α^* (note that the mean buckling load in this case also equals α^*!). Therefore, $\text{Prob}(\Lambda_s/\Lambda_c \geqslant \alpha) = 1$ if $\alpha < \alpha^*$ and $\text{Prob}(\Lambda_s/\Lambda_c > \alpha) = 0$ if $\alpha > \alpha^*$. For a structure with zero mean

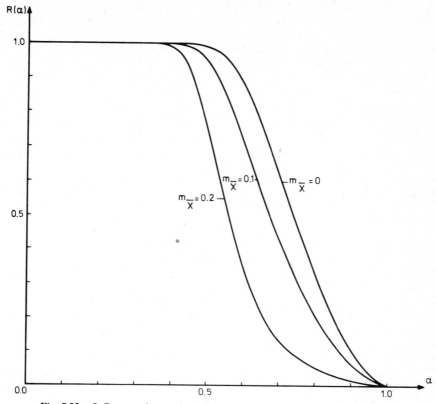

Fig. 5.22.   Influence of mean imperfection on reliability ($\sigma_{\bar{X}} = 0.1, b = -1$).

(see Fig. 5.21), with $\sigma_{\bar{X}}$ approaching zero, we have $\alpha^* = 1$, implying that almost all structures buckle at the classical buckling load. This is as expected, since with $m_{\bar{X}} = 0$ and $\sigma_{\bar{X}} \to 0$, almost all structures tend to become perfect.

The influence of the mean imperfection on the reliability, at constant standard deviation, is illustrated in Figs. 5.22 and 5.23. $\sigma_{\bar{X}}$ is set at 0.1 in Fig. 5.22 and 0.02 in Fig. 5.23. As is seen, for larger values of $m_{\bar{X}}/\sigma_{\bar{X}}$ the reliability curves become steeper and resemble more closely the deterministic case with rectangular "pulse"-type reliability.

5.6   DYNAMIC IMPERFECTION-SENSITIVITY OF A NONLINEAR MODEL STRUCTURE

In the dynamic setting, the central hinge carries a mass M, and the system is under an axial force $\lambda f(t)$. Instead of Eq. (5.45) we have (see Fig. 5.24)

$$\frac{1}{\omega_1^2} \frac{d^2\xi}{dt^2} + \left(1 - \frac{\lambda f(t)}{\lambda_c}\right)\xi + a\xi^2 + b\xi^3 = \frac{\lambda f(t)}{\lambda_c}\bar{\xi} \qquad (5.57)$$

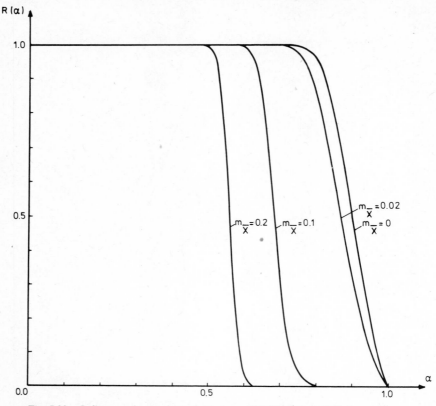

Fig. 5.23. Influence of mean imperfection on reliability ($\sigma_{\bar{X}} = 0.02$, $b = -1$).

where $\omega_1 = \sqrt{k_1/ML}$ is the natural frequency of the mass for $\lambda = 0$. As an example, we have the load represented by the step function of unit magnitude and unlimited duration, defined in Eq. (3.2), $f(t) = U(t)$, which vanishes for $t < 0$ and equals unity for $t \geqslant 0$.

The first integral of Eq. (5.57) subject to the initial conditions $\xi = 0$, $d\xi/dt = 0$ at $t = 0$ (zero displacement and zero velocity at $t = 0$, that is the mass at rest up to that time) is readily found to be

$$\frac{1}{\omega_1^2}\left(\frac{d\xi}{dt}\right)^2 + \left(1 - \frac{\lambda}{\lambda_c}\right)\xi^2 + \tfrac{2}{3}a\xi^3 + \tfrac{1}{2}b\xi^4 = 2\left(\frac{\lambda}{\lambda_c}\right)\bar{\xi}\xi$$

and the corresponding integral curves satisfy

$$\pm \int_0^\xi \left[2\left(\frac{\lambda}{\lambda_c}\right)\bar{\xi}\xi - \left(1 - \frac{\lambda}{\lambda_c}\right)\xi^2 - \tfrac{2}{3}a\xi^3 - \tfrac{1}{2}b\xi^4\right]^{-1/2} d\xi = \omega_1 t$$

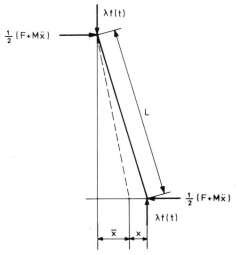

Fig. 5.24.   Derivation of Eq. (5.57).

where the left-hand term may be evaluated in terms of *elliptic integrals*. For sufficiently small λ/λ_c, the motion is periodic, with its amplitude satisfying

$$\left(1 - \frac{\lambda}{\lambda_c}\right)\xi_{max} + \tfrac{2}{3}a\xi_{max}^2 + \tfrac{1}{2}b\xi_{max}^3 = 2\left(\frac{\lambda}{\lambda_c}\right)\bar{\xi} \qquad (5.58)$$

The *dynamic buckling load* λ_d is defined as the maximum value of λ, such that the response $\xi(t)$ still remains finite. At λ_d a finite jump in ξ_{max} is produced by an infinitesimal increase in λ (see Fig. 5.25). For all $\lambda \neq \lambda_d$, the response $\xi(t)$ is periodic; as λ approaches λ_d from below, the period tends to infinity and it takes an infinitely long time for $\xi(t)$ to reach ξ_{max}. The value of λ_d occurs at the first maximum of the relation λ versus ξ_{max}. Thus the dynamic buckling load is defined by the criterion

$$\frac{d\lambda}{d\xi_{max}} = 0 \qquad \lambda = \lambda_d \qquad (5.59)$$

Note that Eq. (5.58) becomes identical with Eq. (5.45) if we make the following formal substitution:

$$\xi \to \xi_{max} \qquad a \to \tfrac{2}{3}a \qquad b \to \tfrac{1}{2}b, \quad \bar{\xi} \to 2\bar{\xi} \qquad (5.60)$$

Analogously, all conclusions in the static case are readily extended to the dynamic one. The structure under a step load buckles for any $b < 0$ (irrespective of the sign of a or $\bar{\xi}$), and does not buckle for $b > 0$ and $a\bar{\xi} > 0$; for $b > 0$

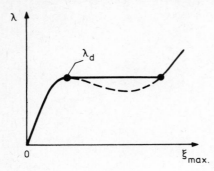

Fig. 5.25. Load versus maximum response; λ_d is the dynamic buckling load. (After Budiansky and Hutchinson)

and $a\bar{\xi} < 0$ it buckles for any $\bar{\xi}$, given the inequality $a^2 \geqslant \frac{27}{8}b$, which is obtainable from its static counterpart $a^2 \geqslant 3b$ by formal substitution (5.60), and also when $a^2 < \frac{27}{8}b$ for $\bar{\xi} \leqslant \bar{\xi}_{cr, d}$, where

$$\bar{\xi}_{cr, d} = -\tfrac{4}{27}\left(\frac{a^3}{b}\right)\frac{1}{\frac{27}{8}b - a^2}$$

The ξ_{max} value associated with $\bar{\xi}_{cr, d}$ equals $-4a/9b$, and consequently the dynamic buckling load is

$$\frac{\lambda_d}{\lambda_c} = \tfrac{4}{9}\frac{a}{b}\left(\tfrac{16}{81}\frac{a^2}{b} - 1\right)\left(2\bar{\xi}_{cr, d} - \tfrac{4}{9}\frac{a}{b}\right)^{-1}$$

At $\bar{\xi} = \bar{\xi}_{cr, d}$ the concept of dynamic buckling is preserved by associating λ_d with the point of inflection in the λ–ξ_{max} curves according to Budiansky (1965) (Fig. 5.26). Comparison of the latter results with their static counterparts shows that the interval $3b < a^2 < \frac{27}{8}b$ is characterized by duality: The structure is statically imperfection-sensitive, but dynamically imperfection-insensitive. In the particular case $b/a^2 = \frac{2}{3}$, considered by Budiansky (1967, pp. 95–96), $\bar{\xi}_{cr, d} = -8/45a$ and $\bar{\xi}_{cr, d} > \bar{\xi}_{cr, s}$, and consequently the interval $\bar{\xi}_{cr, s}$

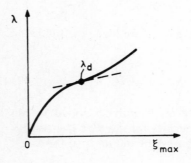

Fig. 5.26. Generalization of dynamic buckling criterion (after Budiansky).

$< \bar{\xi} < \bar{\xi}_{cr, d}$ is similarly characterized by duality, the reverse of the preceding case. For $b/a^2 = \frac{1}{3}$ the structure is statically imperfection-sensitive for any $\bar{\xi}$, but dynamically imperfection-insensitive if $\bar{\xi} > \bar{\xi}_{cr, d} = -\frac{32}{9a}$.

Finally, the relation between the buckling load λ_d and the initial imperfection $\bar{\xi}$ is given by ($b \neq 0$)

$$\left(1 - \frac{\lambda_d}{\lambda_c} - \frac{8}{27}\frac{a^2}{b}\right)^3 = -\frac{27}{8}b\left[\frac{4}{9}\frac{a}{b}\left(1 - \frac{\lambda_d}{\lambda_c}\right) - \frac{64}{729}\frac{a^3}{b^2} + 2\bar{\xi}\frac{\lambda_d}{\lambda_c}\right]^2$$

(5.61)

For vanishing a and $b < 0$, we obtain the relationship

$$\left(1 - \frac{\lambda_d}{\lambda_c}\right)^{3/2} - \frac{3\sqrt{6}}{2}|\bar{\xi}|\sqrt{-b}\,\frac{\lambda_d}{\lambda_c} = 0$$

(5.62)

For the case $b = 0$, $a \neq 0$, and substituting (5.60) in (5.53), we immediately obtain

$$\left(1 - \frac{\lambda_d}{\lambda_c}\right)^2 + \frac{16}{3}a\bar{\xi}\frac{\lambda_d}{\lambda_c} = 0$$

In the case $a < 0$, $b < 0$ with $\bar{\xi}$ such that $\lambda_s/\lambda_c \leqslant 1$ and $\lambda_d/\lambda_c \leqslant 1$, $\bar{\xi}$ is readily eliminated by correlating Eqs. (5.50) and (5.61) for a given structure with a given imperfection. The result relates λ_d to λ_s:

$$\left[-\frac{8}{27b}\left(1 - \frac{\lambda_d}{\lambda_c} - \frac{8}{27}\frac{a^2}{b}\right)^3\right]^{1/2} - \frac{4}{9}\frac{a}{b}\left(1 - \frac{\lambda_d}{\lambda_c}\right) + \frac{64}{729}\frac{a^3}{b^2}$$

$$= 2\frac{\lambda_d}{\lambda_s}\left\{\left[-\frac{4}{27b}\left(1 - \frac{\lambda_s}{\lambda_c} - \frac{a^2}{3b}\right)^3\right]^{1/2} - \frac{a}{3b}\left(1 - \frac{\lambda_s}{\lambda_c}\right) + \frac{2}{27}\frac{a^3}{b^2}\right\}$$

(5.63)

In this form, λ_d/λ_s is no longer directly depended on the imperfection, but via λ_s/λ_c. For vanishing a, we obtain the expression

$$\frac{\lambda_d}{\lambda_s} = \frac{\sqrt{2}}{2}\left(\frac{\lambda_c - \lambda_d}{\lambda_c - \lambda_s}\right)^{3/2}$$

(5.64)

Equations (5.62) and (5.64) are due to Budiansky and Hutchinson.

Recapitualting, our structure is statically imperfection-sensitive and dynamically imperfection-insensitive when

$$3b < a^2 < \frac{27}{8}b, \quad \bar{\xi} > \bar{\xi}_{cr, d}$$

and also when

$$a^2 < 3b, \quad \bar{\xi}_{cr,d} < \bar{\xi} < \bar{\xi}_{cr,s}$$

For this case, following Budiansky, the criterion of the dynamic buckling is generalized, the dynamic buckling load being defined as the point of inflection on the λ–ξ_{max} curve:

$$\frac{d^2\lambda}{d\xi_{max}^2} = \frac{\lambda_c}{(\xi_{max} + 2\bar{\xi})^3}\left[b(\xi_{max} + 2\bar{\xi})^3 - 8b\bar{\xi}^3 + \tfrac{16}{3}a\bar{\xi}^2 - 4\bar{\xi}\right] = 0$$

$$(5.65)$$

(See also Fig. 5.26.) For $b > 0$, $a < 0$, and $\bar{\xi} > 0$, Descartes' rule of signs indicates a single positive root. Eliminating ξ_{max} between Eqs. (5.58) and (5.65), we relate the generalized buckling load and the initial imperfection:

$$\frac{\lambda_d}{\lambda_c}d = (d - 2\bar{\xi})\left[1 + \tfrac{2}{3}a(d - 2\bar{\xi}) + \tfrac{1}{2}b(d - 2\bar{\xi})^2\right] \qquad (5.66)$$

where

$$d = \left(\frac{24b\bar{\xi}^3 - 16a\bar{\xi}^2 + 12\bar{\xi}}{3b}\right)^{1/3}$$

We proceed now to the reliability analysis for a symmetric structure, assuming $\bar{X}$ to be $N(m_{\bar{X}}, \sigma_{\bar{X}}^2)$:

$$R(\alpha) = P\left(\frac{\Lambda_d}{\lambda_c} > \alpha\right) = P(-\bar{\xi}'' < \bar{X} < \bar{\xi}'') \qquad (5.67)$$

where

$$\bar{\xi}'' = \frac{2}{3\sqrt{6}}\frac{(1 - \alpha)^{3/2}}{\alpha\sqrt{-b}}$$

resulting in

$$R(\alpha) = \text{erf}\left[\frac{\bar{\xi}''(\alpha) - m_{\bar{X}}}{\sigma_{\bar{X}}}\right] + \text{erf}\left[\frac{\bar{\xi}''(\alpha) + m_{\bar{X}}}{\sigma_{\bar{X}}}\right] \qquad (5.68)$$

Note that the probability densities of static or dynamic buckling loads are obtainable via the reliability function. Indeed, the unreliability at the load level α is

$$Q(\alpha) = 1 - R(\alpha) = P\left(\frac{\Lambda_s}{\lambda_c} \leqslant \alpha\right) = F_{\Lambda_s/\lambda_c}(\alpha) \qquad (5.69)$$

and therefore

$$f_{\Lambda_s/\lambda_c}\left(\frac{\lambda_s}{\lambda_c}\right) = \frac{dF_{\Lambda_s/\lambda_c}(\alpha)}{d\alpha}\bigg|_{\alpha=\lambda_s/\lambda_c} = -\frac{dR(\alpha)}{d\alpha}\bigg|_{\alpha=\lambda_s/\lambda_c} \tag{5.70}$$

Equation (5.56) yields

$$f_{\Lambda_s/\lambda_c}\left(\frac{\lambda_s}{\lambda_c}\right) = \frac{1}{9}\left(\frac{6}{\pi(-b)}\right)^{1/2}\left(\frac{2 + (\lambda_s/\lambda_c)}{\sigma_{\bar{X}}^2(\lambda_s/\lambda_c)^2}\right)\left(1 - \frac{\lambda_s}{\lambda_c}\right)^{1/2}$$

$$\times \exp\left[-\frac{2(1 - \lambda_s/\lambda_c)^3}{27(\lambda_s/\lambda_c)^2(-b)\sigma_{\bar{X}}^2}\right] \tag{5.71}$$

(see Fig. 5.27). With (5.71) available, we are able, if necessary, to calculate the mean buckling load $E(\Lambda_s)$:

$$E(\Lambda_s) = \int_0^{\lambda_c} \lambda_s f_{\Lambda_s}(\lambda_s) \, d\lambda_s, \tag{5.72}$$

the variance of the buckling load,

$$\mathrm{Var}(\Lambda_s) = \int_0^{\lambda_c} \lambda_s^2 f_{\Lambda_s}(\lambda_s) \, d\lambda_s - \left[E(\Lambda_s)\right]^2 \tag{5.73}$$

and its higher moments.

In particular, in view of the identity

$$f_{\Lambda_s}(\lambda_s) = \frac{1}{\lambda_c} f_{\Lambda_x/\lambda_c}\left(\frac{\lambda_s}{\lambda_c}\right)$$

we obtain for the mean buckling load

$$E(\Lambda_s) = \int_0^{\lambda_s} \lambda_s\left[\frac{1}{\lambda_c} f_{\Lambda_s/\lambda_c}\left(\frac{\lambda_s}{\lambda_c}\right)\right] d\lambda_s$$

or

$$E\left(\frac{\Lambda_s}{\lambda_c}\right) = \int_0^1 \alpha f_{\Lambda_s/\lambda_c}(\alpha) \, d\alpha$$

Further, taking Eq. (5.70) into account, we arrive at

$$E\left(\frac{\Lambda_s}{\lambda_c}\right) = \int_0^1 \alpha\left[-\frac{dR(\alpha)}{d\alpha}\right] d\alpha \tag{5.74}$$

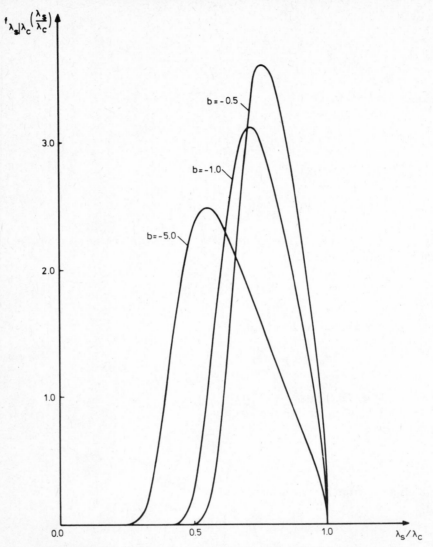

Fig. 5.27. Probability densities of nondimensional static buckling loads ($\sigma_{\bar{X}} = 0.1$).

Integrating by parts,

$$E\left(\frac{\Lambda_s}{\lambda_c}\right) = -\alpha R(\alpha)\Big]_0^1 + \int_0^1 R(\alpha)\, d\alpha \qquad (5.75)$$

However,

$$R(1) = \text{Prob}(\Lambda_s > \lambda_c) = 0 \qquad (5.76)$$

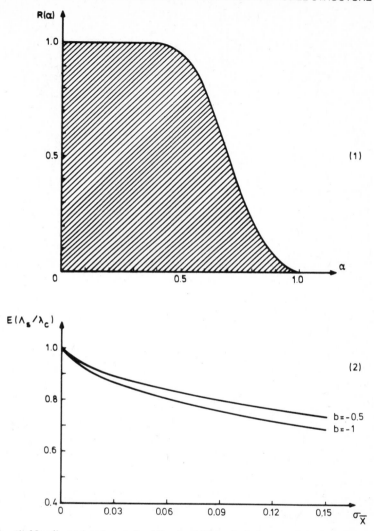

Fig. 5.28. (1) Nondimensional mean buckling load $E(\Lambda_s/\lambda_c)$ equals the area under the reliability curve ($m_{\bar{X}} = 0.05, \sigma_{\bar{X}} = 0.1, b = -1$). (2) $E(\Lambda_s/\lambda_c)$ versus the standard deviation $\sigma_{\bar{X}}$ of initial imperfections.

and finally,

$$E\left(\frac{\Lambda_s}{\lambda_c}\right) = \int_0^1 R(\alpha)\, d\alpha \qquad (5.77)$$

so that the nondimensional mean buckling load equals the area under the reliability curve (see Fig. 5.28).

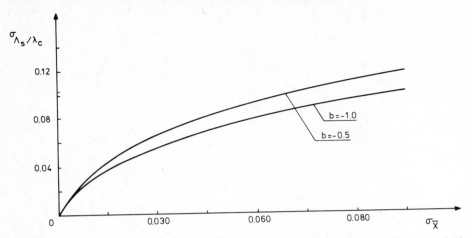

Fig. 5.29. Standard deviation of nondimensional buckling load as a function of the standard deviation of initial imperfection.

The standard deviation of the nondimensional buckling load

$$\sigma_{\Lambda_x/\lambda_c} = \left\{ E\left[\left(\frac{\Lambda_s}{\lambda_c}\right)^2\right] - \left[E\left(\frac{\Lambda_s}{\lambda_c}\right)\right]^2 \right\}^{1/2} \tag{5.78}$$

where

$$E\left[\left(\frac{\Lambda_s}{\lambda_c}\right)^2\right] = \int_0^1 \alpha^2 f_{\Lambda_s/\lambda_c}(\alpha)\, d\alpha$$

is shown in Fig. 5.29. $\sigma_{\Lambda_s/\lambda_c}$ increases with the standard deviation $\sigma_{\bar{X}}$ of the initial imperfections.

With the reliability function available, we can solve the design problem, of determining the allowable load from the requirement

$$R(\alpha_{\text{allow}}) = r \qquad \alpha_{\text{allow}} = \frac{\lambda_{\text{allow}}}{\lambda_c} \tag{5.79}$$

where r is the required reliability. Referring to Eq. (5.56), we obtain the follow transcendental equation for α_{allow}:

$$\text{erf}\left(\frac{\bar{\xi}'(\alpha_{\text{allow}}) - m_{\bar{X}}}{\sigma_{\bar{X}}}\right) + \text{erf}\left(\frac{\bar{\xi}'(\alpha_{\text{allow}}) + m_{\bar{X}}}{\sigma_{\bar{X}}}\right) = r \tag{5.80}$$

For $m_{\bar{X}} = 0$, and bearing in mind the expression for $\bar{\xi}'(\alpha_{\text{allow}})$, Eq. 5.80 reduces to

$$\frac{2}{3\sqrt{3}}\left(\frac{\left(1-\alpha_{\text{allow}}\right)^{3/2}}{\alpha_{\text{allow}}\sqrt{-b}\,\sigma_{\bar{X}}}\right) = \text{erf}^{-1}\left(\frac{r}{2}\right) \tag{5.81}$$

where $\text{erf}^{-1}(\dots)$ is the inverse of $\text{erf}(\dots)$, found from the table in Appendix B. For example, for $r = 0.99$ we have $\text{erf}^{-1}(0.495) = 2.575$. The last equation may be rewritten as

$$\sigma_{\bar{X}} = \frac{2}{3\sqrt{3}}\left(\frac{\left(1-\alpha_{\text{allow}}\right)^{3/2}}{\alpha_{\text{allow}}\sqrt{-b}\,\text{erf}^{-1}(r/2)}\right) \tag{5.82}$$

and $\sigma_{\bar{X}}$ treated as a function of α_{allow}. This function is given in Fig. 5.30 as curve *1*.

For $m_{\bar{X}} \neq 0$, the transcendental equation (5.80) has to be solved numerically. Results are given in Fig. 5.30 as curves *2* and *3*. It is seen that the

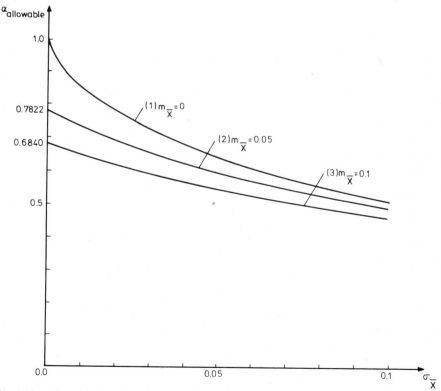

Fig. 5.30. Allowable load corresponding to required reliability $r = 0.99$ as a function of standard deviation of initial imperfections ($b = -1$).

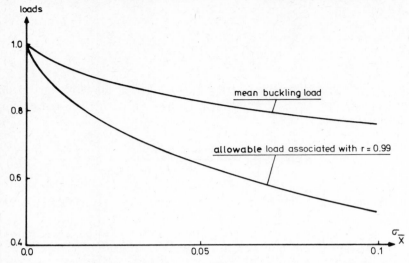

Fig. 5.31.   Mean buckling load exceeding allowable load associated with high reliability.

allowable load decreases as the mean imperfection increases, and that for larger values of $\sigma_{\bar{X}}$ the mean imperfection becomes less significant and the allowable loads associated with different mean imperfections lie closer together.

Note also that for $\sigma_{\bar{X}} = 0$ the allowable load coincides with the mean buckling load irrespective of the required reliability. As is seen from Eq. (5.82), $\sigma_{\bar{X}} = 0$ if $\alpha_{\text{allow}} = 1$ for any r. Figure 5.31 contrasts the allowable load associated with the required reliability $r = 0.99$ and the mean buckling load. As is seen, the allowable loads are much smaller than the mean loads. This implies that the design according to the latter overestimates the load-carrying capacity of the structure, as against α_{allow} associated with the high reliability.

Significantly, Figs. 5.20 and 5.30 apply for the dynamic case as well, albeit for a different b—namely, $b = \frac{1}{2}(-1) = -\frac{1}{2}$, according to the analogy in Eq. (5.60). Comparison of the two cases for the same $b \ (= -1)$ shows (Fig. 5.32) that the dynamic allowable loads are lower than their static counterparts, the analogue of (5.82) being

$$\sigma_{\bar{X}} = \frac{2}{3\sqrt{6}} \left(\frac{(1 - \alpha_{\text{allow}})^{3/2}}{\alpha_{\text{allow}}\sqrt{-b}\, \text{erf}^{-1}(r/2)} \right) \tag{5.83}$$

Comparison of the two analogues shows that for the loads to be equal, the standard deviation of the initial imperfections in the dynamic case should be $1/\sqrt{2}$ that of the static case. In other words, the dynamic allowable load at a specific $\sigma_{\bar{X}}$ is obtainable directly from the static curve by reading the latter at $\sqrt{2}\,\sigma_{\bar{X}}$.

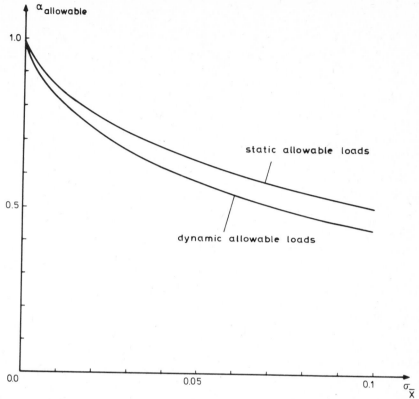

Fig. 5.32. Comparison of allowable loads corresponding to required reliability $r = 0.99$ for static and dynamic cases ($b = -1$, $m_{\bar{X}} = 0$).

5.7 AXIAL IMPACT OF A BAR WITH RANDOM INITIAL IMPERFECTIONS*

We consider now another problem of dynamic buckling, that of the initially imperfect bar under axial impact (Fig. 5.33). In the case under consideration, we use Hoff's definition, which states, "*A structure is in a stable state if admissible finite disturbances of its initial state of static or dynamic equilibrium are followed by displacements whose magnitude remains within allowable bounds during the required lifetime of the structure*" (in contrast to the previous paragraph, where finiteness of the displacements was required for the structure to be considered stable). By virtue of Hoff's definition, we may postulate that a bar with initial imperfections fails (buckles) under axial forces when its dynamic response (deflection, strain, or stress) first reaches an upper-bound

*This section is an extension of the author's paper (1978), from which Figs. 5.35–5.38 are also reproduced.

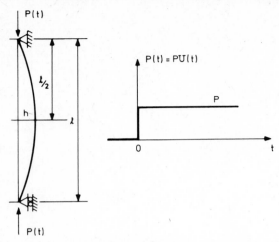

Fig. 5.33. Imperfect bar under axial impact, represented by unit step function in time, $P(t) = PU(t)$.

level Q^+ or a lower-bound level $-Q^-$, Q^+ and Q^- being prescribed positive numbers that represent borderlines between stability and buckling (i.e., safety and failure) (Fig. 5.34).

Consider an ensemble of response histories $y(t)$ in the interval $0 \leqslant t < t^*$, all of them originating under the same initial conditions $t = 0$. Let $R(t; t^*)$ be the probability of $y(t)$ remaining in the same domain throughout the interval $(0, t^*)$. Formally, the reliability is then

$$R(t; t^*) = \text{Prob}\{[y(t) \leqslant Q^+] \cap [y(t) \geqslant -Q^-], \quad 0 \leqslant t \leqslant t^*\} \quad (5.84)$$

The probability of failure (unreliability) is

$$Q(t; t^*) = \text{Prob}\{[y(t) > Q^+] \cup [y(t) < -Q^-], \quad 0 \leqslant t \leqslant t^*\} \quad (5.85)$$

since failure is either the event $\{y(t) > Q^+\}$ or $\{y(t) < -Q^-\}$. Obviously,

$$R(t; t^*) = 1 - Q(t; t^*)$$

which is readily understood if we recall that in these circumstances reliability is the probability of survival up to time t, or that of "being," whereas unreliability is the probability of failure, of "not being." The random event "to be or not to be" has a unity probability.

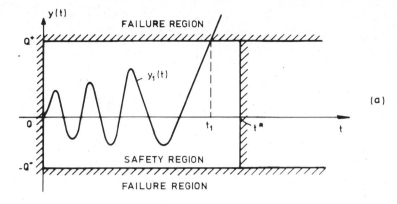

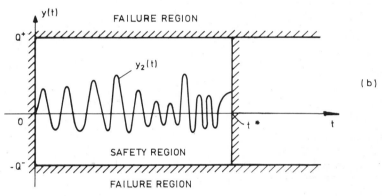

Fig. 5.34. Illustration of Hoff's definition of stability. (a) t_1, time to fail. (b) Structure does not fail.

For t^* tending to infinity, the reliability and the unreliability are functions of t only, and we denote

$$R(t) = \lim_{t^* \to \infty} R(t; t^*) \qquad Q(t) = \lim_{t^* \to \infty} Q(t; t^*)$$

We seek probabilistic information on the random time when failure occurs (i.e., the structure buckles) under suitable initial conditions, in terms of $y(t)$ and perhaps of its derivatives. This problem is very difficult in the general case and is known as the "*first-passage*" (or "*first-excursion*") problem. We give here the *exact* solution to this problem for impact buckling of a bar with random initial imperfections of *given* shape, with the amplitude as a continuous random variable. We formulate the problem as follows:

Given the probability distribution function of the amplitude of random initial imperfection, find the probability of the time required for the response process to move outside the prescribed safe domain (i.e., the first excursion time) being less than the given time, t* being infinity.

Consider first the corresponding deterministic problem. We disregard axial wave propagation and assume uniform compression throughout the bar, whose motion obeys the differential equation

$$EI\frac{\partial^4 y}{\partial x^4} + P\frac{\partial^2 y}{\partial x^2} + \rho A\frac{\partial^2 y}{\partial t^2} = -P\frac{d^2 \bar{y}}{dx^2} \tag{5.86}$$

where x is the axial coordinate; t, time; $\bar{y}(x)$, the initial imperfection (a small perturbation to the perfect, straight shape of the bar); $y(x, t)$, the additional transverse deflection measured from $y_0(x)$ [$y_0(x) + y(x, t)$ being the total deflection of the beam axis from the straight line between the two ends $x = 0$, $x = l$]; E, Young's modulus; I, the moment of inertia; ρ, mass density; A, the cross-sectional area; and P, the applied axial load; with EI and ρA taken as constants.

The differential equation (5.86) is supplemented by the boundary conditions

$$y(x, t) = 0, \quad \frac{\partial^2 y(x, t)}{\partial x^2} = 0, \quad x = 0$$

$$y(x, t) = 0, \quad \frac{\partial^2 y(x, t)}{\partial x^2} = 0, \quad x = l$$

and the initial conditions

$$y(x, t) = 0, \quad \frac{\partial y(x, t)}{\partial t} = 0, \quad t = 0$$

The shape of the imperfection is taken as

$$\bar{y}(x) = H\sin\frac{\pi x}{l} \tag{5.87}$$

We now introduce the nondimensional quantities

$$\xi = \frac{x}{l} \qquad \lambda = \omega_1 t \qquad \alpha = \frac{P}{P_c}$$

$$u(\xi, \lambda) = \frac{y(x, t)}{\Delta} \qquad \bar{u}(\xi) = \frac{\bar{y}(x)}{\Delta} \qquad G = \frac{H}{\Delta}$$

where

$$P_c = \frac{\pi^2 EI}{l^2} \qquad \omega_1 = \left(\frac{\pi}{l}\right)^2 \sqrt{\frac{EI}{\rho A}} \qquad \Delta = \sqrt{\frac{I}{A}}$$

P_c being the *classical (or Euler) buckling load* of a perfect bar; ω_1 its fundamental natural frequency in the absence of axial compression; Δ its radius of gyration; α the nondimensional applied force; $u(\xi, \lambda), \bar{u}(\xi)$ the nondimensional additional and initial displacements, respectively; G the nondimensional amplitude of the initial imperfection; and λ the nondimensional time.

Equations (5.86) and (5.87) become, respectively,

$$\frac{\partial^4 u}{\partial \xi^4} + \pi^2 \alpha \frac{\partial^2 u}{\partial \xi^2} + \pi^4 \frac{\partial^2 u}{\partial \lambda^2} = -\pi^2 \alpha \frac{d^2 \bar{u}}{d\xi^2} \qquad (5.88)$$

$$\bar{u}(\xi) = G \sin \pi \xi \qquad (5.89)$$

Equation (5.88) is based on the conventional strength-of-materials assumptions of uniform geometrical and physical properties, linear elastic behavior, and small displacements, disregarding rotational and axial inertia and shear effects. It is also assumed that the standard deviation of the amplitude of the initial imperfections is smaller than the radius of gyration. Thus the problem may be treated in a probabilistically linear setting.

The boundary conditions are satisfied by setting

$$u(\xi, \lambda) = e(\lambda)\sin \pi \xi \qquad (5.90)$$

Substituting Eqs. (5.89) and (5.90) in (5.88), we obtain the differential equation for $e(\lambda)$

$$\frac{d^2 e(\lambda)}{d\lambda^2} + (1 - \alpha) e(\lambda) = \alpha G \qquad (5.91)$$

with initial condition

$$e(0) = \frac{de(0)}{d\lambda} = 0$$

The solution is given by

$$
e(\lambda) = \begin{cases} \dfrac{G}{1-\beta}(\cosh r\lambda - 1), & \beta < 1 \\ \frac{1}{2}G\lambda^2, & \beta = 1 \\ \dfrac{G}{1-\beta}(\cos r\lambda - 1), & \beta > 1, r = \sqrt{|1-\alpha|}, \beta = \dfrac{1}{\alpha} \end{cases} \tag{5.92}
$$

and the total displacement $v(\xi, \lambda) = V(\lambda)\sin \pi\xi$ by

$$
V(\lambda) = \begin{cases} \dfrac{G}{1-\beta}(\cosh r\lambda - \beta), & \beta < 1 \\ \frac{1}{2}G(\lambda^2 + 2), & \beta = 1 \\ \dfrac{G}{1-\beta}(\cos r\lambda - \beta), & \beta > 1 \end{cases} \tag{5.93}
$$

where $V(\lambda) = e(\lambda) + G$.

We proceed now to buckling under random imperfections. Let G be a continuous random variable with probability distribution function $F_G(g)$. For simplicity, we consider the case of symmetrical bounds, $Q^+ = Q^- \equiv c \, (> 0)$, failure being thus identified with the total displacement reaching, in absolute value, the critical point c.

We seek the probablity $\mathrm{Prob}(T \leqslant t)$ of the first passage time T being equal to or smaller than the given time t; or, in nondimensional form, $\mathrm{Prob}(\Lambda \leqslant \lambda)$ of the nondimensional first passage time $\Lambda = \omega_1 T$ being equal to or smaller than the given nondimensional time $\lambda = \omega_1 t$. Denoting by $\{L\}$ the contingency of buckling being possible [i.e., in the time interval $(0, \infty)$], $\mathrm{Prob}(L)$ is the probability of failure of the system in the infinite time interval. Note, moreover, that if this probability is zero (i.e., buckling cannot occur), then likewise $\mathrm{Prob}(\Lambda \leqslant \lambda) = 0$. This also follows from the formula of overall probability

$$\mathrm{Prob}(\Lambda \leqslant \lambda) = \mathrm{Prob}(\Lambda \leqslant \lambda|L)\mathrm{Prob}(L) + \mathrm{Prob}(\Lambda \leqslant \lambda|\bar{L})\mathrm{Prob}(\bar{L})$$

where $(\Lambda \leqslant \lambda)$ signifies that the first passage time Λ is equal to or smaller than the given time λ; $(\Lambda \leqslant \lambda|L)$ refers to the contingency that the first passage time Λ is equal to or smaller than λ, provided buckling is possible, whereas $(\Lambda \leqslant \lambda|\bar{L})$ refers to the contingency that the first passage time Λ is equal to or smaller than λ, provided buckling is impossible. However,

$$(\Lambda \leqslant \lambda|\bar{L}) = \varnothing \qquad \mathrm{Prob}(\Lambda \leqslant \lambda|\bar{L}) = 0$$

and finally we are left with

$$\mathrm{Prob}(\Lambda \leqslant \lambda) = \mathrm{Prob}(\Lambda \leqslant \lambda|L)\mathrm{Prob}(L) \tag{5.94}$$

Note that for nonzero Prob(L) and

$$\text{Prob}(\Lambda \leqslant \infty | L) = 1 \qquad (5.95)$$

the following equality holds:

$$\text{Prob}(\Lambda \leqslant \infty) = \text{Prob}(L) \qquad (5.96)$$

If the latter probability is unity [i.e., if buckling is certain in the time interval $(0, \infty)$], then also $\text{Prob}(\Lambda \leqslant \infty) = \text{Prob}(\Lambda < \infty) = 1$. In such a case we denote

$$\text{Prob}(\Lambda \leqslant \lambda) = F_\Lambda(\lambda) \qquad \text{Prob}(\Lambda \leqslant \lambda | L) = F_\Lambda(\lambda | L) \qquad (5.97)$$

where $F_\Lambda(\lambda)$ is probability distribution function of Λ.

Note that $F_\Lambda(0)$ does not necessarily vanish. There could be some *nonzero* probability of the amplitude of the initial imperfection having already exceeded the critical point. This probability is

$$F_\Lambda(0) = \text{Prob}(\Lambda \leqslant 0) = \text{Prob}(\Lambda = 0)$$

$$= \text{Prob}(|G| \geqslant c) = 1 - F_G(c) + F_G(-c)$$

If $F_G(c) - F_G(-c) = 1$, then $F_\Lambda(0) = 0$. (Note the analogy with Problem 4.3.)

We now proceed to calculate $F_\Lambda(\lambda)$ for three different cases, in accordance with Eq. (5.92).

Case 1 ($\beta < 1$). In this case, owing to the exponential growth of $u(\frac{1}{2}, \lambda)$ in time, $\{L\}$ is a certain event. The conditional and nonconditional probability distribution functions of the first passage time coincide, $F_\Lambda(\lambda) = F_\Lambda(\lambda | L)$, and Λ satisfies the equation

$$\frac{|G|}{1 - \beta}(\cosh r\Lambda - \beta) = c \qquad (5.98)$$

the probability distribution function sought being

$$F_\Lambda(\lambda) = \text{Prob}\left[\left\{G \leqslant -\frac{c(1 - \beta)}{\cosh(r\lambda) - \beta}\right\} \cup \left\{G \geqslant \frac{c(1 - \beta)}{\cosh(r\lambda) - \beta}\right\}\right]$$

$$= 1 - F_G\left[\frac{c(1 - \beta)}{\cosh(r\lambda) - \beta}\right] + F_G\left[-\frac{c(1 - \beta)}{\cosh(r\lambda) - \beta}\right], \qquad \lambda \geqslant 0$$

$$F_\Lambda(\lambda) = 0, \qquad \lambda < 0$$

$$(5.99)$$

The probability density function is obtained by differentiation:

$$f_\Lambda(\lambda) = \frac{rc(1 - \beta)\sinh(r\lambda)}{[\cosh(r\lambda) - \beta]^2} \left\{ f_G\left[\frac{c(1 - \beta)}{\cosh(r\lambda) - \beta} \right] + f_G\left[-\frac{c(1 - \beta)}{\cosh(r\lambda) - \beta} \right] \right\}$$

$$+ F_\Lambda(0)\delta(\lambda), \quad \lambda \geqslant 0$$

where $\delta(\lambda)$ is the Dirac delta function.

When the initial imperfections are symmetrically distributed in the positive and negative domains, $F_G(-g) = 1 - F_G(g)$, and therefore Eq. (5.99) becomes

$$F_\Lambda(\lambda) = 2 - 2F_G\left[\frac{c(1 - \beta)}{\cosh(r\lambda) - \beta} \right], \qquad\qquad \lambda \geqslant 0$$

$$F_\Lambda(\lambda) = 0, \qquad\qquad \lambda < 0$$

$$f_\Lambda(\lambda) = \frac{2rc(1 - \beta)\sinh(r\lambda)}{[\cosh(r\lambda) - \beta]^2} f_G\left[\frac{c(1 - \beta)}{\cosh(r\lambda) - \beta} \right] + F_\Lambda(0)\delta(\lambda), \quad \lambda \geqslant 0$$

$$(5.100)$$

Case 2 ($\beta = 1$). In this case, $\langle L \rangle$ is again a certain event. The first passage time satisfies the equation

$$\tfrac{1}{2}|G|(\Lambda^2 + 2) = c \qquad\qquad (5.101)$$

The probability distribution of Λ is given by

$$F_\Lambda(\lambda) = \text{Prob}\left[\left\{ G \geqslant \frac{2c}{\lambda^2 + 2} \right\} \cup \left\{ G \leqslant -\frac{2c}{\lambda^2 + 2} \right\} \right]$$

$$= 1 - F_G\left[\frac{2c}{\lambda^2 + 2} \right] + F_G\left[-\frac{2c}{\lambda^2 + 2} \right], \qquad \lambda \geqslant 0 \qquad (5.102)$$

$$F_\Lambda(\lambda) = 0, \qquad\qquad \lambda < 0$$

and the probability density by

$$f_\Lambda(\lambda) = \frac{4c\lambda}{(\lambda^2 + 2)^2} \left\{ f_G\left[\frac{2c}{\lambda^2 + 2} \right] + f_G\left[-\frac{2c}{\lambda^2 + 2} \right] \right\} + F_\Lambda(0)\delta(\lambda), \quad \lambda \geqslant 0$$

For symmetrically distributed initial imperfections, we have

$$F_\Lambda(\lambda) = 2 - 2F_G\left[\frac{2c}{\lambda^2 + 2} \right], \qquad\qquad \lambda \geqslant 0$$

$$F_\Lambda(\lambda) = 0, \qquad\qquad \lambda < 0 \qquad (5.103)$$

$$f_\Lambda(\lambda) = \frac{8c\lambda}{(\lambda^2 + 2)^2} f_G\left[\frac{2c}{\lambda^2 + 2} \right] + F_\Lambda(0)\delta(\lambda), \qquad \lambda \geqslant 0$$

Case 3 ($\beta > 1$).   In this case $\langle L \rangle$ is no longer a certain event; buckling is possible if

$$\max_{\lambda} \max_{\xi} v(\xi, \lambda) = |G| \frac{\beta + 1}{\beta - 1} \tag{5.104}$$

reaches the critical point c. That is,

$$\text{Prob}\langle L \rangle = \text{Prob}\left\{ |G| \frac{\beta + 1}{\beta - 1} \geq c \right\}$$

$$= 1 - F_G \left[c \frac{\beta - 1}{\beta + 1} \right] + F_G \left[-c \frac{\beta - 1}{\beta + 1} \right] \tag{5.105}$$

Note that for $\beta \gg 1$, $\text{Prob}(L) \rightarrow \text{Prob}(\Lambda = 0)$; that is, the probability of the bar buckling at all (i.e., at any time) approaches that of failure at zero time.
   The first passage time satisfies the equation

$$\frac{|G|}{\beta - 1} (\beta - \cos r\Lambda) = c \tag{5.106}$$

and the probability of failure is

$$\text{Prob}(\Lambda \leq \lambda) = \text{Prob}\left[\left\{ G \leq -c \frac{\beta - 1}{\beta - \cos r\lambda} \right\} \cup \left\{ G \geq c \frac{\beta - 1}{\beta - \cos r\lambda} \right\} \right]$$

$$= 1 - F_G \left[\frac{c(\beta - 1)}{\beta - \cos r\lambda} \right] + F_G \left[-\frac{c(\beta - 1)}{\beta - \cos r\lambda} \right], \quad \lambda \in \left[0, \frac{\pi}{r} \right]$$

$$\text{Prob}(\Lambda \leq \lambda) = 0, \qquad\qquad\qquad \lambda < 0$$

$$\text{Prob}(\Lambda \leq \lambda) = \text{Prob}\left(\Lambda \leq \frac{\pi}{r} \right), \quad \lambda > \frac{\pi}{r} \tag{5.107}$$

For symmetrically distributed random imperfections, Eq. (5.107) becomes

$$\text{Prob}(\Lambda \leq \lambda) = 2 - 2F_G \left[\frac{c(\beta - 1)}{\beta - \cos r\lambda} \right], \quad \lambda \geq 0$$

$$\text{Prob}(\Lambda \leq \lambda) = 0, \qquad\qquad\qquad\qquad \lambda < 0 \tag{5.108}$$

The conditional probability distribution function is

$$F_{\Lambda}(\lambda|L) = \frac{1 - F_G\left[\dfrac{c(\beta - 1)}{\beta - \cos r\lambda}\right] + F_G\left[-\dfrac{c(\beta - 1)}{\beta - \cos r\lambda}\right]}{1 - F_G\left[\dfrac{c(\beta - 1)}{\beta + 1}\right] + F_G\left[-\dfrac{c(\beta - 1)}{\beta + 1}\right]}, \quad \lambda \geqslant 0$$

(5.109)

since $F_{\Lambda}(\lambda) = \text{Prob}(\Lambda \leqslant \lambda, L) = \text{Prob}(\Lambda \leqslant \lambda|L)\text{Prob}(L)$.
  The conditional probability density is

$$f_{\Lambda}(\lambda|L) = \frac{1}{\text{Prob}(L)} \frac{dP(\Lambda \leqslant \lambda)}{d\lambda}$$

(5.110)

Note that as $\lambda \to \pi/r$, $F_{\Lambda}(\lambda|L) \to 1$; this is due to the obvious fact that if buckling is possible, it must occur in the time interval $[0, \pi/r]$.
  Note also that in all three cases the functions may be written down in the general form (for $\lambda \geqslant 0$)

$$\text{Prob}(\Lambda \leqslant \lambda) = 1 - F_G[\varphi(\lambda)] + F_G[-\varphi(\lambda)] \tag{5.111}$$

where

$$\varphi(\lambda) = \frac{c(1 - \beta)}{\cosh(r\lambda) - \beta}, \qquad \beta < 1$$

$$\varphi(\lambda) = \frac{2c}{\lambda^2 + 2}, \qquad \beta = 1 \tag{5.112}$$

$$\varphi(\lambda) = \frac{c(\beta - 1)}{\beta - \cos r\lambda}, \qquad \beta > 1$$

Finally, note that for $F_{\Lambda}(0) = 0$, the last terms in the equation for the density functions have to be omitted.
  Consider now some numerical examples. Let G have a uniform distribution in the interval (g_{min}, g_{max}).

$$F_G(g) = \begin{cases} 0, & g \leqslant g_{min} \\ \dfrac{g - g_{min}}{g_{max} - g_{min}}, & g_{min} < g \leqslant g_{max} \\ 1 & g > g_{max} \end{cases} \tag{5.113}$$

For time instances λ satisfying the inequality

$$\varphi(\lambda) \geqslant g_{max} > 0 \tag{5.114}$$

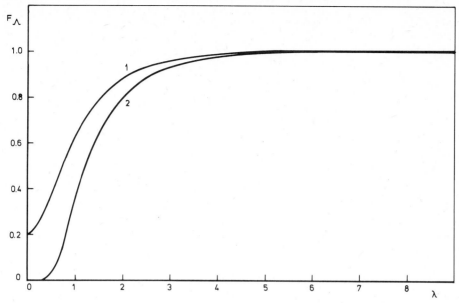

Fig. 5.35. Probability distribution of first passage time ($P = 2P_c$).

then $F_G[\varphi(\lambda)] = 1$ and Eq. (5.111) yields

$$\text{Prob}(\Lambda \leqslant \lambda) = F_G[-\varphi(\lambda)] \qquad (5.115)$$

If, moreover, $-\varphi(\lambda) < g_{min}$, then $\text{Prob}(\Lambda \leqslant \lambda) \equiv 0$. For symmetrically distributed random imperfections $g_{min} = -g_{max}$, this happens for times $\lambda \leqslant \lambda^*$, where

$$\lambda^* = \begin{cases} r^{-1}\cosh^{-1}\left[\dfrac{c(1-\beta)}{g_{max}} + \beta\right], & \beta < 1 \\[2ex] \sqrt{\dfrac{2c}{g_{max}} - 2}\,, & \beta = 1 \\[2ex] r^{-1}\cos^{-1}\left[\beta - \dfrac{c(\beta-1)}{g_{max}}\right], & \beta > 1, c < g_{max} \end{cases} \qquad (5.116)$$

For $\beta > 1$, the probability of buckling may turn out to be zero. From expression (5.104) it may be seen that this is the case, for example, if

$$c > g_{max}\frac{\beta+1}{\beta-1} \quad \text{and} \quad g_{min} = -g_{max} \qquad (5.117)$$

then $\text{Prob}(L) = 0$ and $\text{Prob}(\Lambda \leqslant \lambda) \equiv 0$ for any λ, and the bar remains in the safe region throughout.

The above cases are illustrated in Figs. 5.35–5.37, where $\text{Prob}(\Lambda \leqslant \lambda)$ is plotted against λ. These figures represent, respectively, the cases where the

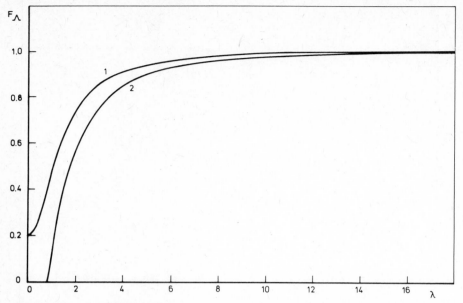

Fig. 5.36.   Probability distribution function of first-passage time ($P = P_c$).

actual load is greater than, equal to, or less than the corresponding classical buckling load of a perfect structure. Particularly, in Fig. 5.35 the actual load was chosen double the classical buckling load, and in Fig. 5.37, it was half. For all calculations the random imperfections were chosen with a symmetric distribution ($g_{max} = -g_{min}$). The curves marked *1* are all associated with $g_{max} = 0.5$, whereas those marked *2* are associated with $g_{max} = 0.3$. The nondimensional critical level c was taken identical for all curves in these figures, the failure being thus identified with the absolute value of the total displacement reaching four-tenths of the bar's radius of gyration. In the cases where $g_{max} > c$ (curves *1*), the probability of the bars buckling at $\Lambda = 0$, Prob($\Lambda = 0$), differs from zero; in particular, Prob($\Lambda = 0$) = 0.2. Thus, under the statistical interpretation of probability, 20% of bars from a large population buckle immediately upon application of the load at $\lambda = 0$. In the cases where $g_{max} < c$ (curves *2*), Prob($\Lambda \leqslant \lambda$) = 0 for $\lambda \leqslant \lambda^*$, λ^* being $\cosh^{-1}$ ($\frac{7}{6}$) = 0.5697 for $P = 2P_c$, $\sqrt{\frac{2}{3}}$ = 0.8165 for $P = P_c$, $\sqrt{2}\cos^{-1}(\frac{2}{3})$ = 1.1895 for $P = 0.5P_c$; that is, below these values of time, we would expect that no bar buckles. For $P < P_c$, Prob(L) $\neq$ 1: For $g_{max} = 0.5$ and $\alpha = 0.5$, Prob(L) = 0.733; that implies in accordance with the statistical interpretation of probability, that, for example, from a series of 1000 bars, the relative number of unbuckled bars (in the time interval $0 \leqslant \lambda < \infty$) would stabilize near (1 − 0.733) × 1000 = 267. For $g_{max} = 0.3$ and $\alpha = 0.5$, Prob(L) = 0.555, indicating once more that the number of unbuckled bars would now stabilize at (1 −

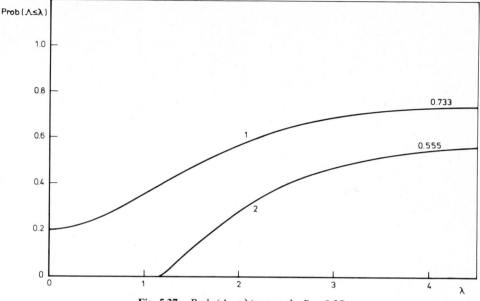

Fig. 5.37.   Prob ($\Lambda \leqslant \lambda$) versus λ, $P = 0.5P_c$.

$0.555) \times 1000 = 445$. For g_{max} less than some critical value, found from Eq. (5.117) as

$$g_{max,cr} = \frac{c(\beta - 1)}{\beta + 1}$$

Prob($\Lambda \leqslant \lambda$) is identically zero. For $\alpha = 0.5$, $g'_{max,cr} = 0.133$, implying that for $g_{max} < 0.133$ a negligible number of bars would buckle at all in a large sample.

Figure 5.38 presents Prob($\Lambda \leqslant \lambda$) as a function of λ for three cases: $P = 2P_c$, $P = P_c$, $P = 0.5P_c$, for specified g_{max}, respectively. As is seen from the figure, the probability of failure λ increases with the actual load for specified λ.

Let us find the mean buckling time, defined as

$$E(\Lambda) = \int_0^\infty \lambda f_\Lambda(\lambda) \, d\lambda \tag{5.118}$$

However, since in general

$$f_\Lambda(\lambda) = \frac{d}{d\lambda} \text{Prob}(\Lambda \leqslant \lambda)$$

$$\text{Prob}(\Lambda \leqslant \lambda) = 1 - \text{Prob}(\Lambda > \lambda) = 1 - R(\lambda)$$

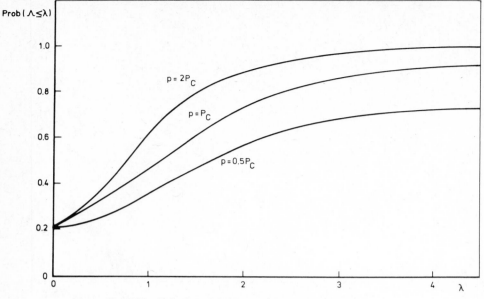

Fig. 5.38.   Influence of load ratio P/P_c on Prob $(\Lambda \leqslant \lambda)$.

where $R(\lambda)$ is the reliability of a structure at time λ, we have

$$E(\Lambda) = \int_0^\infty \lambda \frac{d}{d\lambda} \text{Prob}(\Lambda \leqslant \lambda) \, d\lambda = \int_0^\infty \lambda \frac{d}{d\lambda} [1 - R(\lambda)] \, d\lambda$$

Integration by parts, as in Eq. (5.75), yields

$$E(\Lambda) = -\lambda R(\lambda)]_0^\infty + \int_0^\infty R(\lambda) \, d\lambda \qquad (5.119)$$

Let us demonstrate that the first term is zero:

$$\lambda R(\lambda)]_0^\infty = \lim_{\lambda \to \infty} \lambda R(\lambda) = 0$$

Indeed, for the random variable $\Lambda \geqslant 0$ with finite mathematical expectation, it follows from convergence of the integral $\int_0^\infty \lambda f_\Lambda(\lambda) \, d\lambda$ that

$$\int_M^\infty \lambda f_\Lambda(\lambda) \, d\lambda \to 0 \quad \text{as } M \to \infty$$

However,

$$M \int_M^\infty f_\Lambda(\lambda) \, d\lambda \leqslant \int_M^\infty \lambda f_\Lambda(\lambda) \, d\lambda$$

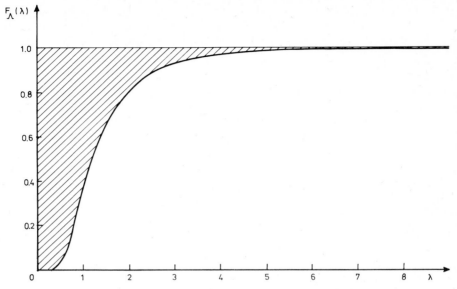

Fig. 5.39. Shaded area equal to mean buckling time of structure ($P = 2P_c$).

Therefore, $M[1 - F_\Lambda(M)] = MR(M) \to 0$ as $M \to \infty$. Consequently,

$$\lim_{\lambda \to \infty} \lambda R(\lambda) = 0 \qquad (5.120)$$

implying that

$$E(\Lambda) = \int_0^\infty R(\lambda) \, d\lambda = \int_0^\infty \left[1 - F_\Lambda(\lambda)\right] d\lambda \qquad (5.121)$$

$E(\Lambda)$ is represented by the shaded area in Fig. 5.39.

For $P < P_c$, however, $\mathrm{Prob}(\Lambda \leq \infty) = \mathrm{Prob}(L) \neq 0$, as we have already seen. In other words, in this case,

$$E(\Lambda) \to \infty \qquad (5.122)$$

This result becomes obvious if we recall that for $P < P_c$ a certain percentage of the bars do not fail, that is, they have an infinite mean buckling time.

With the expressions for the unreliability $\mathrm{Prob}(\Lambda \leq \lambda)$ at hand, we can pose the problem of determining the allowable operation time λ_r for which the reliability $R(\lambda)$ reaches a given level:

$$R(\lambda_r) = r \qquad (5.123)$$

This allowable operation time is clearly finite, whereas the mean buckling time may be infinite, again demonstrating that reliability-based design is superior to mean-behavior based design (compare with Prob. 4.5).

PROBLEMS

5.1. The loads acting on different machine components often have a chi-square distribution with m degrees of freedom, as was shown by Serensen and Bugloff,

$$f_N(n) = \frac{e^{-n/2}n^{\nu-1}}{2^\nu \Gamma(\nu)} U(n), \quad \nu = \frac{m}{2}$$

TABLE 5.1   Values of $\chi^2_{\alpha, \nu}$

ν	$\alpha = .995$	$\alpha = .99$	$\alpha = .975$	$\alpha = .95$	$\alpha = .05$	$\alpha = .025$	$\alpha = .01$	$\alpha = .005$
1	.0000393	.000157	.000982	.00393	3.841	5.024	6.635	7.879
2	.0100	.0201	.0506	.103	5.991	7.378	9.210	10.597
3	.0717	.115	.216	.352	7.815	9.348	11.345	12.838
4	.207	.297	.484	.711	9.488	11.143	13.277	14.860
5	.412	.554	.831	1.145	11.070	12.832	15.086	16.750
6	.676	.872	1.237	1.635	12.592	14.449	16.812	18.548
7	.989	1.239	1.690	2.167	14.067	16.013	18.475	20.278
8	1.344	1.646	2.180	2.733	15.507	17.535	20.090	21.955
9	1.735	2.088	2.700	3.325	16.919	19.023	21.666	23.589
10	2.156	2.558	3.247	3.940	18.307	20.483	23.209	25.188
11	2.603	3.053	3.816	4.575	19.675	21.920	24.725	26.757
12	3.074	3.571	4.404	5.226	21.026	23.337	26.217	28.300
13	3.565	4.107	5.009	5.892	22.362	24.736	27.688	29.819
14	4.075	4.660	5.629	6.571	23.685	26.119	29.141	31.319
15	4.601	5.229	6.262	7.261	24.996	27.488	30.578	32.801
16	5.142	5.812	6.908	7.962	26.296	28.845	32.000	34.267
17	5.697	6.408	7.564	8.672	27.587	30.191	33.409	35.718
18	6.265	7.015	8.231	9.390	28.869	31.526	34.805	37.156
19	6.844	7.633	8.907	10.117	30.144	32.852	36.191	38.582
20	7.434	8.260	9.591	10.851	31.410	34.170	37.566	39.997
21	8.034	8.897	10.283	11.591	32.671	35.479	38.932	41.401
22	8.643	9.542	10.982	12.338	33.924	36.781	40.289	42.796
23	9.260	10.196	11.689	13.091	35.172	38.076	41.638	44.181
24	9.886	10.856	12.401	13.848	36.415	39.364	42.890	45.558
25	10.520	11.524	13.120	14.611	37.652	40.646	44.314	46.928
26	11.160	12.198	13.844	15.379	38.885	41.923	45.642	48.290
27	11.808	12.879	14.573	16.151	40.113	43.194	46.963	49.645
28	12.461	13.565	15.308	16.928	41.337	44.461	48.278	50.993
29	13.121	14.256	16.047	17.708	42.557	45.722	49.588	52.336
30	13.787	14.953	16.791	18.493	43.773	46.979	50.892	53.672

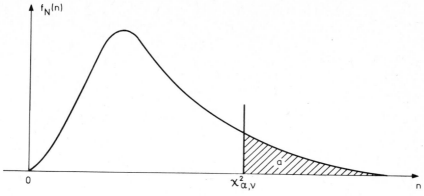

Problem 5.1

Find the reliability of the bar under random tensile force. Table 5.1 contains the values of $\chi^2_{\alpha,\nu}$, where $\int^\infty_{\chi^2_{\alpha,\nu}} f_N(n)\,dn = \alpha$ for $\alpha = 0.995$, 0.99, 0.975, 0.95, 0.05, 0.025, 0.01, 0.005, and $\nu = 1,2,\dots,30$ (see figure).

5.2. Find the reliabilities of the truss structures shown in Figs. 2.9a, c, d, and e, if P is treated as a random variable with given density function $f_P(p)$.

5.3. The truss shown in the figure carries the random normally distributed load P with $E(P) = 10$ kips, $\sigma_P = 5$ kips, $a = 10'$, $\sigma_{\text{allow}} = 24,000$ psi, area $= 1$ in^2. Check whether the reliability exceeds the desired $r = 0.999$.

5.4. A cantilever of rectangular cross section is loaded as shown in the figure. G is a random variable with given $F_G(g)$; σ_Y, the yield stress in the tensile test of the cantilever material, is given. Use the maximum shear-stress criterion to find the reliability of the cantilever. On the concrete numerical example, discuss the change of the reliability estimate under the von Mises criterion.

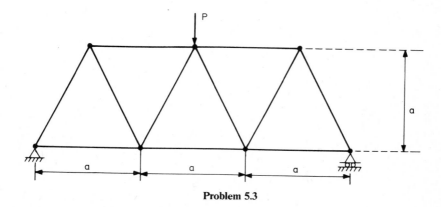

Problem 5.3

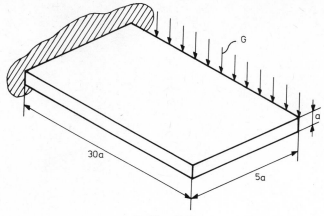

Problem 5.4

5.5. A beam is loaded as shown in the figure. G is a random variable with given probability distribution function $F_G(g)$ and α is a given number. Verify that the extremal bending moment occurs at the section $x = (5\alpha - 1)/4\alpha$ and equals

$$M_z = Ga^2\left[\left(\frac{3 + \alpha}{4}\right)\left(\frac{5\alpha - 1}{4\alpha}\right) - \frac{1}{2}\left(\frac{5\alpha - 1}{4\alpha}\right)^2 - \left(\frac{\alpha - 1}{2}\right)\left(\frac{\alpha - 1}{4\alpha}\right)^2\right]$$

and derive the reliability.

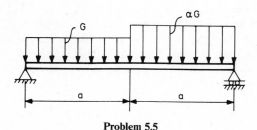

Problem 5.5

5.6. Determine the reliability of the cantilever (see figure) under a given force q applied at the random distance X from the clamped edge. $F_X(x)$ is given. (For a generalization of this problem with both X and q

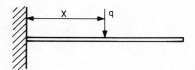

Problem 5.6

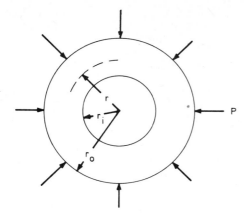

Problem 5.7

random variables, or with random concentrated loads and moments applied at random positions on the beams with different boundary conditions, see the paper by Shukla and Stark.)

5.7. A thick-walled cylinder (see figure) is under external pressure P with a discrete uniform distribution

$$F_P(p) = \tfrac{1}{10} \sum_{i=1}^{10} U(x - p_0 i)$$

that is, P takes on values $p_0, 2p_0, \ldots, 10p_0$ with constant probability $1/10$. For the transverse stresses, the following expressions are valid (see, e.g., Timoshenko and Goodier):

$$\Sigma_r = P\frac{(r_o/r_i)^2 - (r_o/r)^2}{(r_o/r_i)^2 - 1} \qquad \Sigma_\theta = -P\frac{(r_o/r_i)^2 + (r_o/r)^2}{(r_o/r_i)^2 - 1}$$

where r_o and r_i are the outer and inner radii, respectively. Using the von Mises criterion, find r_o/r_i such that the desired reliability is not less than 0.99.

5.8. A rectangular plate, simply supported all around, is under a load Q uniform over its surface, with chi-square distribution (Sec. 4.7)

$$f_Q(q) = \frac{e^{-q/2}q^{\nu-1}}{2^\nu \Gamma(\nu)}, \quad q > 0, \nu = \frac{m}{2}$$

The displacement of the plate under a deterministic uniform load q is (see, e.g., Timoshenko and Woinowski-Krieger):

$$w = \frac{16q}{\pi^2 D} \sum_{m=1}^{\infty} \sum_{n=1}^{\infty} \frac{1}{mn(m^2/a^2 + n^2/b^2)^2} \sin\frac{m\pi x}{a} \sin\frac{n\pi y}{b}$$

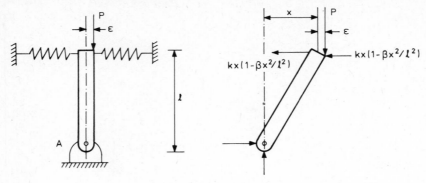

Problem 5.15

where a and b are the sides of the plate; $D = Eh^3/12(1 - v^2)$ is the flexural rigidity; E, the modulus of elasticity; v, Poisson's ratio; and h the thickness of the plate. Find the probability density of the maximum deflection.

5.9. Derive an equation analogous to (5.63) for an asymmetric structure.

5.10. Find the probability density function of the dynamic buckling loads and find the mean dynamic buckling load for the symmetric structure.

5.11. Find the reliability function of the asymmetric structure if the initial imperfection is normally distributed $N(0, \sigma^2)$, in the static setting.

5.12. Repeat 5.11 for the dynamic buckling problem.

5.13. Assume that the initial imperfection has an exponential distribution, with parameter $E(\bar{X})$ given. Find the reliability of the asymmetric structure.

5.14. Plot the nondimensional static buckling λ_s/λ_c versus initial imperfection $\bar{\xi}$ curve for the nonsymmetric structure, according to Eq. (5.50). Find the reliability at the load level λ.

5.15. A rigid weightless bar with a frictionless pin joint at A, constrained by nonlinear springs with $k > 0$, $\beta > 0$, is under an eccentric load P (see figure). The equilibrium equation is

$$P(x + \varepsilon) = 2klx(1 - \beta x^2/l^2)$$

yielding $P_c = 2kl$. Find the expression for the maximum force P_{max} supported by the bar as a function of the eccentricity ε. Assume eccentricity to be a continuous random variable with probability density $f_E(\varepsilon)$. Find the reliability of the structure at load level λ.

5.16. Generalize the results of Sec. 5.7 for the case where the load function is a rectangular impulse $P(t) = P[U(t) - U(t - \tau)]$ with P and τ given positive quantities. (See figure.)

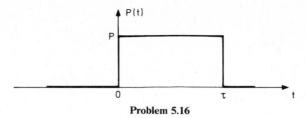

Problem 5.16

5.17. Generalize the results of Sec. 5.7 for the case where failure is considered in the finite time interval $0 \leqslant t \leqslant t^*$ [see Eq. (5.85)].

5.18. Verify that the buckling time Λ in Sec. 5.7 for $P < P_c$ does not represent a random variable. Assign an infinite buckling time to the structure that does not buckle. How does Eq. (5.84) change in these circumstances? Show analytically that for $P < P_c$, $E(\Lambda)$ approaches infinity.

5.19. Modify the results of Sec. 5.7 for the case where the structure possesses viscous damping.

5.20. Consider the load-bearing capacity of an imperfect bar. As can be seen from the figure, a concentric load P produces a bending moment $M_z = -Pw$ and increases the displacement by an amount $w - w_1$. The differential equation of the column is, therefore,

$$EI_z \frac{d^2}{dx^2}(w - w_1) = -Pw \qquad (5.124)$$

The bar is simply supported at its ends and has an initial imperfection

$$w_1 = g \sin \frac{\pi x}{l} \qquad (5.125)$$

Equation (5.124) becomes, upon substitution of Eq. (5.125),

$$\frac{d^2 w}{dx^2} + \frac{Pw}{EI_z} = -a_1 \frac{\pi^2}{l^2} \sin \frac{\pi x}{l}$$

The solution of this equation is

$$w = C_1 \sin\left[\left(\frac{P}{EI_z}\right)^{1/2} x\right] + C_2 \cos\left[\left(\frac{P}{EI_z}\right)^{1/2} x\right] + \frac{a_1}{1 - P/P_c} \sin \frac{\pi x}{L}$$

$$(5.126)$$

where $P_c = \pi^2 EI/l^2$ is the classical or Euler buckling load. The

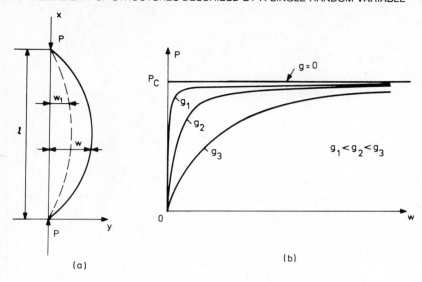

(a)

(b)

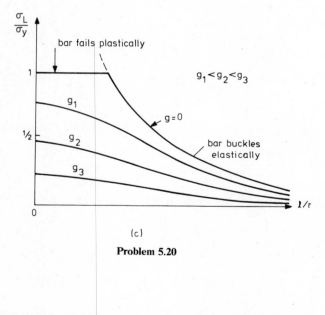

(c)

Problem 5.20

boundary conditions are $w = 0$ at $x = 0, l$, and these yield

$$C_1 \sin \sqrt{\frac{P}{EI_z}}\, l = 0 \qquad C_2 = 0$$

For $P < P_c$, both C_1 and C_2 must be zero, and the total deflection is represented by the last term in Eq. (5.126):

$$w = \frac{1}{1 - P/P_c} g \sin \frac{\pi x}{l} \qquad (5.127)$$

We see that the total deflection becomes increasingly large, as $P \to P_c$. The normal stresses in the bar are given

$$\sigma_x = -\frac{P}{A} - \frac{My}{I_z}, \quad M = -Pw$$

Thus the maximum compressive stress takes place at $x = l/2$ and is given by

$$\sigma_{max} = \frac{P}{A}\left[1 + \frac{gA}{S} \frac{1}{1 - P/P_c}\right] \qquad (5.128)$$

where S is the section modulus ($S = I_z/y_{max}$, y_{max} being the distance from the neutral axis to the point of maximum stress). Denote $P/A = \sigma_{av}$ and $P/P_c = \sigma_{av}/\sigma_c$. Equation (51.27) becomes then

$$\sigma_{max} = \sigma_{av}\left[1 + \frac{gA}{S} \frac{1}{1 - \sigma_{av}/\sigma_c}\right]$$

where $\sigma_c = \pi^2 E/(l/r)^2$, r being the radius of gyration of the cross-sectional area of the bar. The load P_L, for which σ_{max} equals the yield stress σ_y, is the limit load for which the column remains elastic. This load results in the average stress $\sigma_L = P_L/A$, and Eq. (5.127) becomes

$$\sigma_y = \lambda_L\left[1 + \frac{gA}{S} \frac{1}{1 - \sigma_L/\sigma_c}\right]$$

where

$$\left(\frac{\sigma_L}{\sigma_y}\right)^2 - \left[1 + \frac{\sigma_c}{\sigma_y}\left(1 + \frac{gA}{S}\right)\right]\frac{\sigma_L}{\sigma_y} + \frac{\sigma_c}{\sigma_y} = 0 \qquad (5.129)$$

Part (c) of the figure shows σ_L/σ_y as a function of the *slenderness ratio* l/r. Treating the initial imperfection amplitude G as a random variable with gamma distribution (Sec. 4.8),

$$f_G(g) = \frac{1}{\beta^{\alpha+1}\Gamma(\alpha + 1)} g^\alpha e^{-g/\beta} U(g)$$

the average limit stress is also a random variable.

(a) Extract σ_L/σ_y explicitly from Eq. (5.128) as a function of g.

(b) Find $F_\Sigma(\sigma_L)$.

(c) Consider also the special cases of an exponential distribution and of a chi-square distribution with m degrees of freedom. Perform the numerical calculations.

5.21. In problem 5-20, assume G has an exponential distribution. Find the probability of the maximum total displacement $w(l/2)$ taking on values in the interval $[0, 2E(G)]$. Investigate the behavior of this probability as P approaches the classical buckling load P_c.

5.22. Consider now another important case of an initially perfect simply supported bar under eccentric load with eccentricity ε, as shown in part (a) of the accompanying figure. The differential equation for the bar deflection reads

$$EI_z \frac{d^2 w}{dx^2} + Pw = 0 \tag{5.130}$$

The solution is

$$w = C_1 \sin\sqrt{\frac{P}{EI_z}}\, x + C_2 \cos\sqrt{\frac{P}{EI_z}}\, x$$

The integration constants are determined by the boundary conditions $w = \varepsilon$ at $x = \pm l/2$, so that

$$C_1 = 0 \qquad C_2 = \varepsilon \sec\left[\left(\frac{P}{EI_z}\right)^{1/2}\left(\frac{l}{2}\right)\right] \cos\left[\left(\frac{P}{EI_z}\right)^{1/2} x\right]$$

The maximum displacement is reached at $x = 0$

$$w_{max} = \varepsilon \sec\left[\left(\frac{P}{EI_z}\right)^{1/2}\left(\frac{l}{2}\right)\right] = \varepsilon \sec\left[\frac{\pi}{2}\left(\frac{P}{P_c}\right)^{1/2}\right] \tag{5.131}$$

It is seen that the maximum deflection becomes increasingly large as P approaches P_c. The elastic limit is reached at the most stressed point when

$$\sigma_y = \frac{P}{A} + \frac{M_c}{I_z} \tag{5.132}$$

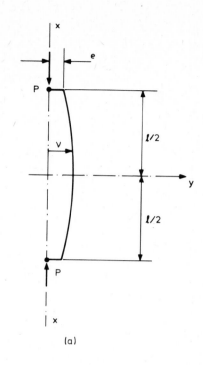

(a)

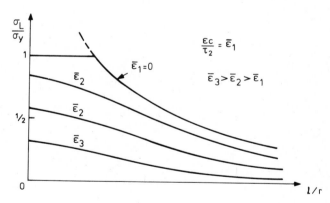

(b)

Problem 5.22

169

or when

$$\sigma_0 = \frac{P}{A} + \frac{P\varepsilon c}{I_z} \sec\left[\frac{\pi}{2}\left(\frac{P}{P_c}\right)^{1/2}\right] \tag{5.133}$$

This equation can be rewritten as

$$\sigma_y = \sigma_L\left\{1 + \frac{\varepsilon c}{r^2}\sec\left[\frac{l}{2r}\left(\frac{\sigma_L}{E}\right)^{1/2}\right]\right\} \tag{5.134}$$

where $\sigma_L = P_L/A$ is the "average" limit stress. Equation (5.134) is referred to as the *secant formula* and is usually plotted in the form of σ_L/σ_y versus l/r for a particular material (with σ_y and E) specified for different values of $\varepsilon c/r^2$, as shown in part (b) of the figure.

Treating the eccentricity as the random variable E, with $F_E(\varepsilon)$ given as exponentially distributed, the average limit stress also is a random variable. Find $F_\Sigma(\sigma_L)$.

5.23. In his (now historic) doctoral thesis in 1945, and in his 1963 paper, Koiter analyzed a sufficiently long cylindrical shell with an axisymmetric initial imperfection, under axial load. He chose an initial imperfection function $w_0(x)$, coconfigurational with the axisymmetric buckling mode of a perfect cylindrical shell, as

$$w_0(x) = gh\sin\frac{\pi i_c x}{L}, \quad i_c = \frac{L}{\pi}\left(\frac{2c}{Rh}\right)^{1/2}, \quad c = [3(1-\nu^2)]^{1/2} \tag{5.135}$$

where μ is the nondimensional initial imperfection magnitude; i_c, the number of half-waves at which the associated perfect shell buckles; L, the shell length; R, the shell radius; and h, the shell thickness. Using his own general nonlinear theory, he derived *inter alia* a relationship between the critical load and the initial imperfection magnitude:

$$(1-\lambda)^2 - \tfrac{3}{2}c|g|\lambda = 0 \tag{5.136}$$

where $\lambda = P_{bif}/P_c$, $P_c = 2\pi Rh\sigma_c$, and $\sigma_c = Eh/Rc$, is the nondimensional buckling load, P_{bif}, the buckling load of an imperfect shell; P_c, the classical buckling load of a perfect shell; E, the modulus of elasticity; and ν, Poisson's ratio. The buckling load P_{bif} was defined as that at which the axisymmetric fundamental equilibrium state bifurcates into a nonsymmetric one. The absolute value of g in Eq. (5.136) stands, since for a sufficiently long shell the sign of the imperfection is immaterial: Positive and negative initial imperfections with equal absolute values cause the same reduction on the buckling load.

Equation (5.136) yields the explicit buckling load-initial imperfection relationship:

$$\lambda = 1 + \tfrac{3}{4}|\xi| - \tfrac{1}{2}\left(6|\xi| + \tfrac{9}{4}|\xi|^2\right)^{1/2}$$

where $\xi = cg$.

(a) Assume X with possible values ξ to be a normally distributed random variable $N(\bar{\xi}, \sigma^2)$.

(b) Find the probability density function of $|X|$. (Consult Example 4.6.)

(c) Find the reliability of the shell at the load level $\bar{\lambda}$.

(d) Assume, after Roorda, $\bar{\xi} = 0.333 \times 10^{-3} R/h$ and $\sigma^2 = 10^{-3}R/h$, and find the stress level at which the system has a given reliability. Compare your result with Roorda's.

5.24. (a) Koiter (1945) also analyzed the imperfection sensitivity of a shell with nonaxisymmetric, periodic imperfections:

$$w_0(x) = gh\left(\cos\frac{i_c \pi x}{L} + 4\cos\frac{i_c \pi x}{2L}\cos\frac{i_c \pi y}{2L}\right) \quad (5.137)$$

(where y is the circumferential coordinate, the remaining notation as in the preceding problem) to arrive, instead of Eq. (5.136), at the equation

$$(1 - \lambda)^2 + 6cg\lambda = 0 \quad (5.138)$$

for the nondimensional buckling load $\lambda = P_{\text{lim}}/P_c$, where P_{lim} is the limit load (as in Sec. 5.5). For the imperfection function (5.138), the limit load exists only at negative values of the imperfection parameter g. For positive g, the origin of the coordinate system may be shifted, and since the shell is sufficiently long, the analysis would be unaffected except that the sign of g would change to yield

$$(1 - \lambda)^2 - 6cg\lambda = 0 \quad (5.139)$$

Combining Eqs. (5.138) and (5.139), we arrive at the final equation

$$(1 - \lambda)^2 - 6c|g|\lambda = 0$$

Perform calculations as in Prob. 5.23. Compare the reliabilities of shells with axisymmetric and nonaxisymmetric imperfections.

(b) Sometimes the imperfection is represented by a local dimple extending over a small region of the shell. A more or less localized imperfection may be represented in the form

$$w_0(x) = gh\left(\cos\frac{i_c\pi x}{L} + 4\cos\frac{i_c\pi x}{2L}\cos\frac{i_c\pi y}{2L}\right)$$

$$\times\exp\left[-\tfrac{1}{2}\frac{\mu^2}{R^2}(x^2 + y^2)\right] \tag{5.140}$$

which is the function in Eq. (5.137) multiplied by an exponentially decaying function. For example, at a distance $x = (4\pi/i_c)R$ or $y = (4\pi/i_c)R$, a complete wavelength of periodic part in Eq. (5.140), the exponential factor reduces to $\exp(-8\pi^2\mu^2/i_c^2)$. At first approximation, the term μ^2/i_c^2 may be neglected with respect to unity. Koiter's analysis (1978) yields then

$$(1 - \lambda)^2 = -4cg\lambda \tag{5.141}$$

Assume again that $X = cG$, with possible values $\xi = cg$, is a normally distributed random variable $N(0, \sigma^2)$, and find the reliability of the shell. Are the localized imperfections as harmful as the periodic ones?

CITED REFERENCES

Budiansky, B., and Hutchinson, J. W., "Dynamic Buckling of Imperfection Sensitive Structures," in H. Görtler, Ed., *Proc. Eleventh Intern. Congr. Appl. Mech.*, 1964, pp. 636–651.

Budiansky, B., "Dynamic Buckling of Elastic Structures: Criteria and Estimates," in G. Herrmann, Ed., *Dynamic Stability of Structures*, Pergamon Press, New York, 1967, pp. 83–106.

Elishakoff, I., "Axial Impact Buckling of Column With Random Initial Imperfections," *ASME J. Appl. Mech.* **45**, 361–365 (1978).

Elishakoff, I., "Remarks on the Static and Dynamic Imperfection-Sensitivity of Nonsymmetric Structures," *ASME J. Appl. Mech.*, **46**, 111–115 (1980).

Hoff, N. J., "Dynamic Stability of Structures" (Keynote Address), in G. Herrmann, Ed., *Dynamic Stability of Structures* Pergamon Press, New York, 1965, pp. 7–44.

Koiter, W. T., "On the Stability of Elastic Equilibrium" (in Dutch), Ph.D. thesis, Delft Univ. Technology, H. J. Paris, Amsterdam; English translations: (a) NASA-TTF-10, 833, 1967, (b) AFFDL-TR-70-20, 1970 (translated by E. Riks).

Koiter, W. T., "The Effect of Axisymmetric Imperfections on the Buckling of Cylindrical Shells under Axial Compression," *Proc. Kon. Ned. Akad. Wet., Amsterdam*, Ser. B., **6**, 265–279 (1963) (also, Lockheed Missiles and Space Co., Rep. 6-90-63-86, Palo Alto, CA, Aug. 1963).

Koiter, W. T., "The Influence of More or Less Localized Imperfections on the Buckling of Circular Cylindrical Shells Under Axial Compression," in *Complex Analysis and Its Applications* (Dedicated to the 70th Birthday of Academician I. N. Vekua,), Acad. U.S.S.R. Sci., "Nauka" Publ. House, Moscow, 1978, pp. 242–244.

Lomakin, V. A., "Strength and Stiffness Calculations of the Beam Bent Under a Random Load," *Mechanics of Solids*, Faraday Press, New York, 1966, No. 4, pp. 162–164.

Ogata, K., *Modern Control Engineering*, Prentice-Hall, Englewood Cliffs, NJ, 1970.

Roorda, J., "Buckling of Shells: an Old Idea With a New Twist," *J. Eng. Mech. Div., Proc. ASCE*, **98** (EM3), 531–538 (1972).

Serensen, S. B., and Bugloff, E. G., "On the Probabilistic Representations of the Varying Loading of Machine Details," *Vestn. Mashinostr.*, 1960, No. 10 (in Russian).

Shukla, D. K., and Stark, R. M., "Statics of Random Beams," *J. Eng. Mech. Div., Proc. ASCE*, **98** (EM6), 1487–1497 (1972).

Timoshenko, S. P., and Goodier, J. N., *Theory of Elasticity*, 3rd ed., McGraw-Hill, New York, 1970.

Timoshenko, S. P., and Woinowski-Krieger, S., *Theory of Plates and Shells*, McGraw-Hill, New York, 1959.

RECOMMENDED FURTHER READING

Augusti, G., and Baratta, A., "Reliability of Slender Columns: Comparison of Different Approximations," in B. Budiansky, Ed., *Buckling of Structures*, Springer Verlag, Berlin, 1976, pp. 183–198.

Bolotin, V. V., "Statistical Methods in the Nonlinear Theory of Elastic Shells," NASA TTF-85, 1962.

Fraser, W. B., "Buckling of a Structure With Random Imperfections," Ph.D. Thesis, Div. Eng. Appl. Phys., Harvard Univ., Cambridge, MA, 1965.

Hansen, J. S., and Roorda, J., "Reliability of Imperfection Sensitive Structures," in S. T. Ariaratnam and H. H. E. Leipholz, Eds., *Stochastic Problems in Mechanics, Proc. Symp. Stochastic Prob. Mech.*, Waterloo Univ. Press, Waterloo, Ont., 1973, pp. 229–242.

Konishi, I., and Takaoka, N., "Some Comments on the Reliability Analysis of Civil Engineering Structures with Special Reference to Compression Members", in T. Moan and M. Shinozuka, Eds., *Structural Safety and Reliability*, Elsevier, Amsterdam, 1981, pp. 341–357.

Miller, R. K., and Hedgepeth, J. M., "The Buckling of Lattice Columns with Stochastic Imperfections," *Int. J. Solids Structures*, **15**, 13–24 (1979).

Perry, S. H., and Chilver, A. H., "The Statistical Variation of the Buckling Strength of Columns", *Proc. Inst. Civ. Engrs.*, Part 2, **61**, 109–125, (1976).

Thompson, J. M. T., "Towards a General Statistical Theory of Imperfection-Sensitivity in Elastic Post-Buckling," *J. Mech. Phys. Solids*, **15**, 413–417 (1967).

chapter **6**

Two or More Random Variables

With the aid of the theory of a single random variable, we were able to calculate the reliability of a structure characterized by such a variable, for example that of a bar under a random tensile force, with strength treated as a deterministic quantity. We considered also the "reverse" example, where the force was assumed to be given and the strength considered as a random variable. In practice, however, both force and strength are random. Moreover, the beam may have to bear several concentrated loads rather than a single one, and so on. In these circumstances the theory must be extended to the case of multiple random variables.

We begin with a pair of such variables and then proceed to the general multiple case.

6.1 JOINT DISTRIBUTION FUNCTION OF TWO RANDOM VARIABLES

Let X and Y be random variables as defined in Sec. 3.1. The *joint distribution function* of X and Y, denoted by $F_{XY}(x, y)$, is defined as the probability of the intersection of two random events $\{X \leqslant x\}$ and $\{Y \leqslant y\}$:

$$F_{XY}(x, y) = P\{(X \leqslant x) \cap (Y \leqslant y)\} = P(X \leqslant x, Y \leqslant y) \qquad (6.1)$$

treated as a function of two variables x and y. $F_{XY}(x, y)$ is sometimes called the *joint cumulative or bivariate distribution function* of X and Y. If $F_{XY}(x, y)$ is the joint (cumulative) distribution function of X and Y, then the distribution functions $F_X(x)$ and $F_Y(y)$ are called marginal distribution functions. While knowledge of the marginal distribution functions does not imply that of the joint distribution function, the opposite is correct: We can readily find the marginal probability functions from the joint probability function. Indeed, consider

$$F_{XY}(x, \infty) = P(X \leqslant x, Y \leqslant \infty) = P(X \leqslant x) = F_X(x) \qquad (6.2)$$

since $\{Y \leqslant \infty\}$ is a certain event and $(X \leqslant x, Y \leqslant \infty) = (X \leqslant x)$.
   Similarly,

$$F_{XY}(\infty, y) = P(X \leqslant \infty, Y \leqslant y) = P(Y \leqslant y) = F_Y(y) \qquad (6.3)$$

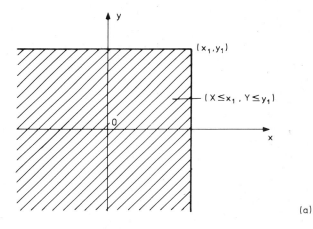

(a)

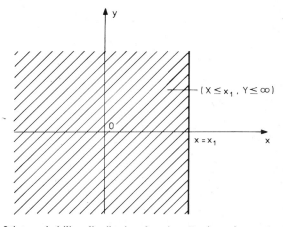

(b)

Fig. 6.1. (a) Joint probability distribution function $F_{XY}(x, y_1)$ equals the probability of the random point falling within the shaded area. (b, c) The marginal distribution functions $F_X(x_1)$ and $F_Y(y_1)$, respectively, equal the probability of the random point falling within shaded areas.

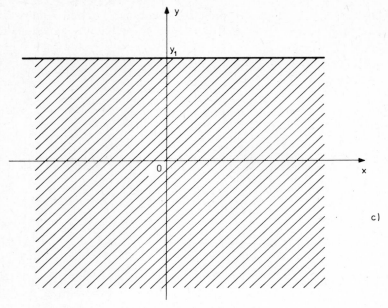

Fig. 6.1. (*Continued*)

Geometrically, the joint probability distribution function $F_{XY}(x_1, y_1)$ represents the probability of a random point (a point with random coordinates X
and Y) falling within the quadrant bounded by the lines $x = x_1$ and $y = y_1$,
intersecting at (x_1, y_1) as shown in Fig. 6.1a. The marginal probability density
$F_X(x_1)$ is the probability of it falling to the left of $x = x_1$ (Fig. 6.1b), and
$F_Y(y_1)$ that of it falling below $y = y_1$ (Fig. 6.1c). Falling *anywhere* in the xy
plane is a certain event; hence,

$$F_{XY}(\infty, \infty) = 1 \tag{6.4}$$

The joint probability distribution function approaches zero as either or both of
its arguments, x, y approach $-\infty$:

$$F_{XY}(-\infty, y) = F_{XY}(x, -\infty) = F_{XY}(-\infty, -\infty) = 0 \tag{6.5}$$

Indeed, since

$$\{X \leqslant -\infty, Y \leqslant y\} \subset \{X \leqslant -\infty\}$$

we have

$$P(X \leqslant -\infty, Y \leqslant y) \leqslant \text{Prob}\{X \leqslant -\infty\} = 0$$

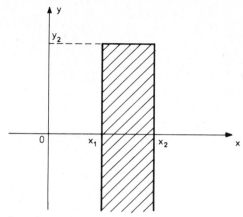

Fig. 6.2. Probability of random point falling within the shaded area equals the difference $F_{XY}(x_2, y_2) - F_{XY}(x_1, y_2)$.

which leads us to the first of the equalities in Eq. (6.5), the other equalities are arrived at in an analogous manner. Geometrically, Eq. (6.5) is quite clear: As one of the arguments, x or y, approaches minus infinity, the probability of the random point falling within the shaded quadrant approaches zero.

Now suppose $x_1 < x_2$. We have, then, for fixed $y = y_2$,

$$P(x_1 < X \leqslant x_2, Y \leqslant y_2) = P(X \leqslant x_2, Y \leqslant y_2) - P(X \leqslant x_1, Y \leqslant y_2)$$

which is illustrated in Fig. 6.2. In view of the definition (6.1), we have

$$P(x_1 < X \leqslant x_2, Y \leqslant y_2) = F_{XY}(x_2, y_2) - F_{XY}(x_1, y_2) \qquad (6.6)$$

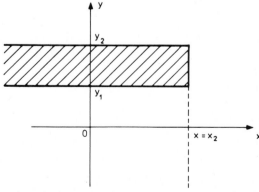

Fig. 6.3. Probability of random point falling within the shaded area equals the difference $F_{XY}(x_2, y_2) - F_{XY}(x_2, y_1)$.

Since the left-hand term of Eq. (6.6) is a probability, it is nonnegative, implying that the joint probability distribution function is a nondecreasing function of x. Similarly, it is a nondecreasing function of y; indeed, if $y_1 < y_2$ we have, for fixed $x = x_2$ (see Fig. 6.3),

$$P(X \leqslant x_2, y_1 < Y < y_2) = P(X \leqslant x_2, Y \leqslant y_2) - P(X \leqslant x_2, Y \leqslant y_1)$$

The rectangle in Fig. 6.4 may be considered as the difference of two half-strips,

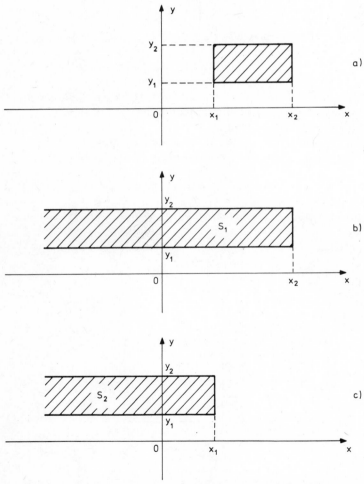

Fig. 6.4. (a) Probability of a point falling within the rectangle equals the difference of those of it falling within half-strip S_1 (b) and half-strip S_2 (c), as per Eq. (6.7).

so that

$$P(x_1 < X \leqslant x_2, y_1 < Y \leqslant y_2)$$

$$= [F_{XY}(x_2, y_2) - F_{XY}(x_2, y_1)] - [F_{XY}(x_1, y_2) - F_{XY}(x_1, y_1)]$$

$$= F_{XY}(x_2, y_2) - F_{XY}(x_1, y_2) - F_{XY}(x_2, y_1) + F_{XY}(x_1, y_1) \quad (6.7)$$

The joint probability distribution function $F_{XY}(x, y)$ is continuous on the right in either argument; that is,

$$\lim_{0 < \varepsilon \to 0} F_{XY}(x + \varepsilon, y) = \lim_{0 < \varepsilon \to 0} F_{XY}(x, y + \varepsilon) = F_{XY}(x, y) \quad (6.8)$$

6.2   JOINT DENSITY FUNCTION OF TWO RANDOM VARIABLES

Consider now a pair of continuous random variables X and Y. We define as their *joint density function* the limiting probability of a random point with coordinates (X, Y) falling within the elementary rectangle with vertices (x, y), $(x + \Delta x, y)$, $(x, y + \Delta y)$, $(x + \Delta x, y + \Delta y)$ as its area $\Delta x \, \Delta y$ approaches zero, and denote it by $f_{XY}(x, y)$:

$$f_{XY}(x, y) = \lim_{\substack{\Delta x \to 0 \\ \Delta y \to 0}} \frac{P(x < X \leqslant x + \Delta x, y < Y \leqslant y + \Delta y)}{\Delta x \, \Delta y}$$

$$= \lim_{\substack{\Delta x \to 0 \\ \Delta y \to 0}} \frac{1}{\Delta x \, \Delta y} [F_{XY}(x + \Delta x, y + \Delta y) - F_{XY}(x + \Delta x, y)$$

$$- F_{XY}(x, y + \Delta y) + F_{XY}(x, y)] \quad (6.9)$$

Since

$$P(x < X \leqslant x + dx, y < Y \leqslant y + dy) \simeq f_{XY}(x, y) \, dx \, dy \quad (6.10)$$

we conclude that the joint probability density is nonnegative for all x and y. Another conclusion is that the probability of a random point on the plane, with coordinates X, Y, falling within the rectangle $(x_1 < X \leqslant x_2, y_1 < Y \leqslant y_2)$ equals

$$P(x_1 < X \leqslant x_2, y_1 < Y \leqslant y_2) = \int_{x_1}^{x_2} \int_{y_1}^{y_2} f_{XY}(x, y) \, dx \, dy \quad (6.11)$$

Moreover, the probability of the random point falling within the area A is

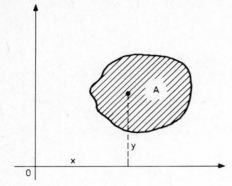

Fig. 6.5. Probability of a point falling within the area A equals the integral of $f_{XY}(x, y)$ over that area.

given by the formula

$$P\{(X, Y) \in A\} = \iint_A f_{XY}(x, y) \, dx \, dy \tag{6.12}$$

(see Fig. 6.5), that is, it equals the *volume* under $f_{XY}(x, y)$ over the domain A. In particular, the joint distribution function $F_{XY}(x, y)$ of random variables X and Y equals the probability of the coordinates of the random point falling within the quadrant with vertex (x, y). Therefore,

$$F_{XY}(x, y) = \int_{-\infty}^{x} \int_{-\infty}^{y} f_{XY}(\xi, \eta) \, d\xi \, d\eta \tag{6.13}$$

which yields the formula for $f_{XY}(x, y)$, provided $F_{XY}(x, y)$ is known,

$$f_{XY}(x, y) = \frac{\partial^2 F_{XY}(x, y)}{\partial x \, \partial y} \tag{6.14}$$

which actually coincides with Eq. (6.9).

By virtue of Eq. (6.4), Eq. (61.3) yields

$$\int_{-\infty}^{\infty} \int_{-\infty}^{\infty} f_{XY}(x, y) \, dx \, dy = 1 \tag{6.15}$$

The marginal probability distribution functions are obtainable from Eq. (6.13), given $f_{XY}(x, y)$:

$$F_X(x) = F_{XY}(x, \infty) = \int_{-\infty}^{x} d\xi \int_{-\infty}^{\infty} f_{XY}(\xi, \eta) \, d\eta$$

$$F_Y(y) = F_{XY}(\infty, y) = \int_{-\infty}^{y} d\eta \int_{-\infty}^{\infty} f_{XY}(\xi, \eta) \, d\xi \tag{6.16}$$

which, in turn, yield the marginal probability densities:

$$f_X(x) = \frac{dF_X(x)}{dx} = \int_{-\infty}^{\infty} f_{XY}(x, y) \, dy$$

$$f_Y(y) = \frac{dF_Y(y)}{dy} = \int_{-\infty}^{\infty} f_{XY}(x, y) \, dx \qquad (6.17)$$

Example 6.1

A pair of random variables X and Y have a jointly uniform distribution in the rectangle $|x| \leqslant 1, |y| \leqslant 1$ if their joint probability density is

$$f_{XY}(x, y) = \begin{cases} \frac{1}{4}, & |x| \leqslant 1, |y| \leqslant 1 \\ 0, & \text{otherwise} \end{cases} \qquad (6.18)$$

We seek the joint distribution function $F_{XY}(x, y)$ and the marginal density functions $f_X(x)$ and $f_Y(y)$.

Substituting Eq. (6.18) in Eq. (6.13), we have

$$F_{XY}(x, y) = \begin{cases} \frac{1}{4}(x + 1)(y + 1), & |x| \leqslant 1, |y| \leqslant 1 \\ \frac{1}{2}(x + 1), & |x| \leqslant 1, y > 1 \\ \frac{1}{2}(y + 1), & x > 1, |y| \leqslant 1 \\ 0, & x < -1, y < -1 \\ 1, & x > 1, y > 1 \end{cases}$$

The marginal density functions are obtained by applying (6.17):

$$f_X(x) = \begin{cases} \frac{1}{2}, & |x| \leqslant 1 \\ 0, & \text{otherwise} \end{cases}$$

$$f_Y(y) = \begin{cases} \frac{1}{2}, & |y| \leqslant 1 \\ 0, & \text{otherwise} \end{cases}$$

implying that each random coordinate has a uniform distribution. As can be concluded from Prob. 6.1, however, this is not always the case.

Example 6.2

Given the joint probability density of two random variables $f_{XY}(x, y)$, the following expressions can be written down for the probabilities of various

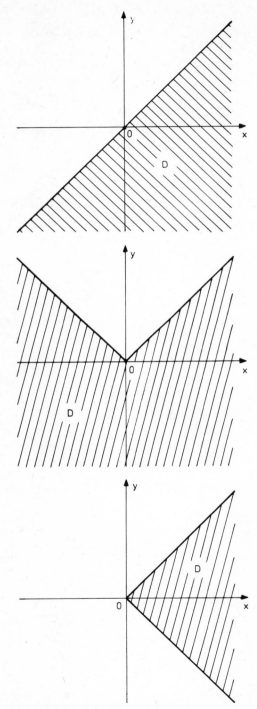

Fig. 6.6. Integration domains D in Example 6.2.

events:

(a) $P(X > Y) = \int_{-\infty}^{\infty} dx \int_{-\infty}^{x} f_{XY}(x, y)\, dy$

(b) $P(|X| > Y) = \int_{-\infty}^{\infty} dx \int_{-\infty}^{|x|} f_{XY}(x, y)\, dy$

(c) $P(X > |Y|) = \int_{0}^{\infty} dx \int_{-x}^{x} f_{XY}(x, y)\, dy$

These situations are illustrated in Fig. 6.6a, b, and c, respectively.

6.3 CONDITIONAL PROBABILITY DISTRIBUTION AND DENSITY FUNCTIONS

In Sec. 2.6, we defined *conditional probability*, the probability of occurrence of a random event A under the hypothesis that B has taken place, as

$$P(A|B) = \frac{P(AB)}{P(B)} \tag{6.19}$$

provided $P(B) \neq 0$. Now let X and Y be a pair of random variables. Denoting

$$A = \{X \leqslant x\} \qquad B = \{y_1 < Y \leqslant y_2\}$$

then from Eq. (6.19) we obtain

$$P(X \leqslant x | y_1 < Y \leqslant y_2) = \frac{P(X \leqslant x, y_1 < Y \leqslant y_2)}{P(y_1 < Y \leqslant y_2)} \tag{6.20}$$

We call $F_X(x|Y \leqslant y)$ the *conditional probability distribution function* of X; in accordance with Sec. 3.7,

$$F_X(x|y_1 < Y \leqslant y_2) = \frac{F_{XY}(x, y_2) - F_{XY}(x, y_1)}{F_Y(y_2) - F_Y(y_1)} \tag{6.21}$$

or, in terms of the probability densities,

$$F_X(x|y_1 < Y \leqslant y_2) = \frac{\displaystyle\int_{-\infty}^{x} dx \int_{y_1}^{y_2} f_{XY}(x, y)\, dy}{\displaystyle\int_{y_1}^{y_2} f_Y(y)\, dy} \tag{6.22}$$

Similarly,

$$F_Y(y|x_1 < X \leqslant x_2) = \frac{F_{XY}(x_2, y) - F_{XY}(x_1, y)}{F_X(x_2) - F_X(x_1)} \tag{6.23}$$

or

$$F_Y(y|x_1 < X \leqslant x_2) = \frac{\int_{-\infty}^{y} dy \int_{x_1}^{x_2} f_{XY}(x, y)\, dx}{\int_{x_1}^{x_2} f_X(x)\, dx}$$

The conditional probability density function $f_X(x|y_1 < Y \leqslant y_2)$ is obtained by differentiating Eq. (6.21) with respect to x:

$$f_X(x|y_1 < Y \leqslant y_2) = \frac{d}{dx} F_X(x|y_1 < Y \leqslant y_2)$$

$$= \frac{1}{F_Y(y_2) - F_Y(y_1)} \left[\frac{\partial F_{XY}(x, y_2)}{\partial x} - \frac{\partial F_{XY}(x, y_1)}{\partial x} \right]$$

$$(6.24)$$

or, in terms of density functions,

$$f_X(x|y_1 < Y \leqslant y_2) = \frac{\int_{y_1}^{y_2} f_{XY}(x, y)\, dy}{\int_{y_1}^{y_2} f_Y(y)\, dy}$$

The conditional probability density $f_Y(y|x_1 < X \leqslant x_2)$ is found in a similar fashion:

$$f_Y(y|x_1 < X \leqslant x_2) = \frac{1}{F_X(x_2) - F_X(x_1)} \left[\frac{\partial F_{XY}(x_2, y)}{\partial y} - \frac{\partial F_{XY}(x_1, y)}{\partial y} \right]$$

or

$$f_Y(y|x_1 < X \leqslant x_2) = \frac{\int_{x_1}^{x_2} f_{XY}(x, y)\, dx}{\int_{x_1}^{x_2} f_X(x)\, dx}$$

Next, we proceed to define the conditional probability distribution function of

X under the hypothesis that $Y = y$ as the limit, in view of Eq. (6.21):

$$F_X(x|Y = y) = \lim_{\Delta y \to 0} F_X(x|y < Y \leqslant y + \Delta y)$$

$$= \lim_{\Delta y \to 0} \frac{F_{XY}(x, y + \Delta y) - F_{XY}(x, y)}{F_Y(y + \Delta y) - F_Y(y)}$$

$$= \frac{1}{f_Y(y)} \frac{\partial F_{XY}(x, y)}{\partial y} \tag{6.25}$$

Differentiating this equation with respect to x, we obtain the conditional probability density function

$$f_X(x|Y = y) = \frac{1}{f_Y(y)} \frac{\partial^2 F_{XY}}{\partial x \, \partial y} = \frac{f_{XY}(x, y)}{f_Y(y)}$$

which we denote by $f_X(x|y)$, that is,

$$f_X(x|y) = \frac{f_{XY}(x, y)}{f_Y(y)} \tag{6.26}$$

Similarly,

$$f_Y(y|x) = \frac{f_{XY}(x, y)}{f_X(x)} \tag{6.27}$$

Comparing Eqs. (6.26) and (6.27), we conclude that the two conditional probability densities are interrelated:

$$f_{XY}(x, y) = f_X(x|y)f_Y(y) = f_Y(y|x)f_X(x) \tag{6.28}$$

We call two random variables *dependent* if the events $\{X \leqslant x\}$ and $\{Y \leqslant y\}$ are dependent for at least one pair of x and y. We call them *independent* in the opposite case, that is, if $\{X \leqslant x\}$ and $\{Y \leqslant y\}$ are independent for any x and y:

$$F_{XY}(x, y) = P(X \leqslant x, Y \leqslant y) = P(X \leqslant x)P(Y \leqslant y) = F_X(x)F_Y(y) \tag{6.29}$$

In other words, the joint probability distribution function of a pair of independent random variables equals the product of the marginal probability distribution functions. Differentiating Eq. (6.29) once with respect x and once with respect y, we obtain

$$f_{XY}(x, y) = f_X(x) f_Y(y) \qquad (6.30)$$

Comparing Eqs. (6.28) and (6.30), we conclude that for independent random variables the conditional probability densities equal their unconditional counterparts:

$$f_X(x|y) = f_X(x) \qquad f_Y(y|x) = f_Y(y) \qquad (6.31)$$

Example 6.3

We seek the conditional probability densities $f_X(x|y)$ and $f_Y(y|x)$ for a pair of random variables with jointly uniform distribution, as per Example 6.1. Via Eq. (6.28) we obtain

$$f_X(x|y) = \frac{f_{XY}(x, y)}{f_Y(y)} = \begin{cases} \frac{1}{2}, & |x| \leqslant 1, |y| \leqslant 1 \\ 0, & \text{otherwise} \end{cases}$$

$$f_Y(y|x) = \frac{f_{XY}(x, y)}{f_X(x)} = \begin{cases} \frac{1}{2}, & |x| \leqslant 1, |y| \leqslant 1 \\ 0, & \text{otherwise} \end{cases}$$

X and Y are thus independent, since the conditional probability densities coincide with their unconditional, marginal, counterparts.

6.4   MULTIDIMENSIONAL RANDOM VECTOR

The pair of random variables defined in Sec. 6.1 is often referred to as a two-dimensional random vector. Indeed, they may be viewed as coordinates of a random point in a plane, whose radius vector from the origin of the coordinate system may in turn be treated as a two-dimensional vector with coordinates X and Y. Analogously, a triplet of random variables may be viewed as coordinates of a random point in space, consitituting a three-dimensional random vector. In general, a set of random variables $X_1, X_2, \ldots, X_n$ may be viewed as an n-dimensional random vector $\{X\}$, the joint distribution function of its components being

$$F_{\{X\}}(x_1, x_2, \ldots, x_n) = P(X_1 \leqslant x_1, X_2 \leqslant x_2, \ldots, X_n \leqslant x_n) \qquad (6.32)$$

The joint density function $f_{\{X\}}(x_1, x_2, \ldots, x_n)$ is defined as the following limit:

$$f_{\{X\}}(x_1, x_2, \ldots, x_n) = \lim_{\substack{\Delta x_i \to 0 \\ i = 1, 2, \ldots, n}} \frac{P\left[\cap_{i=1}^{n}(x_i < X_i \leqslant x_i + \Delta x_i)\right]}{\Delta x_1 \Delta x_2 \cdots \Delta x_n}$$

If any of the arguments of $F_{(X)}(x_1, x_2, \ldots, x_n)$ equals $-\infty$, the distribution function is zero. If $x_1 = x_2 = \cdots = x_n = +\infty$, then $F_{(X)}(+\infty, +\infty, \ldots, +\infty) = 1$. If $m < n$ arguments of $F_{(X)}$ equal $+\infty$, we obtain the distribution function of the remaining $n - m$ random variables, for example,

$$F_{(X)_{(n)}}(+\infty, x_2, x_3, \ldots, x_n) = F_{(X)_{(n-1)}}(x_2, x_3, \ldots, x_n)$$

The distribution function is a nondecreasing function of each of the arguments. The distribution function and the joint density function are interrelated as follows:

$$F_{(X)}(x_1, x_2, \ldots, x_n) = \int_{-\infty}^{x_1} \cdots \int_{-\infty}^{x_n} f_{(X)}(\xi_1, \xi_2, \ldots, \xi_n)\, d\xi_1 \cdots d\xi_n \quad (6.33)$$

$$f_{(X)}(x_1, x_2, \ldots, x_n) = \frac{\partial^n F_{(X)}(x_1, x_2, \ldots, x_n)}{\partial x_1\, \partial x_2 \cdots \partial x_n} \quad (6.34)$$

Integrating $f_{(X)}(x_1, x_2, \ldots, x_n)$ with respect to some of the variables, we obtain the joint density of the remaining ones, for example,

$$f_{X_1 X_2}(x_1, x_2) = \int_{-\infty}^{\infty} f_{X_1 X_2 X_3}(x_1, x_2, \xi_3)\, d\xi_3$$

$$f_{X_1}(x_1) = \int_{-\infty}^{\infty} \int_{-\infty}^{\infty} f_{X_1 X_2 X_3}(x_1, \xi_2, \xi_3)\, d\xi_2\, d\xi_3$$

The conditional probability density of the random variables $X_1, X_2, \ldots, X_k$, under the hypothesis that $X_{k+1} = x_{k+1}$, $X_{k+2} = x_{k+2}, \ldots, X_n = x_n$, is, in complete analogy with Eq. (6.26),

$$f(x_1, x_2, \ldots, x_k | x_{k+1}, x_{k+2}, \ldots, x_n) = \frac{f(x_1, x_2, \ldots, x_k, x_{k+1}, \ldots, x_n)}{f(x_{k+1}, x_{k+2}, \ldots, x_n)}$$

$$(6.35)$$

6.5 FUNCTIONS OF RANDOM VARIABLES

Given an n-dimensional random vector $\{X\}$ with components $X_1, X_2, \ldots, X_n$ and joint probability density $f_{(X)}(x_1, x_2, \ldots, x_n)$, we seek the joint probability $f_{(Y)}(y_1, y_2, \ldots, y_m)$ of an m-dimensional random vector $\{Y\}$ with components

$Y_1, Y_2, \ldots, Y_m$, representing the given function of $X_1, X_2, \ldots, X_n$:

$$Y_1 = \varphi_1(X_1, X_2, \ldots, X_n)$$
$$Y_2 = \varphi_2(X_1, X_2, \ldots, X_n)$$
$$\vdots$$
$$Y_m = \varphi_m(X_1, X_2, \ldots, X_n)$$

(6.36)

The joint distribution function $F_{(Y)}(y_1, y_2, \ldots, y_m)$ reads

$$F_{(Y)}(y_1, y_2, \ldots, y_m) = P(Y_1 \leqslant y_1, Y_2 \leqslant y_2, \ldots, Y_m \leqslant y_m)$$

$$= P[\varphi_1(X_1, X_2, \ldots, X_n) \leqslant y_1, \varphi_2(X_1, X_2, \ldots, X_n)$$

$$\leqslant y_2, \ldots, \varphi_m(X_1, X_2, \ldots, X_n) \leqslant y_m]$$ (6.37)

So that

$$F_{(Y)}(y_1, y_2, \ldots, y_m) = \iint_D \ldots \int f_{(X)}(x_1, x_2, \ldots, x_n) \, dx_1 \, dx_2 \, \cdots \, dx_n$$

(6.38)

where the integration domain D is defined by the following inequalities

$$\varphi_1(x_1, x_2, \ldots, x_n) \leqslant y_1$$
$$\varphi_2(x_1, x_2, \ldots, x_n) \leqslant y_2$$
$$\vdots$$
$$\varphi_m(x_1, x_2, \ldots, x_n) \leqslant y_m$$

(6.39)

Note that in the case of a single function of a random variable the integration domain D is shown in Eq. (4.58).

The joint probability density $f_{(Y)}(y_1, y_2, \ldots, y_m)$ is obtained from $F_{(Y)}$ by differentiating with respect to $y_1, y_2, \ldots, y_m$:

$$f_{(Y)}(y_1, y_2, \ldots, y_m) = \frac{\partial^m F_{(X)}(y_1, y_2, \ldots, y_m)}{\partial y_1 \, \partial y_2 \, \cdots \, \partial y_m}.$$ (6.40)

In the particular case $m = 1$,

$$Y = \varphi(X_1, X_2, \ldots, X_n)$$ (6.41)

we have

$$F_Y(y) = \iint_D \ldots \int f_{(X)}(x_1, x_2, \ldots, x_n) \, dx_1 \, dx_2 \, \cdots \, dx_n$$ (6.42)

where D is defined by the inequality

$$\varphi(x_1, x_2, \ldots, x_n) \leqslant y$$

and

$$f_Y(y) = \frac{dF_Y(y)}{dy}$$

In the general case, we first find the conditional probability of random vector $\{X\}$ under the hypothesis $X_1 = x_1, X_2 = x_2, \ldots, X_n = x_n$. For any values of $x_1, x_2, \ldots, x_n$ of the random variables $X_1, X_2, \ldots, X_n$, each of the random variables $Y_1, Y_2, \ldots, Y_m$ has a single value given by Eqs. (6.36). We have

$$f_{\{Y\}|\{X\}}(y_1, \ldots, y_m | x_1, x_2, \ldots, x_n) = \prod_{j=1}^{m} \delta[y_j - \varphi_j(x_1, \ldots, x_n)] \quad (6.43)$$

where components of the vector $\{Y\}|\{X\}$ represent the causally distributed random variables (see Sec. 4.1). The joint probability density of the random vectors $\{X\}$ and $\{Y\}$ is, in accordance with Eq. (6.35),

$$f_{\{X\}\{Y\}}(x_1, \ldots, x_n, y_1, \ldots, y_m) = f_{\{X\}}(x_1, \ldots, x_n) \prod_{j=1}^{m} \delta[y_j - \varphi_j(x_1, \ldots, x_n)]$$

$$(6.44)$$

Integrating with respect to $x_1, \ldots, x_n$ we obtain the desired probability density of vector $\{Y\}$:

$$f_{\{Y\}}(y_1, \ldots, y_m) = \int_{-\infty}^{\infty} \cdots \int_{-\infty}^{\infty} f_{\{X\}}(\xi_1, \ldots, \xi_n)$$

$$\times \prod_{j=1}^{m} \delta[y_j - \varphi_j(\xi_1, \ldots, \xi_n)] \, d\xi_1 \ldots d\xi_n \quad (6.45)$$

The joint distribution function $F_{\{Y\}}(y_1, y_2, \ldots, y_m)$ is obtained by integrating Eq. (6.45):

$$F_{\{Y\}}(y_1, y_2, \ldots, y_m) = \int_{-\infty}^{y_1} \int_{-\infty}^{y_2} \cdots \int_{-\infty}^{y_m} f_{\{Y\}}(\eta_1, \eta_2, \ldots, \eta_m) \, d\eta_1 \, d\eta_2 \cdots d\eta_m$$

$$(6.46)$$

For a scalar function Y of the vector $\{X\}$, $m = 1$, we have

$$f_Y(y) = \int_{-\infty}^{\infty} \cdots \int_{-\infty}^{\infty} f_{\{X\}}(\xi_1, \ldots, \xi_n) \delta[y - \varphi(\xi_1, \ldots, \xi_n)] \, d\xi_1 \cdots d\xi_n$$

$$(6.47)$$

and

$$F_Y(y) = \int_{-\infty}^{y} f_Y(y) \, dy \qquad (6.48)$$

Example 6.4

Consider the case $m = 2$, $n = 1$, with φ_1 and φ_2 the linear functions, so that

$$Y_1 = \alpha_1 X + \beta_1$$

$$Y_2 = \alpha_2 X + \beta_2$$

where $\alpha_1, \alpha_2, \beta_1, \beta_2$ are given real numbers, $\alpha_1 \neq 0$, $\alpha_2 \neq 0$. Then

$$f_{\langle Y \rangle}(y_1, y_2) = \int_{-\infty}^{\infty} f_X(x)\delta(y_1 - \alpha_1 x - \beta_1)\delta(y_2 - \alpha_2 x - \beta_2) \, dx$$

we introduce a new variable $\xi = \alpha_1 x$ and obtain

$$f_{\langle Y \rangle}(y_1, y_2) = \frac{1}{\alpha_1} \int_{-\infty \, \text{sign} \, \alpha_1}^{+\infty \, \text{sign} \, \alpha_1} f_X\left(\frac{\xi}{\alpha_1}\right)\delta(y_1 - \xi - \beta_1)\delta\left(y_2 - \frac{\alpha_2}{\alpha_1}\xi - \beta_2\right) d\xi$$

$$= \frac{1}{|\alpha_1|} \int_{-\infty}^{\infty} f_X\left(\frac{\xi}{\alpha_1}\right)\delta(y_1 - \xi - \beta_1)\delta\left(y_2 - \frac{\alpha_2}{\alpha_1}\xi - \beta_2\right) d\xi$$

$$= \frac{1}{|\alpha_1|} f_X\left(\frac{y_1 - \beta_1}{\alpha_1}\right)\delta\left(y_2 - \frac{\alpha_2}{\alpha_1}y_1 + \frac{\alpha_2\beta_1}{\alpha_1} - \beta_2\right)$$

or

$$f_{\langle Y \rangle}(y_1, y_2) = \frac{1}{|\alpha_2|} f_X\left(\frac{y_2 - \beta_2}{\alpha_2}\right)\delta\left(y_1 - \frac{\alpha_1}{\alpha_2}y_2 + \frac{\alpha_1\beta_2}{\alpha_2} - \beta_1\right)$$

If, $\beta_1 = \beta_2 = 0$, and $\alpha_1 = \alpha_2 = 1$, we have

$$f_{\langle Y \rangle}(y_1, y_2) = f_X(y_1)\delta(y_1 - y_2) = f_X(y_2)\delta(y_2 - y_1)$$

which could be anticipated, since in this case $Y_1 = Y_2 = X$.
In the case of the functions

$$Y_1 = \alpha_1 |G| \qquad Y_2 = \alpha_2 |G|$$

where G is a continuous random variable with probability density $f_G(g)$, the above equations yield

$$f_{(Y)}(y_1, y_2) = \frac{1}{|\alpha_1|} f_{|G|}\left(\frac{y_1}{\alpha_1}\right) \delta\left(y_2 - \frac{\alpha_2}{\alpha_1} y_1\right)$$

$$f_{(Y)}(y_1, y_2) = \frac{1}{|\alpha_2|} f_{|G|}\left(\frac{y_2}{\alpha_2}\right) \delta\left(y_1 - \frac{\alpha_1}{\alpha_2} y_2\right) \tag{6.49}$$

We can apply these results to Example 5.1, where G was a random force applied at the middle section of a beam, simply supported at its ends, and

$$\alpha_1 = \frac{l^2}{8S\sigma_{\text{allow}}} \qquad \alpha_2 = \frac{5l^4}{384EIw_{\text{allow}}}$$

with the reliability defined as

$$R = P(\alpha_1|G| \leqslant 1, \alpha_2|G| \leqslant 1) = P(Y_1 \leqslant 1, Y_2 \leqslant 1)$$

which, since $\alpha_1, \alpha_2 > 0$, may be rewritten as

$$R = \int_0^1 \int_0^1 f_{Y_1Y_2}(y_1, y_2)\, dy_1\, dy_2 = \int_0^1 \int_0^1 \frac{1}{\alpha_1} f_{|G|}\left(\frac{y_1}{\alpha_1}\right) \delta\left(y_2 - \frac{\alpha_2}{\alpha_1} y_1\right) dy_1\, dy_2$$

Now, for $\alpha_1 > \alpha_2$ we have

$$\int_0^1 \delta\left(y_2 - \frac{\alpha_2}{\alpha_1} y_1\right) dy_2 = 1$$

and therefore

$$R = \int_0^1 \frac{1}{\alpha_1} f_{|G|}\left(\frac{y_1}{\alpha_1}\right) dy_1 = F_{|G|}\left(\frac{1}{\alpha_1}\right)$$

For $\alpha_1 < \alpha_2$, however, we have

$$R = \int_0^1 dy_2 \int_0^1 \frac{1}{\alpha_1} f_{|G|}\left(\frac{y_1}{\alpha_1}\right) \delta\left(y_2 - \frac{\alpha_2}{\alpha_1} y_1\right) dy_1$$

$$= \frac{1}{\alpha_1} \frac{\alpha_1}{\alpha_2} \int_0^1 f_{|G|}\left(\frac{1}{\alpha_1} \frac{\alpha_1}{\alpha_2} y_2\right) dy_2 = \frac{1}{\alpha_2} \int_0^1 f_{|G|}\left(\frac{y_2}{\alpha_2}\right) dy_2 = F_{|G|}\left(\frac{1}{\alpha_2}\right)$$

and finally, for $\alpha_1 = \alpha_2 = \alpha$,

$$R = F_{|G|}\left(\frac{1}{\alpha}\right)$$

These results coincide with those obtained in Example 5.1.

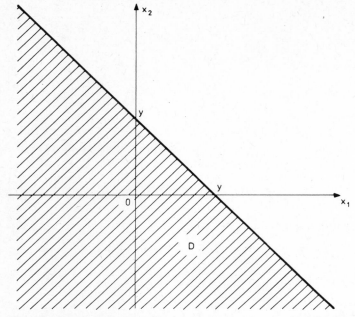

Fig. 6.7. Integration domain D for the sum of a pair of random variables.

Example 6.5

We seek the distribution function of a random variable Y representing the sum of a pair of random variables X_1 and X_2 with $f_{(X)}(x_1, x_2)$ given. In this case the domain D of the $x_1 x_2$ plane is determined by the equation $x_1 + x_2 \leqslant y$ and is shown in Fig. 6.7. For the distribution function $F_Y(y)$ we obtain

$$F_Y(y) = \int_{-\infty}^{\infty} dx_2 \int_{-\infty}^{y-x_2} f_{(X)}(x_1, x_2)\, dx_1$$

or

$$F_Y(y) = \int_{-\infty}^{\infty} dx_1 \int_{-\infty}^{y-x_1} f_{(X)}(x_1, x_2)\, dx_2$$

Differentiating these equations with respect to y, we have

$$f_Y(y) = \frac{dF_Y(y)}{dy} = \int_{-\infty}^{\infty} f_{(X)}(y - x_2, x_2)\, dx_2 = \int_{-\infty}^{\infty} f_{(X)}(x_1, y - x_1)\, dx_1$$

$$(6.50)$$

In the particular case where the components X_1 and X_2 of $\{X\}$ are independent, that is,

$$f_{\langle X\rangle}(x_1, x_2) = f_{X_1}(x_1)f_{X_2}(x_2)$$

and

$$F_Y(y) = \int_{-\infty}^{\infty} F_{X_1}(y - x_2)f_{X_2}(x_2)\, dx_2 = \int_{-\infty}^{\infty} F_{X_2}(y - x_1)f_{X_1}(x_1)\, dx_1$$

$$(6.51)$$

which in turn yields $f_Y(y)$ in the form

$$f_Y(y) = \int_{-\infty}^{\infty} f_{X_1}(y - x_2)f_{X_2}(x_2)\, dx_2 = \int_{-\infty}^{\infty} f_{X_2}(y - x_1)f_{X_1}(x_1)\, dx_1$$

$$(6.52)$$

This result follows also from Eq. (6.45), which now reads

$$f_Y(y) = \int_{-\infty}^{\infty}\int_{-\infty}^{\infty} f_{\langle X\rangle}(\xi_1, \xi_2)\delta(y - \xi_1 - \xi_2)\, d\xi_1\, d\xi_2$$

The integral in Eq. (6.52) is called the *convolution* of $f_{X_1}(y)$ and $f_{X_2}(y)$ and is denoted by

$$f_Y(y) = f_{X_1}(y) * f_{X_2}(y)$$

Thus, the probability density of the sum of a pair of independent random variables is represented by the convolution of the density functions of the component random variables.

Example 6.6

Let now

$$Y = X_1 X_2$$

The integration domain D (Fig. 6.8) is determined by the part of the $x_1 x_2$ plane where

$$x_1 x_2 \leqslant y$$

The distribution function $F_Y(y)$ is obtained by

$$F_Y(y) = \int_{-\infty}^{0} dx_1 \int_{y/x_1}^{+\infty} f_{\langle X\rangle}(x_1, x_2)\, dx_2 + \int_{0}^{\infty} dx_1 \int_{-\infty}^{y/x_1} f_{\langle X\rangle}(x_1, x_2)\, dx_2$$

$$(6.53)$$

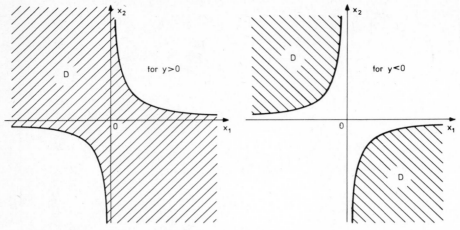

Fig. 6.8. Integration domain D for the product of a pair of random variables.

Differentiation yields

$$f_Y(y) = \int_0^\infty \frac{1}{x_1} f_{(X)}\left(x_1, \frac{y}{x_1}\right) dx_1 - \int_{-\infty}^0 \frac{1}{x_1} f_{(X)}\left(x_1, \frac{y}{x_1}\right) dx_1 \quad (6.54)$$

In the particular case where X_1 and X_2 are independent, we have

$$F_Y(y) = \int_{-\infty}^0 f_{X_1}(x_1)\left[1 - F_{X_2}\left(\frac{y}{x_1}\right)\right] dx_1 + \int_0^\infty f_{X_1}(x_1) F_{X_2}\left(\frac{y}{x_1}\right) dx_1$$

$$f_Y(y) = \int_0^\infty \frac{1}{x_1} f_{X_1}(x_1) f_{X_2}\left(\frac{y}{x_1}\right) dx_1 - \int_{-\infty}^0 \frac{1}{x_1} f_{X_1}(x_1) f_{X_2}\left(\frac{y}{x_1}\right) dx_1 \quad (6.55)$$

Example 6.7

Let $Y = X_1/X_2$. The relevant integration domain D is shown in Fig. 6.9. We have

$$F_Y(y) = \int_0^\infty dx_2 \int_{-\infty}^{yx_2} f_{(X)}(x_1, x_2) \, dx_1 + \int_{-\infty}^0 dx_2 \int_{yx_2}^\infty f_{(X)}(x_1, x_2) \, dx_1 \quad (6.56)$$

For the case where X_1 and X_2 are independent, we obtain, respectively,

$$F_Y(y) = \int_0^\infty F_{X_1}(yx_2) f_{X_2}(x_2) \, dx_2 + \int_{-\infty}^0 \left[1 - F_{X_1}(yx_2)\right] f_{X_2}(x_2) \, dx_2$$

$$f_Y(y) = \int_0^\infty x_2 f_{X_1}(yx_2) f_{X_2}(x_2) \, dx_2 - \int_{-\infty}^0 x_2 f_{X_1}(yx_2) f_{X_2}(x_2) \, dx_2 \quad (6.57)$$

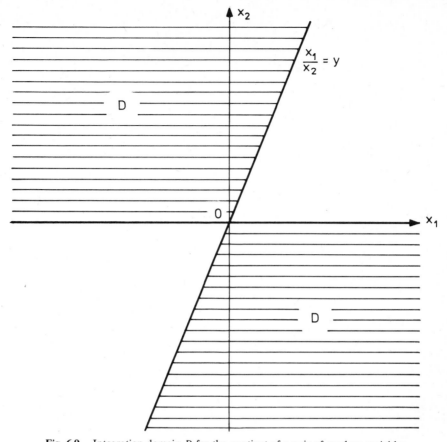

Fig. 6.9. Integration domain D for the quotient of a pair of random variables.

Example 6.8

We seek the distribution function $F_Y(y)$ of a random variable Y representing the radius vector of a random point in the plane whose components are X_1 and X_2 with the given joint density function $f_{\langle X \rangle}(x_1, x_2)$.

Here $Y = \sqrt{X_1^2 + X_2^2}$, and the integration domain D, defined by $x_1^2 + x_2^2 \leqslant y^2$ is a circle. For $F_Y(y)$, $y > 0$, we have

$$F_Y(y) = \int_D \int f_{\langle X \rangle}(x_1, x_2)\, dx_1\, dx_2$$

Resorting to polar coordinates, $r = \sqrt{x_1^2 + x_2^2}$ and $\theta = \tan^{-1}(x_2/x_1)$, we obtain

$$F_Y(y) = \int_0^{2\pi} \int_0^y f_{\langle X \rangle}(r\cos\theta, r\sin\theta)\, r\, dr\, d\theta \tag{6.58}$$

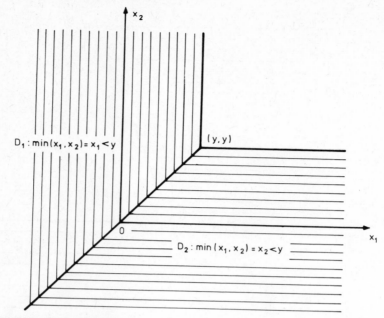

Fig. 6.10.   Shaded domain represents $D = D_1 + D_2$ for $Y = \min(X_1, X_2)$.

The probability density function is

$$f_Y(y) = \frac{dF_Y(y)}{dy} = \int_0^{2\pi} f_{(X)}(y\cos\theta, \, y\sin\theta)\, y\, d\theta \qquad (6.59)$$

Example 6.9

Let $Y = \min(X_1, X_2)$. The integration domain (Fig. 6.10) is defined by

$$\min(x_1, x_2) \leqslant y$$

or

$$
\begin{aligned}
x_1 &\leqslant y, & x_1 &< x_2 \\
x_2 &\leqslant y, & x_2 &< x_1 \\
x &\leqslant y, & x_1 &= x_2 = x
\end{aligned}
$$

Integration over D_1 and D_2 yields

$$F_Y(y) = F_{X_1}(y) + F_{X_2}(y) - F_{X_1 X_2}(y, y) \qquad (6.60)$$

and differentiation yields

$$f_Y(y) = f_{X_1}(y) + f_{X_2}(y) - \frac{d}{dy} F_{X_1 X_2}(y, y)$$

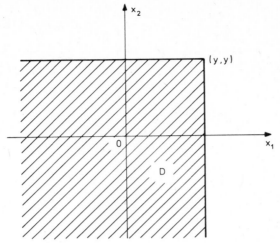

Fig. 6.11. Shaded domain represents D for $Y = \max(X_1, X_2)$.

If X_1 and X_2 are independent random variables, this equation becomes

$$f_Y(y) = f_{X_1}(y) + f_{X_2}(y) - \frac{d}{dy} F_{X_1}(y) F_{X_2}(y)$$

$$= f_{X_1}(y) + f_{X_2}(y) - F_{X_1}(y) f_{X_2}(y) - f_{X_1}(y) F_{X_2}(y)$$

$$= f_{X_1}(y) \big[1 - F_{X_2}(y) \big] + f_{X_2}(y) \big[1 - F_{X_1}(y) \big] \qquad (6.61)$$

Example 6.10

Let $Y = \max(X_1, X_2)$. The integration domain (Fig. 6.11) is defined by

$$\max(x_1, x_2) \leqslant y$$

We immediately observe that

$$F_Y(y) = F_{X_1 X_2}(y, y) \qquad (6.62)$$

as could be anticipated. Indeed, for the maximum of X_1 and X_2 not to exceed y, neither of them must do so: $F_Y(y) = P\{X_1 \leqslant y, X_2 \leqslant y\}$, leading to Eq.

(6.62). Differentiation with respect to y gives

$$f_Y(y) = \frac{dF_Y(y)}{dy} = \frac{d}{dy}\left\{\int_{-\infty}^{y_1}\left[\int_{-\infty}^{y_2} f_{X_1 X_2}(x_1, x_2)\, dx_2\right] dx_1\right\}$$

$$= \frac{\partial F_Y(y)}{\partial y_1}\frac{dy_1}{dy} + \frac{\partial F_Y(y)}{\partial y_2}\frac{dy_2}{dy}$$

$$= \int_{-\infty}^{y} f_{X_1 X_2}(x_1, x_2)\, dx_2 + \int_{-\infty}^{y} f_{X_1 X_2}(x_1, x_2)\, dx_1 \qquad (6.63)$$

If X_1 and X_2 are independent, we obtain

$$f_Y(y) = \frac{d}{dy} F_{X_1}(y) F_{X_2}(y) = F_{X_1}(y) f_{X_2}(y) + F_{X_2}(y) f_{X_1}(y)$$

6.6 EXPECTED VALUES, MOMENTS, COVARIANCE

Let $\{X\} = (X_1, X_2, \ldots, X_n)$ be an n-dimensional random continuous vector with joint probability density $f_{(X)}(x_1, x_2, \ldots, x_n)$, and $g(x_1, x_2, \ldots, x_n)$ a given function. The *mathematical expectation* of the function $g(X_1, X_2, \ldots, X_n)$ of $\{X\}$ is defined as

$$E[g(X_1, X_2, \ldots, X_n)] = \int_{-\infty}^{\infty}\int_{-\infty}^{\infty}\cdots\int_{-\infty}^{\infty} g(x_1, x_2, \ldots, x_n)$$

$$\times f_{(X)}(x_1, x_2, \ldots, x_n)\, dx_1\, dx_2\, \cdots\, dx_n \qquad (6.64)$$

It is readily verified that if $g(X_1, X_2, \ldots, X_n)$ depends on only one of the arguments, Eqs. (6.64) and (3.27) yield the same result. Indeed, if $g(X_1, X_2, \ldots, X_n) = g(X_1)$, then

$$E[g(X_1)] = \int_{-\infty}^{\infty}\int_{-\infty}^{\infty}\cdots\int_{-\infty}^{\infty} g(x_1) f_{(X)}(x_1, x_2, \ldots, x_n)\, dx_1\, dx_2\, \cdots\, dx_n$$

$$= \int_{-\infty}^{\infty} g(x_1)\, dx_1 \int_{-\infty}^{\infty}\cdots\int_{-\infty}^{\infty} f_{(X)}(x_1, x_2, \ldots, x_n)\, dx_2\, \cdots\, dx_n$$

The inner integral, however, equals $f_{X_1}(x_1)$:

$$\int_{-\infty}^{\infty}\cdots\int_{-\infty}^{\infty} f_{(X)}(x_1, x_2, \ldots, x_n)\, dx_2\, \cdots\, dx_n = f_{X_1}(x_1)$$

so that Eq. (6.64) may be rewritten as

$$E[g(X_1)] = \int_{-\infty}^{\infty} g(x_1) f_{X_1}(x_1)\, dx_1$$

which is the desired result.

It is also readily shown that

$$\text{Var}(X_i) = E(X_i^2) - [E(X_i)]^2$$

Letting, in Eq. (6.64), $g(x_1, x_2, \ldots, x_n) = x_i^k x_j^r$, we obtain the $(k + r)$th *moment* of the random variables X_i and X_j:

$$m_{kr} = E(X_i^k X_j^r) = \int_{-\infty}^{\infty} \int_{-\infty}^{\infty} x_i^k x_j^r f_{X_i X_j}(x_i, x_j) \, dx_i \, dx_j \qquad (6.65)$$

Denoting $X_i = X$, $X_j = Y$, we obtain

$$m_{kr} = E(X^k Y^r) = \int_{-\infty}^{\infty} \int_{-\infty}^{\infty} x^k y^r f_{XY}(x, y) \, dx \, dy$$

The joint central moments μ_{kr} are defined by

$$\mu_{kr} = E\{[X - E(X)]^k [Y - E(Y)]^r\}$$

$$= \int_{-\infty}^{\infty} \int_{-\infty}^{\infty} [x - E(X)]^k [y - E(Y)]^r f_{XY}(x, y) \, dx \, dy \qquad (6.66)$$

while

$$\mu_{k0} = E\{[X - E(X)]^k\} \qquad \mu_{0r} = E\{[Y - E(Y)]^r\}$$

are the appropriate central moments of the components X and Y, respectively. In particular,

$$\mu_{20} = \text{Var}(X) \qquad \mu_{02} = \text{Var}(Y)$$

The second central moment μ_{11} is denoted by $\text{Cov}(X, Y)$:

$$\text{Cov}(X, Y) = E\{[X - E(X)][Y - E(Y)]\}$$

$$= \int_{-\infty}^{\infty} \int_{-\infty}^{\infty} [x - E(X)][y - E(Y)] f_{XY}(x, y) \, dx \, dy \qquad (6.67)$$

and is referred to as the covariance of X and Y.

The ratio

$$r_{XY} = \frac{\text{Cov}(X, Y)}{\sqrt{\text{Var}(X)\text{Var}(Y)}} \qquad (6.68)$$

is called the *correlation coefficient*.

It is readily shown that

$$\text{Cov}(X, Y) = E(XY) - E(X)E(Y) \tag{6.69}$$

Indeed, from Eq. (6.67) we have

$$\text{Cov}(X, Y) = E(XY) - E(X)E(Y) - E(Y)E(X) + E(X)E(Y)$$

which leaves us with Eq. (6.69).

We say that two random variables X and Y are *uncorrelated* if their covariance (and therefore their correlation coefficient also) is zero. We immediately observe from Eq. (6.69) that for uncorrelated random variables

$$E(XY) = E(X)E(Y) \tag{6.70}$$

That is, the mathematical expectation of the product of two uncorrelated random variables equals the product of their mathematical expectations.

We prove now that independent random variables, that is, those possessing the property $f_{XY}(x, y) = f_X(x)f_Y(y)$, are also uncorrelated. Indeed, calculation of the covariance of independent random variables yields

$$\text{Cov}(X, Y) = \int_{-\infty}^{\infty} \int_{-\infty}^{\infty} [x - E(X)][y - E(Y)]f_{XY}(x, y)\, dx\, dy$$

$$= \int_{-\infty}^{\infty} \int_{-\infty}^{\infty} [x - E(X)][y - E(Y)]f_X(x)f_Y(y)\, dx\, dy$$

$$= \left\{ \int_{-\infty}^{\infty} [x - E(X)]f_X(x)\, dx \right\} \left\{ \int_{-\infty}^{\infty} [y - E(Y)]f_Y(y)\, dy \right\}$$

$$\tag{6.71}$$

Each of these integrals is zero, since the central moment of first order of a random variable vanishes. Thus, independent random variables are uncorrelated, while correlated random variables are dependent. The opposite, by contrast, is not necessarily valid; indeed, dependent random variables may be either correlated or uncorrelated. For example, if $E(X) = E(Y) = 0$ and $f_{XY}(x, y) = f_{XY}(-x, -y)$, then $\mu_{11} = 0$, although X and Y may be dependent; such is the case in Prob. 6.1, where the covariance is zero, since the distribution $f_{XY}(x, y)$ is symmetric. The conclusion is that independence is a stronger property than uncorrelatedness.

We now introduce the *Cauchy-Schwarz inequality*, namely,

$$|\text{Cov}(X, Y)| \leqslant \sigma_X \sigma_Y \tag{6.72}$$

where

$$\sigma_X = \sqrt{\text{Var}(X)} \qquad \sigma_Y = \sqrt{\text{Var}(Y)}$$

are the standard deviations of the random variables X and Y, respectively. The above may be rewritten as

$$|r_{XY}| \leqslant 1 \qquad (6.73)$$

To prove it, we introduce new random variables

$$Z = \frac{X - E(X)}{\sigma_X} \pm \frac{Y - E(Y)}{\sigma_Y}$$

Since $\text{Var}(Z) \geqslant 0$, we have

$$E\left\{\left[\frac{X - E(X)}{\sigma_X} \pm \frac{Y - E(Y)}{\sigma_Y}\right]^2\right\} = 2(1 \pm r_{XY}) > 0$$

which is equivalent to the sought property (6.73). The equality sign in Eq. (6.73) is obtained when

$$E\left\{\left[\frac{X - E(X)}{\sigma_X} + \frac{Y - E(Y)}{\sigma_Y}\right]^2\right\} = 0, \quad r_{XY} = -1$$

$$E\left\{\left[\frac{X - E(X)}{\sigma_X} - \frac{Y - E(Y)}{\sigma_Y}\right]^2\right\} = 0, \quad r_{XY} = 1$$

Signifying that the following equalities hold:

$$\frac{X - E(X)}{\sigma_X} + \frac{Y - E(Y)}{\sigma_Y} = 0, \quad r_{XY} = -1$$

$$\frac{X - E(x)}{\sigma_X} - \frac{Y - E(Y)}{\sigma_Y} = 0, \quad r_{XY} = 1 \qquad (6.74)$$

In other words, there is a linear functional dependence between absolutely, fully correlated random variables (defined as those with either $r_{XY} = -1$ or $r_{XY} = 1$). Therefore the magnitude of the correlation coefficient is a measure of the degree of *linear* dependence between two random variables. In fact, there may exist a nonlinear functional relationship between two uncorrelated random variables. This happens, for example, when $Y = X^2$, if in addition $f_X(x)$ is an even function. When, for example, X is normally distributed, $N(0, \sigma_X^2)$, then

$$E[XY] - E[X]E[Y] = E[X^3] - E[X]E[X^2] = 0 \qquad (6.75)$$

since all odd moments vanish.

Example 6.11

Given n random variables $X_1, X_2, \ldots, X_n$ with known $f_{(X)}(x_1, x_2, \ldots, x_n)$. We construct their sum

$$Y = X_1 + X_2 + \cdots X_n$$

and seek $E(Y)$ and Var(Y).

Applying Eq. (6.64), we find

$$E(Y) = E(X_1 + X_2 + \cdots + X_n)$$

$$= \int_{-\infty}^{\infty} \int_{-\infty}^{\infty} \cdots \int_{-\infty}^{\infty} (x_1 + x_2 + \cdots + x_n)$$

$$\times f_{(X)}(x_1, x_2, \ldots, x_n)\, dx_1\, dx_2 \cdots dx_n$$

$$= E(X_1) + E(X_2) + \cdots + E(X_n) \tag{6.76}$$

that is, the mathematical expectation of the sum of a set of random variables equals the sum of mathematical expectations of the component variables (irrespective of whether they are correlated or uncorrelated).

The variance of Y is

$$\mathrm{Var}(Y) = E\{[Y - E(Y)]^2\}$$

$$= E\left\{ \sum_{j=1}^{n} [X_j - E(X_j)] \right\}^2$$

$$= \sum_{j=1}^{n} \sum_{k=1}^{n} E\{[X_j - E(X_j)][X_k - E(X_k)]\}$$

$$= \sum_{j=1}^{n} \mathrm{Var}(X_j) + \sum_{\substack{j=1 \\ j \neq k}}^{n} \sum_{k=1}^{n} \mathrm{Cov}(X_j, X_k)$$

$$= \sum_{j=1}^{n} \sigma_{X_j}^2 + \sum_{\substack{j=1 \\ j \neq k}}^{n} \sum_{k=1}^{n} r_{X_j X_k} \sigma_{X_j} \sigma_{X_k} \tag{6.77}$$

where σ_{X_j} and σ_{X_k} are the standard deviations of X_j and X_k, respectively. Equation (6.77) indicates that the variance of the sum of the random variables equals the sum of all variances and covariances of the component variables. If the components $X_1, X_2, \ldots, X_n$ of a random variable X are uncorrelated, all correlation coefficients $r_{X_j X_k}$ vanish and

$$\mathrm{Var}\left(\sum_{j=1}^{n} X_j \right) = \sum_{j=1}^{n} \mathrm{Var}(X_j) \qquad (6.78)$$

that is, the variance of the sum of *uncorrelated* random variables equals the sum of variances of the components. Obviously, Eq. (6.78) holds for *independent* random variables.

In the case of a pair of random variables $X_1 = X$, $X_2 = Y$, Eqs. (6.76) and (6.77) become

$$E(X + Y) = E(X) + E(Y)$$

$$\sigma_{X+Y}^2 = \sigma_X^2 + \sigma_Y^2 + 2r_{XY}\sigma_X\sigma_Y \qquad (6.79)$$

where σ_{X+Y} is the standard deviation of $X + Y$.

For absolutely positively correlated X and Y, $r_{XY} = +1$, and

$$\sigma_{X+Y} = \sigma_X + \sigma_Y \qquad (6.80)$$

and if $\sigma_X = \sigma_Y$, we have $\sigma_{X+Y} = 2\sigma_X = 2\sigma_Y$.

For absolutely negatively correlated X and Y, $r_{XY} = -1$, and

$$\sigma_{X+Y} = \sigma_X - \sigma_Y \qquad (6.81)$$

and if $\sigma_X = \sigma_Y$ we have $\sigma_{X+Y} = 0$.

If X and Y are uncorrelated, we have from Eq. (6.79),

$$\sigma_{X+Y} = \sqrt{\sigma_X^2 + \sigma_Y^2} \qquad (6.82)$$

For the difference of two random variables, we similarly have

$$E(X - Y) = E(X) - E(Y)$$

$$\sigma_{X-Y}^2 = \sigma_X^2 + \sigma_Y^2 - 2r_{XY}\sigma_X\sigma_Y \qquad (6.83)$$

For absolutely positively correlated X and Y, we have

$$\sigma_{X-Y} = |\sigma_X - \sigma_Y| \qquad (6.84)$$

whereas for absolutely negatively correlated X and Y

$$\sigma_{X-Y} = \sigma_X + \sigma_Y \qquad (6.85)$$

For uncorrelated X and Y

$$\sigma_{X-Y}^2 = \sigma_{X+Y}^2 = \sigma_X^2 + \sigma_Y^2 \qquad (6.86)$$

The central moments of second order of an n-dimensional random vector $\{X\}$ with components $X_1, X_2, \ldots, X_n$ represent the variances $\mathrm{Var}(X_1)$, $\mathrm{Var}(X_2), \ldots, \mathrm{Var}(X_n)$ and all covariances $\mathrm{Cov}(X_1, X_2), \ldots, \mathrm{Cov}(X_1, X_n)$, $\ldots, \mathrm{Cov}(X_{n-1}, X_n)$, their number being n^2. They may be treated as elements of a matrix $[V]$:

$$[V] = \begin{bmatrix} \mathrm{Var}(X_1) & \mathrm{Cov}(X_1, X_2) & \cdots & \mathrm{Cov}(X_1, X_n) \\ \mathrm{Cov}(X_2, X_1) & \mathrm{Var}(X_2) & \cdots & \mathrm{Cov}(X_2, X_n) \\ \vdots & & & \\ \mathrm{Cov}(X_n, X_1) & \mathrm{Cov}(X_n, X_2) & \cdots & \mathrm{Var}(X_n) \end{bmatrix} \quad (6.87)$$

This is referred to as the *variance-covariance matrix*, and is symmetric, since $\mathrm{Cov}(X_i, X_j) = \mathrm{Cov}(X_j, X_i)$ according to the definition (6.67).

For a pair of random variables $X_1 \equiv X$ and $X_2 \equiv Y$, the variance-covariance matrix reads

$$[V] = \begin{bmatrix} \mathrm{Var}(X) & \mathrm{Cov}(X, Y) \\ \mathrm{Cov}(Y, X) & \mathrm{Var}(Y) \end{bmatrix}$$

The Cauchy-Schwarz inequality (6.72) implies that the determinant of $[V]$ is nonnegative:

$$\det[V] > 0 \quad (6.88)$$

Since $\mathrm{Var}(X)$ is also nonnegative, we conclude that $[V]_{2 \times 2}$ is likewise *nonnegative* or *positive-semidefinite*. We will show that this property is possessed by the variance-covariance matrix of any n-dimensional random vector. To this end, consider the random variable

$$Y = \alpha_1 X_1 + \alpha_2 X_2 + \cdots + \alpha_n X_n$$

in which $\alpha_1, \alpha_2, \ldots, \alpha_n$ are real. The mathematical expectation and variance of Y are, respectively,

$$E(Y) = \sum_{j=1}^{n} \alpha_j E(X_j)$$

$$\mathrm{Var}(Y) = \sum_{j=1}^{n} \sum_{k=1}^{n} \alpha_j \alpha_k \mathrm{Cov}(X_j, X_k) \quad (6.89)$$

Now the right-hand term of Eq. (6.89.2) is nonnegative, irrespective of the values of $\alpha_1, \alpha_2, \ldots, \alpha_n$. Thus the matrix $[\mathrm{Cov}(X_j, X_k)]$ is nonnegative-definite. This implies, according to the Sylvester's theorem (see e.g., Chetaev) that all principal minor determinants associated with matrix $[\text{Cov}(X_j, X_k)]$ are non-negative:

$$\text{Var}(X_1) \geqslant 0 \qquad \begin{vmatrix} \text{Var}(X_1) & \text{Cov}(X_1, X_2) \\ \text{Cov}(X_1, X_2) & \text{Var}(X_2) \end{vmatrix} \geqslant 0$$

$$\begin{vmatrix} \text{Var}(X_1) & \text{Cov}(X_1, X_2) & \text{Cov}(X_1, X_3) \\ \text{Cov}(X_1, X_2) & \text{Var}(X_2) & \text{Cov}(X_2, X_3) \\ \text{Cov}(X_1, X_3) & \text{Cov}(X_2, X_3) & \text{Var}(X_3) \end{vmatrix} \geqslant 0 \qquad (6.90)$$

etc.

Example 6.12

A box contains a red, b white, and c blue balls. A ball is picked at random from the box. The random variables X, Y, and Z denote the following events:

$$X = \begin{cases} 1, & \text{if a red ball is picked} \\ 0, & \text{otherwise} \end{cases}$$

$$Y = \begin{cases} 1, & \text{if a white ball is picked} \\ 0, & \text{otherwise} \end{cases}$$

$$Z = \begin{cases} 1, & \text{if a blue ball is picked} \\ 0, & \text{otherwise} \end{cases}$$

We seek the variance-covariance matrix formed by the random variables X, Y, Z.
Obviously,

$$P[X = 1] = \frac{a}{a + b + c} \qquad P[X = 0] = \frac{b + c}{a + b + c}$$

$$P[Y = 1] = \frac{b}{a + b + c} \qquad P[Y = 0] = \frac{a + c}{a + b + c}$$

$$P[Z = 1] = \frac{c}{a + b + c} \qquad P[Z = 0] = \frac{a + b}{a + b + c}$$

The mathematical expectations are found from Eq. (3.21):

$$E(X) = \sum_{i=0}^{1} x_i P[X = x_i] = 0 \cdot P[X = 0] + 1 \cdot P[X = 1] = \frac{a}{a + b + c}$$

with $x_0 = 0$ and $x_1 = 1$. Analogously,

$$E(Y) = P[Y = 1] = \frac{b}{a + b + c}$$

$$E(Z) = P[Z = 1] = \frac{c}{a + b + c}$$

Since $E(X^2) = E(X)$, we have for the variance of X

$$\text{Var}(X) = E(X^2) - E^2(X) = \frac{a}{a + b + c} - \frac{a^2}{(a + b + c)^2}$$

$$= \frac{a(b + c)}{(a + b + c)^2}$$

and the variances of Y and Z are, respectively,

$$\text{Var}(Y) = \frac{b(a + c)}{(a + b + c)^2} \qquad \text{Var}(Z) = \frac{c(a + b)}{(a + b + c)^2}$$

In order to find $\text{Cov}(X, Y)$, we first calcuate the following probabilities:

$$P[(X = 1) \cap (Y = 0)] = P[(X = 1) \cap (Y = 0) \cap (Z = 1)]$$

$$+ P[(X = 1) \cap (Y = 0) \cap (Z = 0)]$$

by the formula of overall probability, (2.25). However,

$$P[(X = 1) \cap (Y = 0) \cap (Z = 1)] = 0$$

as the probability of an impossible event, and

$$P[(X = 1) \cap (Y = 0) \cap (Z = 0)] = \frac{a}{a + b + c}$$

Therefore,

$$P[(X = 1) \cap (Y = 0)] = \frac{a}{a + b + c}$$

In a similar manner,

$$E[(X = 1) \cap (Y = 1)] = 0$$

$$E[(X = 0) \cap (Y = 1)] = \frac{b}{a + b + c}$$

$$E[(X = 0) \cap (Y = 0)] = \frac{c}{a + b + c}$$

so that the covariance becomes

$$\text{Cov}(X, Y) = \sum_{i=0}^{1} \sum_{j=0}^{1} [x_1 - E(X)][y_j - E(Y)] P[(X = x_i) \cap (Y = y_j)]$$

$$= \left(0 - \frac{a}{a + b + c}\right)\left(0 - \frac{b}{a + b + c}\right)\left(\frac{c}{a + b + c}\right)$$

$$+ \left(0 - \frac{a}{a + b + c}\right)\left(1 - \frac{b}{a + b + c}\right)\left(\frac{b}{a + b + c}\right)$$

$$+ \left(1 - \frac{a}{a + b + c}\right)\left(0 - \frac{b}{a + b + c}\right)\left(\frac{a}{a + b + c}\right)$$

$$+ \left(1 - \frac{a}{a + b + c}\right)\left(1 - \frac{b}{a + b + c}\right) \cdot 0$$

$$= - \frac{ab}{(a + b + c)^2}$$

Analogously,

$$\text{Cov}(X, Z) = - \frac{ac}{(a + b + c)^2} \qquad \text{Cov}(Y, Z) = - \frac{bc}{(a + b + c)^2}$$

The correlation coefficients, defined in Eq. (6.68), are

$$r_{XY} = - \sqrt{\frac{ab}{(a + c)(b + c)}} \qquad r_{XZ} = - \sqrt{\frac{ac}{(a + b)(b + c)}}$$

$$r_{YZ} = - \sqrt{\frac{bc}{(a + b)(a + c)}}$$

with all of them equal to $-\frac{1}{2}$ for $a = b = c$. Finally, the variance-covariance matrix is

$$[V] = \frac{1}{(a+b+c)^2} \begin{bmatrix} a(b+c) & -ab & -ac \\ -ab & b(a+c) & -bc \\ -ac & -bc & c(a+b) \end{bmatrix}$$

Let us check the nonnegativeness property of the variance-covariance matrix. Indeed, $a(b+c) > 0$, and

$$\begin{vmatrix} a(b+c) & -ab \\ -ab & b(a+c) \end{vmatrix} = abc(a+b+c) > 0$$

$$\begin{vmatrix} a(b+c) & -ab & -ac \\ -ab & b(a+c) & -bc \\ -ac & -bc & c(a+b) \end{vmatrix} = 0$$

since the elements of the first row are the sums of the corresponding elements of the other two rows, taken with a minus sign.

Note that the variance of the sum $X + Y + Z$ is, in accordance with Eq. (6.77),

$$\text{Var}(X + Y + Z) = \text{Var}(X) + \text{Var}(Y) + \text{Var}(Z)$$

$$+ 2[\text{Cov}(X, Y) + \text{Cov}(X, Z) + \text{Cov}(Y, Z)] = 0$$

which is explained by the fact that $X + Y + Z$ is a certain event: either a red, a white, or a blue ball will be picked from the box during the experiment. If we neglect the covariances between the random variables X, Y, and Z, the approximate value of the variance could be

$$\widetilde{\text{Var}}(X + Y + Z) \simeq \text{Var}(X) + \text{Var}(Y) + \text{Var}(Z)$$

$$= \frac{2(ab + bc + ac)}{(a+b+c)^2}$$

The relative percentage error with respect to the exact value of the variance is defined as

$$\eta = \frac{\text{Var}(X + Y + Z) - \widetilde{\text{Var}}(X + Y + Z)}{\text{Var}(X + Y + Z)} \times 100\%$$

with infinity as its upper limit. The conclusion is that the error induced by assumption of uncorrelatedness may be very large.

6.7 APPROXIMATE EVALUATION OF MOMENTS OF FUNCTIONS

As can be seen from Eq. (6.89), determination of moments of linear functions of random variables is a straightforward task, whereas that of moments of nonlinear ones is often rather cumbersome and approximation may be preferable. Consider, for example, the mathematical expectation of the function g of random variables $X_1, X_2, \ldots, X_n$ given in Eq. (6.64). In order to evaluate the integral there, we resort to the *Laplace approximation*. $f_{(X)}(x_1, x_2, \ldots, x_n)$ takes on significant values in an interval containing the point with coordinates $E(X_1), E(X_2), \ldots, E(X_n)$; if $g(x_1, x_2, \ldots, x_n)$ is a slowly varying function in this interval (see Fig. 6.12 for the case $n = 1$), it may be withdrawn outside the integral sign for $x_1 = E(X_1), x_2 = E(X_2), \ldots, x_n = E(X_n)$ to yield

$$E[g(X_1, X_2, \ldots, X_n)] \simeq g[E(X_1), E(X_2), \ldots, E(X_n)]$$

$$\times \int_{-\infty}^{\infty} \cdots \int_{-\infty}^{\infty} f_{(X)}(x_1, x_2, \ldots, x_n) \, dx_1 \, dx_2, \ldots, dx_n$$

$$= g[E(X_1), E(X_2), \ldots, E(X_n)]$$

This estimate may be improved by expanding $g(x_1, x_2, \ldots, x_n)$ in series about the point $E(X_1), E(X_2), \ldots, E(X_n)$:

$$g(x_1, x_2, \ldots, x_n) = g[E(X_1), E(X_2), \ldots, E(X_n)]$$

$$+ \sum_{j=1}^{n} \left(\frac{\partial g}{\partial x_j} \right)_{x_j = E(X_j)} [x_j - E(X_j)]$$

$$+ \frac{1}{2} \sum_{j=1}^{n} \sum_{k=1}^{n} \left(\frac{\partial^2 g}{\partial x_j \, \partial x_k} \right)_{x_j = E(X_j)} [x_j - E(X_j)][x_k - E(X_k)] + \cdots$$

Substituting this in Eq. (6.64) and noting that $E[X_j - E(X_j)] = 0$, we obtain

$$E[g(X_1, X_2, \ldots, X_n)] \simeq g[E(X_1), E(X_2), \ldots, E(X_n)]$$

$$+ \tfrac{1}{2} \sum_{j=1}^{n} \sum_{k=1}^{n} \left(\frac{\partial^2 g}{\partial x_j \, \partial x_k} \right)_{x_j = E(X_j)} \mathrm{Cov}(X_j, X_k)$$

$$(6.91)$$

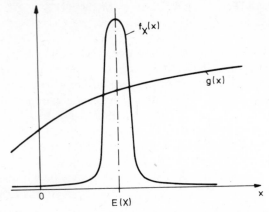

Fig. 6.12. To the Laplace approximation of the moments of the function of a random variable: $f_x(x)$ has a peak at $E(X)$, and $g(x)$ is a "smooth" function.

For the variance of $g(X_1, X_2, \ldots, X_n)$ we obtain in a similar manner

$$\mathrm{Var}\big[g(X_1, X_2, \ldots, X_n)\big] \simeq \sum_{j=1}^{n} \sum_{k=1}^{n} \left(\frac{\partial g}{\partial x_j}\right)_{x_j = E(X_j)} \left(\frac{\partial g}{\partial x_k}\right)_{x_k = E(X_k)} \mathrm{Cov}(X_j, X_k)$$

(6.92)

Formulas (6.91) and (6.92) are readily extended to the n functions g_j of n random variables $X_1, X_2, \ldots, X_n$.

Example 6.11

Application of the above formulas for the mathematical expectation and variance of the quotient of two random variables $X_1 = X$ and $X_2 = Y$ yields

$$E\left(\frac{X}{Y}\right) = \frac{E(X)}{E(Y)} - \frac{1}{E^2(Y)}\,\mathrm{Cov}(X, Y) + \frac{E(X)}{E^3(Y)}\,\mathrm{Var}(Y)$$

$$\mathrm{Var}\left(\frac{X}{Y}\right) \simeq \left[\frac{\mathrm{Var}(X)}{E^2(X)} + \frac{\mathrm{Var}(Y)}{E^2(Y)} - \frac{2\,\mathrm{Cov}(X, Y)}{E(X)E(Y)}\right]\left[\frac{E(X)}{E(Y)}\right]^2 \quad (6.93)$$

Note that if the standard deviations are of the same order of magnitude as the mathematical expectations, then $f_{(X)}(x_1, x_2, \ldots, x_n)$ takes on significant values outside the interval containing the point $E(X_1), E(X_2), \ldots, E(X_n)$, and approximations (6.91) and (6.92) will be too rough. If, for example, X and Y are two independent exponentially distributed random variables

$$f_X(x) = e^{-x}U(x)$$

$$f_Y(y) = e^{-y}U(y)$$

then, according to Eq. (6.57) with $Z = X/Y$

$$f_Z(z) = \int_0^\infty y f_X(zy) f_Y(y)\, dy - \int_{-\infty}^0 y f_X(zy) f_Y(y)\, dy$$

The second term vanishes, since X and Y take on positive values and

$$f_Z(z) = \int_0^\infty y e^{-zy} e^{-y}\, dy = \int_0^\infty y e^{-(z+1)y}\, dy = \frac{1}{(1+z)^2}$$

The mean $E(Z)$ is infinite

$$E(Z) = \int_0^\infty z f_Z(z)\, dz = \int_0^\infty \frac{z\, dz}{(1+z)^2} \rightarrow \infty$$

whereas formula (6.93.1) yields 2.

6.8 JOINT CHARACTERISTIC FUNCTION

The *joint characteristic function* of the components $X_1, X_2, \ldots, X_n$ of an n-dimensional random vector $\{X\}$ is defined as the following mathematical expectation:

$$M_{(X)}(\theta_1, \theta_2, \ldots, \theta_n) = E\{\exp[i(\theta_1 X_1 + \theta_2 X_2 + \cdots + \theta_n X_n)]\}$$

$$= \int_{-\infty}^\infty \int_{-\infty}^\infty \cdots \int_{-\infty}^\infty f_{(X)}(x_1, x_2, \ldots, x_n)$$

$$\times \exp[i(\theta_1 x_1 + \theta_2 x_2 + \cdots + \theta_n x_n)]\, dx_1\, dx_2 \cdots dx_n$$

$$(6.94)$$

and represents the generalization of the characteristic function $M_X(\theta)$ of a single random variable, defined in Eq. (3.40). In Eq. (6.94), $\theta_1, \theta_2, \ldots, \theta_n$ are the arguments of the joint characteristic function. Since the joint probability density function $f_{(X)}$ is nonnegative and integrable, it can be represented as a Fourier integral. The inverse Fourier transform of the characteristic function equals the joint probability density function:

$$f_{\langle X \rangle}(x_1, x_2, \ldots, x_n) = \frac{1}{(2\pi)^n} \int_{-\infty}^{\infty} \int_{-\infty}^{\infty} \cdots \int_{-\infty}^{\infty} M_{\langle X \rangle}(\theta_1, \theta_2, \ldots, \theta_n)$$

$$\times \exp\left[-i(\theta_1 x_1 + \theta_2 x_2 + \cdots + \theta_n x_n)\right] d\theta_1 \, d\theta_2 \cdots d\theta_n$$

$$(6.95)$$

The marginal characteristic functions are obtained from the joint characteristic function. For example,

$$M_{X_1}(\theta_1) = E\left[e^{i\theta_1 X_1}\right] = M_{\langle X \rangle}(\theta_1, 0, \ldots, 0)$$

$$M_{X_1 X_2}(\theta_1, \theta_2) = E\left[e^{i(\theta_1 X_1 + \theta_2 X_2)}\right] = M_{\langle X \rangle}(\theta_1, \theta_2, 0, \ldots, 0)$$

etc.

Using the joint characteristic function, we readily obtain the moments of any order of the random variables $X_1, X_2, \ldots, X_n$. For example,

$$E\left(X_1^{k_1} X_2^{k_2} \cdots X_n^{k_n}\right)$$

$$= (-i)^{k_1 + k_2 + \cdots + k_n} \left[\frac{\partial^{k_1 + k_2 + \cdots + k_n} M_{\langle X \rangle}(\theta_1, \theta_2, \ldots, \theta_n)}{\partial \theta_1^{k_1} \partial \theta_2^{k_2} \cdots \partial \theta_n^{k_n}}\right]_{\theta_1 = \theta_2 = \cdots = \theta_n = 0}$$

$$(6.96)$$

Note that $M_{\langle X \rangle}(0, 0, \ldots, 0) = 1$; indeed,

$$M_{\langle X \rangle}(0, 0, \ldots, 0) = \int_{-\infty}^{\infty} \int_{-\infty}^{\infty} \cdots \int_{-\infty}^{\infty} f_{\langle X \rangle}(x_1, x_2, \ldots, x_n) \, dx_1 \, dx_2 \cdots dx_n = 1$$

Another important property of the joint characteristic function is that for independent random variables $X_1, X_2, \ldots, X_n$ it equals the product of the marginal characteristic functions

$$M_{\langle X \rangle}(\theta_1, \theta_2, \ldots, \theta_n) = \prod_{j=1}^{n} M_{X_j}(\theta_j) \qquad (6.97)$$

Since in that case

$$f_{\langle X \rangle}(x_1, x_2, \ldots, x_n) = \prod_{j=1}^{n} f_{X_j}(x_j)$$

we have

$$M_{\langle X\rangle}(\theta_1, \theta_2, \ldots, \theta_n) = \int_{-\infty}^{\infty} \int_{-\infty}^{\infty} \cdots \int_{-\infty}^{\infty} f_{\langle X\rangle}(x_1, x_2, \ldots, x_n)$$

$$\times \exp[i(\theta_1 x_1 + \theta_2 x_2 + \cdots + \theta_n x_n)] \, dx_1 \, dx_2 \cdots dx_n$$

$$= \prod_{j=1}^{n} \int_{-\infty}^{\infty} f_{X_j}(x_j) \exp(i\theta_j x_j) \, dx_j$$

and since

$$\int_{-\infty}^{\infty} f_{X_j}(x_j) \exp(i\theta_j x_j) \, dx_j = M_{X_j}(\theta_j)$$

we obtain Eq. (6.97).

Distinction is again possible between the independent and uncorrelated random variables. Consider, for example, two random variables, X_1 and X_2; then, from (6.96) we have, if X_1 and X_2 are independent,

$$E(X_1^{k_1} X_2^{k_2}) = (-i)^{k_1 + k_2} \left[\frac{\partial^{k_1 + k_2} M_{\langle X\rangle}(\theta_1, \theta_2)}{\partial\theta_1^{k_1} \partial\theta_2^{k_2}} \right]_{\theta_1 = \theta_2 = 0}$$

But since for independent random variables Eq. (6.97) is valid,

$$E(X_1^{k_1} X_2^{k_2}) = (-i)^{k_1 + k_2} \left[\frac{\partial^{k_1} M_{X_1}(\theta_1)}{\partial\theta_1^{k_1}} \right]_{\theta_1 = 0} \left[\frac{\partial^{k_2} M_{X_2}(\theta_2)}{\partial\theta_2^{k_2}} \right]_{\theta_2 = 0}$$

$$= E(X_1^{k_1}) E(X_2^{k_2}) \tag{6.98}$$

whereas uncorrelatedness implies only that

$$E(X_1 X_2) = E(X_1) E(X_2)$$

If we substitute

$$\theta_1 = \theta_2 = \cdots = \theta_n = \theta$$

in Eq. (6.94), we arrive at the characteristic function of the sum $Y = X_1 + X_2 + \cdots + X_n$ of random variables:

$$M_Y(\theta) = E[\exp(i\theta Y)] = E\{\exp[i\theta(X_1 + X_2 + \cdots + X_n)]\}$$

$$= M_{\langle X\rangle}(\theta, \theta, \ldots, \theta) \tag{6.99}$$

If $X, X_2, \ldots, X_n$ are independent, then combining Eqs. (6.97) and (6.99) we obtain

$$M_Y(\theta) = \prod_{j=1}^{n} M_{X_j}(\theta)$$

(6.100)

where $M_{X_j}(\theta)$ is the marginal characteristic function of X_j. Thus, the characteristic function of the sum of independent random variables equals the product of characteristic functions of the constitutents. In the particular case where all variables have an identical characteristic function

$$M_{X_j}(\theta) = M(\theta), \quad j = 1, 2, \dots, n$$

we have

$$M_Y(\theta) = [M(\theta)]^n$$

(6.101)

Example 6.12

Consider n independent random variables, each with a normal distribution $N(a_j, \sigma_j^2), j = 1, 2, \dots, n$. The characteristic function of their sum is, in accordance with Eqs. (6.100) and (4.23),

$$M_Y(\theta) = \prod_{j=1}^{n} \exp\left(i a_j \theta - \frac{\sigma_j^2 \theta^2}{2} \right) = \exp\left(i\theta \sum_{j=1}^{n} a_j - \frac{\theta^2}{2} \sum_{j=1}^{n} \sigma_j^2 \right)$$

Denoting

$$a = \sum_{j=1}^{n} a_j \qquad \sigma^2 = \sum_{j=1}^{n} \sigma_j^2$$

(6.102)

Then

$$M_Y(\theta) = \exp\left(i a \theta - \frac{\sigma^2 \theta^2}{2} \right)$$

(6.103)

Comparing Eqs. (4.23) and (6.103), we notice that the characteristic function of the sum of *independent* normal variables is also normally distributed, $N(a, \sigma^2)$, where a and σ^2 are defined by Eq. (6.102). The mathematical expectation and variance of this sum equal the respective sums of those of the constituents. These results are particular cases of Eqs. (6.76) and (6.78), respectively. What is new is that the sum of n normally distributed variables turns out to be normally distributed as well.

Example 6.13

Let $X_1, X_2, \dots, X_n$ be independent random variables, identically normally distributed, $N(a, \sigma^2)$. We wish to find $f_Y(y)$ of

$$Y = \sum_{j=1}^{n} (X_j - a)^2$$

We first determine the characteristic function $M_Y(y)$. Since $X_1, X_2, \ldots, X_n$ are independent, $(X_1 - a)^2, (X_2 - a)^2, \ldots, (X_n - a)^2$ are also independent; moreover, since all X_j have identical distributions, so have all $(X_j - a)^2$-s. Denote

$$Z_j = (X_j - a)^2$$

Since X_j is $N(a, \sigma^2)$, $X_j - a$ is $N(0, \sigma^2)$. Now, according to Eq. (4.62), we have

$$f_Z(z_j) = \frac{1}{\sigma(2\pi z_j)^{1/2}} \exp\left(-\frac{z_j}{2\sigma^2}\right) U(z_j)$$

The characteristic function of Z_j (with the j's dropped) is

$$M_Z(\theta) = \int_0^\infty \frac{e^{i\theta z}}{\sigma\sqrt{2\pi z}} \exp\left(-\frac{z}{2\sigma^2}\right) dz = (1 - i2\sigma^2\theta)^{-1/2}$$

Now, since all Z_j's have identical characteristic functions, by Eq. (6.101) we have

$$M_Y(\theta) = [M_Z(\theta)]^n = (1 - i2\sigma^2\theta)^{-n/2}$$

and the probability density function of Y

$$f_Y(y) = \frac{1}{2\pi} \int_{-\infty}^\infty M_Y(\theta) e^{-iy\theta} d\theta$$

reads

$$f_Y(y) = \frac{1}{2\sigma^2\Gamma(n/2)} \left(\frac{y}{2\sigma^2}\right)^{n/2-1} \exp\left(-\frac{y}{2\sigma^2}\right) U(y)$$

That is, the random variable Y/σ^2 has a χ^2 (chi-square) distribution (see Eq. 4.9).

6.9   PAIR OF JOINTLY NORMAL RANDOM VARIABLES

A pair of random variables X and Y are said to be *jointly normal*, or to have *a bivariate normal distribution* if their joint probability density reads

$$f_{XY}(x, y) = \frac{1}{2\pi\sigma_1\sigma_2(1 - r^2)^{1/2}}$$

$$\times \exp\left\{-\frac{1}{2(1 - r^2)}\left[\left(\frac{x - a}{\sigma_1}\right)^2 - 2r\frac{x - a}{\sigma_1}\frac{y - b}{\sigma_2} + \left(\frac{y - b}{\sigma_2}\right)^2\right]\right\}$$

$$(6.104)$$

This density depends on five parameters a, b, σ_1, σ_2 and r, the significance of which will be shown further on. The (cumulative) joint distribution function reads

$$F_{XY}(x, y) = \frac{1}{2\pi\sigma_1\sigma_2(1 - r^2)^{1/2}}$$

$$\times \int_{-\infty}^{x}\int_{-\infty}^{y} \exp\left\{-\frac{1}{2(1 - r^2)}\left[\left(\frac{\xi - a}{\sigma_1}\right)^2 - 2r\frac{\xi - a}{\sigma_1}\frac{\eta - b}{\sigma_2}\right.\right.$$

$$\left.\left. + \left(\frac{\eta - b}{\sigma_2}\right)^2\right]\right\} d\xi\, d\eta \qquad (6.105)$$

We find the marginal probability densities $f_X(x)$ and $f_Y(y)$ as

$$f_X(x) = \int_{-\infty}^{\infty} f_{XY}(x, y)\, dy, \qquad f_Y(y) = \int_{-\infty}^{\infty} f_{XY}(x, y)\, dx$$

For $f_X(x)$ we have

$$f_X(x) = \frac{1}{2\pi\sigma_1\sigma_2(1 - r^2)^{1/2}}$$

$$\times \int_{-\infty}^{\infty} \exp\left\{-\frac{1}{2(1 - r^2)}\left[\left(\frac{x - a}{\sigma_1}\right)^2 - 2r\frac{x - a}{\sigma_1}\frac{y - b}{\sigma_2}\right.\right.$$

$$\left.\left. + \left(\frac{y - b}{\sigma_2}\right)^2\right]\right\} dy$$

Denoting

$$\frac{y - b}{\sigma_2} = \eta$$

then

$$f_X(x) = \frac{1}{2\pi\sigma_1(1 - r^2)^{1/2}} \int_{-\infty}^{\infty} \exp\left[-\tfrac{1}{2}(A\eta^2 - 2B\eta + C)\right] d\eta$$

where

$$A = \frac{1}{1 - r^2} \qquad B = \frac{r(x - a)}{\sigma_1(1 - r^2)} \qquad C = \frac{(x - a)^2}{\sigma_1^2(1 - r^2)}$$

Using the formula (A2) derived in Appendix A, we obtain

$$f_X(x) = \frac{1}{\sigma_1 [2\pi(1 - r^2)]^{1/2} \sqrt{A}} \exp\left(-\frac{AC - B^2}{2A}\right)$$

or, finally,

$$f_X(x) = \frac{1}{\sigma_1 \sqrt{2\pi}} \exp\left[-\frac{(x - a)^2}{2\sigma_1^2}\right]$$

That is, X has a normal distribution with $a = E(X)$ and $\sigma_1 = \sqrt{\text{Var}(X)} = \sigma_X$. Similarly, Y has a normal distribution with $b = E(Y)$ and $\sigma_2 = \sqrt{\text{Var}(Y)} = \sigma_Y$.

Let us show now the significance of r. To do this, we calculate the covariance

$$\text{Cov}(X, Y) = \int_{-\infty}^{\infty} \int_{-\infty}^{\infty} (x - a)(y - b) f_{XY}(x, y) \, dx \, dy$$

Applying a change of variables,

$$\frac{x - a}{\sigma_1} = \xi, \qquad \frac{y - b}{\sigma_2} = r\xi + (1 - r^2)^{1/2}\eta \qquad (6.106)$$

and the incremental area $dx \, dy$ transforms into

$$dx \, dy = |J| \, d\xi \, d\eta$$

where J is the *Jacobian* of the transformation (6.106),

$$J = \frac{\partial(x, y)}{\partial(\xi, \eta)} = \begin{vmatrix} \dfrac{\partial x}{\partial \xi} & \dfrac{\partial x}{\partial \eta} \\ \dfrac{\partial y}{\partial \xi} & \dfrac{\partial y}{\partial \eta} \end{vmatrix} = \begin{vmatrix} \sigma_1 & 0 \\ r\sigma_2 & (1 - r^2)^{1/2}\sigma_2 \end{vmatrix} = \sigma_1\sigma_2(1 - r^2)^{1/2}$$

so that

$$\text{Cov}(X, Y) = \frac{\sigma_1\sigma_2}{2\pi} \int_{-\infty}^{\infty} \int_{-\infty}^{\infty} \xi\left[r\xi + (1 - r^2)^{1/2}\eta\right] \exp\left[-\frac{1}{2}(\xi^2 + \eta^2)\right] d\xi \, d\eta$$

or

$$\text{Cov}(X, Y) = r\sigma_1\sigma_2 \left[\frac{1}{\sqrt{2\pi}} \int_{-\infty}^{\infty} \xi^2 e^{-(1/2)\xi^2} \, d\xi\right]\left[\frac{1}{\sqrt{2\pi}} \int_{-\infty}^{\infty} e^{-(1/2)\eta^2} \, d\eta\right]$$

$$+ (1 - r^2)^{1/2}\sigma_1\sigma_2 \left[\frac{1}{\sqrt{2\pi}} \int_{-\infty}^{\infty} \xi e^{-(1/2)\xi^2} \, d\xi\right]$$

$$\times \left[\frac{1}{\sqrt{2\pi}} \int_{-\infty}^{\infty} \eta e^{-(1/2)\eta^2} \, d\eta\right]$$

and

$$\text{Cov}(X, Y) = \sigma_1 \sigma_2 r$$

implying that r is the correlation coefficient of X and Y.

If X and Y are uncorrelated, their joint probability density becomes

$$f_{XY}(x, y) = \frac{1}{2\pi\sigma_1\sigma_2} \exp\left[-\frac{(x-a)^2}{2\sigma_1^2}\right] \exp\left[-\frac{(y-b)^2}{2\sigma_2^2}\right] = f_X(x)f_Y(y)$$

that is, they are then also independent.

The conditional probability densities $f_Y(y|x)$ and $f_X(x|y)$ are obtained from Eqs. (6.26) and (6.27):

$$f_X(x|y) = \frac{1}{\sigma_1[2\pi(1-r^2)]^{1/2}} \exp\left\{-\frac{1}{2(1-r^2)}\left[\frac{x-a}{\sigma_1} - r\frac{y-b}{\sigma_2}\right]^2\right\}$$

$$f_Y(y|x) = \frac{1}{\sigma_2[2\pi(1-r^2)]^{1/2}} \exp\left\{-\frac{1}{2(1-r^2)}\left[\frac{y-b}{\sigma_2} - r\frac{x-a}{\sigma_1}\right]^2\right\}$$

$$(6.107)$$

Note that the joint probability density of two normally distributed random variables reaches its maximum at (a, b):

$$f_{XY}(a, b) = \frac{1}{2\pi\sigma_1\sigma_2[1-r^2]^{1/2}}$$

The density (6.104) is constant along the ellipse

$$Q(x, y) = \frac{(x-a)^2}{\sigma_1^2} - 2r\frac{(x-a)(y-b)}{\sigma_1\sigma_2} + \frac{(y-b)^2}{\sigma_2^2} = c^2 \qquad (6.108)$$

with center (a, b). The probability of the point with random coordinates X, Y, having jointly normal distribution, falling within the ellipse of constant density is

$$P(c) = \frac{1}{2\pi\sigma_1\sigma_2(1-r^2)^{1/2}} \int\int_{A(c)} \exp\left[-\frac{1}{2(1-r^2)}Q(x, y)\right] dx\, dy$$

$$(6.109)$$

where $A(c)$ is the region bounded by the ellipse (6.107) in the xy plane. The integral in Eq. (6.109) is calculated by transformation to polar coordinates. The result is

$$P(c) = 1 - \exp\left[-\frac{c^2}{2(1 - r^2)}\right] \tag{6.110}$$

We next determine the joint characteristic function of X and Y:

$$M_{XY}(\theta_1, \theta_2) = \int_{-\infty}^{\infty}\int_{-\infty}^{\infty} f_{XY}(x, y)\exp\{i(\theta_1 x + \theta_2 y)\}\, dx\, dy$$

or

$$M_{XY}(\theta_1, \theta_2) = \exp\{i(\theta_1 a + \theta_2 b)\}$$

$$\times \int_{-\infty}^{\infty}\int_{-\infty}^{\infty} f_{XY}(x, y)\exp\{i[\theta_1(x - a) + \theta_2(y - b)]\}\, dx\, dy$$

Making use again of the substitution (6.106), we have

$$M_{XY}(\theta_1, \theta_2) = \exp\{i(\theta_1 a + \theta_2 b)\}$$

$$\times \frac{1}{2\pi}\int_{-\infty}^{\infty}\int_{-\infty}^{\infty} \exp\left\{i\left[\theta_1\sigma_1\xi + \theta_2\sigma_2\left(r\xi + (1 - r^2)^{1/2}\eta\right)\right]\right.$$

$$\left. - \tfrac{1}{2}(\xi^2 + \eta^2)\right\}\, d\xi\, d\eta$$

$$= \frac{\exp\{i(\theta_1 a + \theta_2 b)\}}{2\pi}$$

$$\times \left[\int_{-\infty}^{\infty} \exp\{-\tfrac{1}{2}[\xi^2 - 2i(\theta_1\sigma_1 + \theta_2\sigma_2 r)\xi]\}\, d\xi\right]$$

$$\times \left[\int_{-\infty}^{\infty} \exp\{-\tfrac{1}{2}[\eta^2 - 2i\theta_2\sigma_2(1 - r^2)^{1/2}\eta]\}\, d\eta\right] \tag{6.111}$$

Using formula (A2) in Appendix A, we evaluate these integrals as follows:

$$\int_{-\infty}^{\infty} \exp\{-\tfrac{1}{2}[\xi^2 - 2i(\theta_1\sigma_1 + \theta_2\sigma_2 r)\xi]\}\, d\xi = \sqrt{2\pi}\, \exp\left[-\tfrac{1}{2}(\theta_1\sigma_1 + \theta_2\sigma_2 r)^2\right]$$

$$\int_{-\infty}^{\infty} \exp\{-\tfrac{1}{2}[\eta^2 - 2i\theta_2\sigma_2\sqrt{1 - r^2}\,\eta]\}\, d\eta = \sqrt{2\pi}\, \exp\left[-\tfrac{1}{2}\theta_2^2\sigma_2^2(1 - r^2)\right]$$

Substituting the latter equations in Eq. (6.111), we obtain finally

$$M_{XY}(\theta_1, \theta_2) = \exp\left[i(\theta_1 a + \theta_2 b) - \tfrac{1}{2}(\theta_1^2\sigma_1^2 + 2\theta_1\theta_2\sigma_1\sigma_2 r + \theta_2^2\sigma_2^2)\right]$$

$$\tag{6.112}$$

Using the joint characteristic function $M_{XY}(\theta_1, \theta_2)$, we readily find the characteristic function of the sum or difference of two random variables $X \pm Y$. We first note that the characteristic function of the difference is obtained from Eq. (6.94) by setting $\theta_1 = \theta$, $\theta_2 = -\theta$, $\theta_3 = \cdots = \theta_n = 0$, that is,

$$M_{X-Y}(\theta) = E\{\exp[i\theta(X - Y)]\} = M_{XY}(\theta, -\theta) \tag{6.113}$$

Thus, comparison of Eqs. (6.99) and 6.113) yields

$$M_{X \pm Y}(\theta) = M_{XY}(\theta, \pm\theta)$$

and

$$M_{X \pm Y}(\theta) = \exp\{i\theta(a \pm b) - \tfrac{1}{2}\theta^2(\sigma_1^2 \pm 2r\sigma_1\sigma_2 + \sigma_2^2)\}$$

implying that

$$E(X \pm Y) = a \pm b$$

$$\sigma_{X \pm Y}^2 = \sigma_1^2 \pm 2r\sigma_1\sigma_2 + \sigma_2^2 \tag{6.114}$$

These results are again particular cases of Eqs. (6.76) and (6.77), respectively. What is new is that the sum or difference of a pair of jointly normally distributed *dependent* variables turns out to be normally distributed as well.

6.10 SEVERAL JOINTLY NORMAL RANDOM VARIABLES

Note first that the probability density function of a pair of jointly normal random variables, given in Eq. (6.104), can be rewritten as

$$f_{X_1 X_2}(x_1, x_2) = \frac{1}{2\pi(\det[V])^{1/2}} \exp\left(-\tfrac{1}{2}\{x - m\}^T[V]^{-1}\{x - m\}\right)$$

where $X \equiv X_1$, $Y \equiv X_2$, $x \equiv x_1$, $y \equiv x_2$ and

$$\{x\} = \begin{Bmatrix} x_1 \\ x_2 \end{Bmatrix} \qquad \{m\} = \begin{Bmatrix} m_1 \\ m_2 \end{Bmatrix}$$

$$[V] = E\left[\{x - m\}\{x - m\}^T\right] = \begin{bmatrix} \sigma_1^2 & r\sigma_1\sigma_2 \\ r\sigma_1\sigma_2 & \sigma_2^2 \end{bmatrix}$$

where $[V]$ is the variance-covariance matrix, and T indicates transpose. Accordingly,

$$(\det[V])^{1/2} = \sigma_1\sigma_2(1 - r^2)^{1/2}$$

$$[V]^{-1} = \frac{1}{\sigma_1^2\sigma_2^2(1 - r^2)}\begin{bmatrix} \sigma_2^2 & -r\sigma_1\sigma_2 \\ -r\sigma_1\sigma_2 & \sigma_1^2 \end{bmatrix}$$

$$= \frac{1}{1 - r^2}\begin{bmatrix} \dfrac{1}{\sigma_1^2} & -\dfrac{r}{\sigma_1\sigma_2} \\ -\dfrac{r}{\sigma_1\sigma_2} & \dfrac{1}{\sigma_2^2} \end{bmatrix}$$

hence,

$$\{x - m\}^T[V]^{-1}\{x - m\}$$

$$= \frac{1}{1 - r^2}\left[\frac{(x_1 - m_1)^2}{\sigma_1^2} - 2r\frac{(x_1 - m_1)(x_2 - m_2)}{\sigma_1\sigma_2} + \frac{(x_2 - m_2)^2}{\sigma_2^2}\right]$$

which then yields Eq. (6.104).

In the general case of an n-dimensional random vector $\{X\}$, its components $X_1, X_2, \ldots, X_n$ are said to be jointly normal if their joint probability density is given by

$$f_{\langle X\rangle}(x_1, x_2, \ldots, x_n) = \frac{1}{\langle(2\pi)^n\det[V]\rangle^{1/2}}\exp\left(-\tfrac{1}{2}\{x - m\}^T[V]^{-1}\{x - m\}\right)$$

$$(6.115)$$

where

$$\{x\}^T = [x_1 \quad x_2 \quad \cdots \quad x_n]$$

$$\{m\}^T = [m_1 \quad m_2 \quad \cdots \quad m_n]$$

and

$$[V] = E\big[\{x - m\}\{x - m\}^T\big] = \begin{bmatrix} v_{11} & v_{12} & \cdots & v_{1n} \\ v_{21} & v_{22} & \cdots & v_{2n} \\ \vdots & & & \\ v_{n1} & v_{n2} & \cdots & v_{nn} \end{bmatrix}$$

is the variance-covariance matrix. The density is characterized by $n + (n^2 - n)/2 + n = (n^2 + 3n)/2$ parameters: n mathematical expectations $m_1, m_2, \ldots, m_n$, and n variances $v_{11}, v_{22}, \ldots, v_{nn}$, and $(n^2 - n)/2$ covariances v_{jk} ($j < k$). If the random variables are uncorrelated, then all the covariances vanish, $[V]$ becomes diagonal

$$[V]^{-1} = \begin{bmatrix} \dfrac{1}{v_{11}} & & & \text{zeros} \\ & \dfrac{1}{v_{22}} & & \\ & & \ddots & \\ \text{zeros} & & & \dfrac{1}{v_{nn}} \end{bmatrix}$$

and

$$f_{\langle X \rangle}(x_1, x_2, \ldots, x_n) = \prod_{j=1}^{n} f_{X_j}(x_j)$$

where

$$f_{X_j}(x_j) = \frac{1}{(2\pi v_{jj})^{1/2}} \exp\left[-\frac{(x_j - m_j)^2}{2v_{jj}} \right]$$

We have thus reached the important conclusion that the n uncorrelated jointly normal random variables are independent.

The joint characteristic function of the random normal vector $\{X\}$ is (proof omitted)

$$M_{\langle X \rangle}(\theta_1, \theta_2, \ldots, \theta_n) = \exp\big(i\{m\}^T\{\theta\} - \tfrac{1}{2}\{\theta\}^T[V]\{\theta\} \big)$$

$$= \exp\left(i \sum_{j=1}^{n} m_j \theta_j - \frac{1}{2} \sum_{j=1}^{n} \sum_{k=1}^{n} v_{jk} \theta_j \theta_k \right) \qquad (6.116)$$

and

$$\{\theta\}^T = [\theta_1 \quad \theta_2 \quad \cdots \quad \theta_n]$$

Example 6.14

Consider n dependent jointly normal variables. By virtue of Eqs. (6.99) and (6.116), the characteristic function of their sum Y is

$$M_Y(\theta) = M_{\langle X \rangle}(\theta, \theta, \dots, \theta) = \exp\left\langle i\theta \sum_{j=1}^{n} m_j - \frac{\theta^2}{2} \sum_{j=1}^{n} \sum_{k=1}^{n} \sigma_j \sigma_k r_{jk} \right\rangle \quad (6.117)$$

where r_{jk} is the correlation coefficient of the random variables X_j and X_k,

$$r_{jk} = \frac{v_{jk}}{(v_{jj} v_{kk})^{1/2}}$$

Equation (6.117) indicates that the sum of the jointly normal random variables has a normal distribution with mean and variance, respectively,

$$a = \sum_{j=1}^{n} m_j \qquad \sigma^2 = \sum_{j=1}^{n} \sum_{k=1}^{n} \sigma_j \sigma_k r_{jk} \quad (6.118)$$

Consider now the random variables $X_1 - m_1, X_2 - m_2, \dots, X_n - m_n$. Their characteristic function is

$$\tilde{M}_{\langle X \rangle}(\theta_1, \theta_2, \dots, \theta_n) = E\langle \exp[i(X_1 - m_1)\theta_1 + i(X_2 - m_2)\theta_2$$

$$+ \cdots + i(x_n - m_n)\theta_n]\rangle$$

$$= \int_{-\infty}^{\infty} \int_{-\infty}^{\infty} \cdots \int_{-\infty}^{\infty} \exp\{i(x_1 - m_1)\theta_1 + i(x_2 - m_2)\theta_2$$

$$+ \cdots + (x_n - m_n)\theta_n\}$$

$$\times f_{\langle X \rangle}(x_1, x_2, \dots, x_n) \, dx_1 \, dx_2 \cdots dx_n$$

$$= \exp\{-i(m_1\theta_1 + m_2\theta_2 + \cdots + m_n\theta_n)\}$$

$$\times M_{\langle X \rangle}(\theta_1, \theta_2, \dots, \theta_n) \quad (6.119)$$

so that the joint characteristic function of $X_1 - m_1, X_2 - m_2, \dots, X_n - m_n$, where $X_1, X_2, \ldots, X_n$ are jointly normal, is

$$\tilde{M}_{\langle X \rangle}(\theta_1, \theta_2, \ldots, \theta_n) = \exp\left(-\tfrac{1}{2} \sum_{j=1}^{n} \sum_{k=1}^{n} v_{jk}\theta_j\theta_k\right) \qquad (6.120)$$

With Eq. (6.120) available, we readily find the central moment of different orders:

$$E\left[(X_1 - m_1)^{k_1}(X_2 - m_2)^{k_2} \cdots (X_n - m_n)^{k_n}\right]$$

$$= \frac{1}{i^k} \left.\frac{\partial^k \tilde{M}_{\langle X \rangle}(\theta_1, \theta_2, \ldots, \theta_n)}{\partial\theta_1^{k_1}\,\partial\theta_2^{k_2} \cdots \partial\theta_n^{k_n}}\right|_{\theta_1 = \theta_2 = \cdots = 0} \qquad (6.121)$$

where $k = k_1 + k_2 + \cdots + k_n$.

6.11   FUNCTIONS OF RANDOM VARIABLES

The general case of m functions $Y_1, Y_2, \ldots, Y_m$ of n random variables $X_1, X_2, \ldots, X_n$ was already considered in Sec. 6.9. In particular, the joint probability density $f_{\langle Y \rangle}(y_1, y_2, \ldots, y_m)$ of derived random variables is expressed by Eq. (6.45) in terms of the density $f_{\langle X \rangle}(x_1, x_2, \ldots, x_m)$ of initial random variables. Consider now the important particular case where $n = m$ and the functions $\varphi_1, \varphi_2, \ldots, \varphi_m$ in Eq. (6.36) define one-to-one mapping. This restriction ensures existence of transverse transformation, and we may therefore express $\{X\}$ as a function of $\{Y\}$ by the formula

$$\{X\} = \varphi^{-1}(\{Y\}) \equiv h(\{Y\})$$

or, in component notation

$$\begin{aligned}
x_1 &= h_1(y_1, y_2, \ldots, y_n) \\
x_2 &= h_2(y_1, y_2, \ldots, y_n) \\
&\;\;\vdots \\
x_n &= h_n(y_1, y_2, \ldots, y_n)
\end{aligned} \qquad (6.122)$$

Let us use Eqs. (6.122) for transformation of the variables in Eq. (6.45):

$$\xi_i = h_i(\eta_1, \eta_2, \ldots, \eta_n), \quad i = 1, 2, \ldots, n \qquad [\eta_i = \varphi_i(\xi_1, \xi_2, \ldots, \xi_n)]$$

It follows then, from a well-known result on coordinate transformation, that

$$d\xi_1\, d\xi_2 \cdots d\xi_n = |J|\, d\eta_1\, d\eta_2 \cdots d\eta_n$$

where J is the Jacobian of the transformation

$$J = \frac{\partial(\xi_1, \xi_2, \dots, \xi_n)}{\partial(\eta_1, \eta_2, \dots, \eta_n)} = \frac{\partial(h_1, h_2, \dots, h_n)}{\partial(\eta_1, \eta_2, \dots, \eta_n)}$$

$$= \begin{vmatrix} \dfrac{\partial h_1}{\partial \eta_1} & \dfrac{\partial h_1}{\partial \eta_2} & \cdots & \dfrac{\partial h_1}{\partial \eta_n} \\[2mm] \dfrac{\partial h_2}{\partial \eta_1} & \dfrac{\partial h_2}{\partial \eta_2} & \cdots & \dfrac{\partial h_2}{\partial \eta_n} \\[2mm] \vdots & & & \\[2mm] \dfrac{\partial h_n}{\partial \eta_1} & \dfrac{\partial h_n}{\partial \eta_2} & \cdots & \dfrac{\partial h_n}{\partial \eta_n} \end{vmatrix}$$

and we obtain, instead of Eq. (6.45),

$$f_{(Y)}(y_1, y_2, \dots, y_n) = \int_{-\infty}^{\infty} \int_{-\infty}^{\infty} \cdots$$

$$\times \int_{-\infty}^{\infty} f_{(X)}[h_1(\eta_1, \eta_2, \dots, \eta_n), \dots, h_n(\eta_1, \eta_2, \dots, \eta_n)]$$

$$\times \prod_{j=1}^{n} \delta(y_j - \eta_j) \left| \frac{\partial(h_1, h_2, \dots, h_n)}{\partial(\eta_1, \eta_2, \dots, \eta_n)} \right| d\eta_1 \, d\eta_2 \cdots d\eta_n$$

Integration of the above is preformed using the basic property (3.14) of Dirac's delta function. The result reads:

$$f_{(Y)}(y_1, y_2, \dots, y_n) = f_{(X)}[h_1(y_1, y_2, \dots, y_n), \dots, h_n(y_1, y_2, \dots, y_n)]$$

$$\times \left| \frac{\partial(h_1, h_2, \dots, h_n)}{\partial(y_1, y_2, \dots, y_n)} \right| \tag{6.123}$$

where

$$\frac{\partial(h_1, h_2, \dots, h_n)}{\partial(y_1, y_2, \dots, y_n)} = \det\left[\frac{\partial h_j(y_1, y_2, \dots, y_n)}{\partial y_k} \right] \tag{6.124}$$

Example 6.15

Consider the transformation

$$\{Y\} = [A]\{X\} \tag{6.125}$$

where the square matrix $[A]$ is *nonsingular*; in these circumstances $[A]^{-1}$ exists and

$$\{X\} = [A]^{-1}\{Y\} \tag{6.126}$$

It is readily shown that the Jacobian of the transformation, J, is the determinant of matrix $[A]^{-1}$, and we obtain

$$f_{\langle Y\rangle}(y_1, y_2, \ldots, y_n) = f_{\langle X\rangle}[B_{11}y_1 + B_{12}y_2 + \cdots + B_{1n}y_n, \ldots, B_{n1}y_1$$

$$+ B_{n2}y_2 + \cdots + B_{nn}y_n]|\det[B]|$$

$$[B] = [A]^{-1} \tag{6.127}$$

The mathematical expectation vector $E(\langle Y\rangle)$, with elements $E(Y_j)$,

$$\{E(Y)\}^T = [E(Y_1)\ E(Y_2) \cdots E(Y_n)] \tag{6.128}$$

or the variance-covariance matrix of vector Y:

$$[W] = \left[\mathrm{Cov}(Y_j, Y_k)\right]_{n\times n} \tag{6.129}$$

may be derived directly from (6.125); in fact,

$$E(\langle Y\rangle) = E([A]\langle X\rangle) = [A]E(\langle X\rangle) \tag{6.130}$$

since $[A]$ is a matrix of constants. Also,

$$[W] = E\left[\langle\{Y\} - E(\langle Y\rangle)\rangle\langle\{Y\} - E(\langle Y\rangle)\rangle^T\right]$$

$$= E\left[\langle[A]\{X\} - [A]E(\langle X\rangle)\rangle\langle[A]\{X\} - [A]E(\langle X\rangle)\rangle^T\right]$$

$$= E\left([A]\langle\{X\} - E(\langle X\rangle)\rangle\{[A]\langle\{X\} - E(\langle X\rangle)\rangle\}^T\right)$$

$$= [A]\left[E\langle\{X\} - E(\langle X\rangle)\rangle\langle\{X\} - E(\langle X\rangle)\rangle^T\right][A]^T$$

$$= [A][V][A]^T \tag{6.131}$$

where $[V]$ is the variance-covariance matrix of $\{X\}$,

$$[V] = \left[\mathrm{Cov}(X_j, X_k)\right]_{n\times n} \tag{6.132}$$

Now let $\{X\}$ be a normal random vector with density (6.115) and let $\{Y\}$ be determined by the linear transformation (6.122). Equation (6.123) yields

$$f_{\langle Y\rangle}(y_1, y_2, \ldots, y_n) = \frac{1}{\left\langle (2\pi)^n \det[V]\right\rangle^{1/2}} \exp\left\langle -\tfrac{1}{2}\left([A]^{-1}\{y\} - \{m\}\right)^T [V]^{-1}\right.$$

$$\left. \times \left([A]^{-1}\{y\} - \{m\}\right)\right\rangle \det[A]^{-1}$$

$$= \frac{\det[A]^{-1}}{\left\langle (2\pi)^r \det[V]\right\rangle^{1/2}}$$

$$\times \exp\left\langle -\tfrac{1}{2}(\{y\} - [A]\{m\})^T [A]^{-1T}[V]^{-1}\right\rangle$$

$$\times \left\langle [A]^{-1}(\{y\} - [A]\{m\})\right\rangle$$

But

$$[A]^{-1T}[V]^{-1}[A]^{-1} = [W]^{-1}$$

and

$$\frac{\det[A]^{-1}}{(\det[V])^{1/2}} = \frac{1}{(\det[W])^{1/2}}$$

and recalling (6.130) we have

$$f_{\langle Y\rangle}(y_1, y_2, \ldots, y_n) = \frac{1}{\left\langle (2n)^n \det[W]\right\rangle^{1/2}}$$

$$\times \exp\left\langle -\tfrac{1}{2}(\{y\} - \{\mu\})^T [W]^{-1}(\{y\} - \{\mu\})\right\rangle \quad (6.133)$$

that is, the linear transformation of the jointly normal random variables yields jointly normal random variables with

$$\{\mu\} = [A]\{m\}, \qquad [W] = [A][V][A]^T \qquad (6.134)$$

The mathematical expectation vector and the variance-covariance matrix of the derived random vector $\{Y\}$ were already given by Eqs. (6.130) and (6.131). The property of normality is obtained by using Eq. (6.123).

Example 6.16

Let the matrix $[A]$ in Eq. (6.125) be

$$[A] = \begin{bmatrix} \cos\theta & \sin\theta \\ -\sin\theta & \cos\theta \end{bmatrix} \qquad (6.135)$$

Here transformation as per Eq. (6.125) is replaced with rotation of the axes (Fig. 6.13) through the angle θ. $[A]^{-1}$ is

$$[A]^{-1} = \begin{bmatrix} \cos\theta & -\sin\theta \\ \sin\theta & \cos\theta \end{bmatrix}$$

and Eq. (6.127) becomes

$$f_{(Y)}(y_1, y_2) = f_{(X)}(y_1\cos\theta - y_2\sin\theta,\; y_1\sin\theta + y_2\cos\theta)$$

If X_1 and X_2 were jointly normal with zero means,

$$f_{X_1 X_2}(x_1, x_2) = \frac{1}{2\pi\sigma_1\sigma_2\sqrt{1 - r^2}} \exp\left\langle -\frac{1}{2(1 - r^2)}\left(\frac{x_1^2}{\sigma_1^2} - 2r\frac{x_1 x_2}{\sigma_1\sigma_2} + \frac{x_2^2}{\sigma_2^2}\right)\right\rangle$$

then

$$f_{Y_1 Y_2}(y_1, y_2) = \frac{1}{2\pi\sigma_1\sigma_2\sqrt{1 - r^2}} \exp\left\langle -\frac{1}{2(1 - r^2)}\left(Py_1^2 - 2Qy_1 y_2 + Ry_2^2\right)\right\rangle$$

where

$$P = \frac{\cos^2\theta}{\sigma_1^2} - r\frac{\sin 2\theta}{\sigma_1\sigma_2} + \frac{\sin^2\theta}{\sigma_2^2}$$

$$Q = \frac{\sin 2\theta}{2\sigma_1^2} + r\frac{\cos 2\theta}{\sigma_1\sigma_2} - \frac{\sin 2\theta}{2\sigma_2^2}$$

$$R = \frac{\sin^2\theta}{\sigma_1^2} + r\frac{\sin 2\theta}{\sigma_1\sigma_2} + \frac{\cos^2\theta}{\sigma_2^2}$$

Note that for an angle satisfying

$$\tan 2\theta = \frac{2r\sigma_1\sigma_2}{\sigma_1^2 - \sigma_2^2} \tag{6.136}$$

we have $Q = 0$ and

$$f_{Y_1 Y_2}(y_1, y_2) = \frac{1}{2\pi\sigma_1\sigma_2\sqrt{1 - r^2}} \exp\left\langle -\frac{1}{2(1 - r^2)}\left(Py_1^2 + Ry_2^2\right)\right\rangle$$

implying that linear transformation with the $[A]$ of Eq. (6.135) yields independent normal variables.

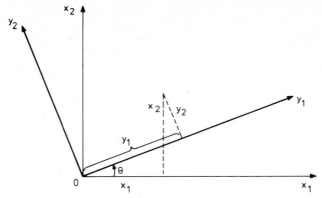

Fig. 6.13. Linear transformation according to Eq. (6.135), with θ as per (6.136), makes Y_1 and Y_2 independent.

This result can be extended to multidimensional random normal variables. We saw in Example 6.15 that the linear transformation $\{Y\} = [A]\{X\}$ of the random normal vector X similarly yields the random normal vector $\{Y\}$ with mean $m_{\langle Y \rangle} = [A]E(\{X\})$ and variance-covariance matrix $[W] = [A][V][A]^T$, where $E(\{X\})$ and $[V]$ are the mean and the variance-covariance matrix of the initial vector $\{X\}$. However, due to the symmetry of $[V]$, it is always possible to find a square matrix $[B]$ such that

$$[B][V][B]^T = \lceil \Lambda \rfloor \qquad [B]^T = [B]^{-1} \qquad (6.137)$$

Here $\lceil \Lambda \rfloor$ is a diagonal matrix with diagonal elements representing the eigenvalues of $[V]$:

$$\lceil \Lambda \rfloor_{jk} = \lambda_j \delta_{jk}$$

where δ_{jk} is the *Kronecker delta*

$$\delta_{jk} = \begin{cases} 1, & j = k \\ 0, & j \neq k \end{cases} \qquad (6.138)$$

Since $[V]$ is nonnegative, the eigenvalues $\lambda_1, \lambda_2, \ldots, \lambda_n$ are also nonnegative, and if it is nonsingular (that is, if $\det[V] \neq 0$), then $\lambda_1, \lambda_2, \ldots, \lambda_n$ are positive.

A matrix $[B]$ satisfying the conditions given in (6.137) is said to be *orthogonal*. Thus if the transformation $[A]$ is chosen to be $[B]$, then the probability density function of $\{Y\}$ is, by virtue of Eq. (6.115),

$$f_{\langle Y\rangle}(y_1, y_2, \ldots, y_n) = \frac{1}{\langle (2\pi)^n \det[W]\rangle^{1/2}}$$

$$\times \exp\left\langle -\tfrac{1}{2}(\{y\} - \{m_Y\})^T [W]^{-1}(\{y\} - \{m_Y\})\right\rangle$$

with

$$[W] = [B][V][B]^T = \lceil \Lambda \rfloor$$

Then $[W]^{-1} = \lceil \Lambda \rfloor^{-1}$, where $\lceil \Lambda \rfloor^{-1}$ is also a diagonal matrix and the diagonal elements of $\lceil \Lambda \rfloor^{-1}$ are the reciprocals of appropriate elements of $\lceil \Lambda \rfloor$:

$$\lceil \Lambda^{-1} \rfloor_{jk} = \lambda_{jj}^{-1} \delta_{jk}$$

Furthermore, $\det[W] = \det\lceil \Lambda \rfloor$, and $f_{\langle Y\rangle}$ becomes

$$f_{\langle Y\rangle}(y_1, y_2, \ldots, y_n) = \frac{1}{\langle (2\pi)^n \det\lceil \Lambda \rfloor\rangle^{1/2}}$$

$$\times \exp\left\langle -\tfrac{1}{2}(\{y\} - \{m_Y\})^T \lceil \Lambda \rfloor^{-1}(\{y\} - \{m_Y\})\right\rangle$$

which in scalar notation reads

$$f_{\langle Y\rangle}(y_1, y_2, \ldots, y_n) = \frac{1}{\langle (2\pi)^n \lambda_1 \lambda_2 \cdots \lambda_n\rangle^{1/2}}$$

$$\times \exp\left\langle -\tfrac{1}{2} \sum_{j=1}^n \frac{1}{\lambda_j}(y_j - E(Y_j))^2\right\rangle$$

or

$$f_{\langle Y\rangle}(y_1, y_2, \ldots, y_n) = \prod_{j=1}^n \frac{1}{\sqrt{2\pi\lambda_j}} \exp\left\langle -\frac{(y_j - E(Y_j))^2}{2\lambda_j}\right\rangle = \prod_{j=1}^n f_{Y_j}(y_j)$$

$$(6.139)$$

The conclusion is that jointly normal *dependent* random variables can be linearly transformed into jointly normal *independent* variables. Note that the eigenvalues λ_j of the matrix $\lceil \Lambda \rfloor$ represents the variances of the derived random variables Y_j.

Now consider the remaining case where $[V]$ is singular. Its rank r is then less than n, and $n - r$ eigenvalues denoted $\lambda_1, \lambda_2, \dots, \lambda_{n-r}$ of the matrix $\lceil \Lambda \rfloor$ are zeros, implying that $Y_1, Y_2, \dots, Y_{n-r}$ have zero variances and take on values equal to their mathematical expectations $E(Y_1), E(Y_2), \dots, E(Y_{n-r})$ with probability unity. Thus $Y_1, Y_2, \dots, Y_{n-r}$ have a causal distribution (see Sec. 4.1)

$$f_{Y_1 Y_2 \cdots Y_{n-r}}(y_1, y_2, \dots, y_{n-r}) = \prod_{j=1}^{n-r} \delta\big(y_j - E(Y_j)\big)$$

and instead of Eq. (6.139) we have

$$f_{(Y)}(y_1, y_2, \dots, y_{n-r}, y_{n-r+1}, \dots, y_n) = \prod_{j=1}^{n-r} \delta\big(y_j - E(y_j)\big) \prod_{j=n-r+1}^{n} \frac{1}{\sqrt{2\pi\lambda_j}}$$

$$\times \exp\left\langle -\frac{\big(y_j - E(Y_j)\big)^2}{2\lambda_j} \right\rangle \quad (6.140)$$

6.12   COMPLEX RANDOM VARIABLES

Let X and Y be random variables as defined in Sec. 3.1; that is, let real numbers $X(\omega)$ and $Y(\omega)$ correspond to each outcome $\omega \in \Omega$. We say that Z is a *complex random variable* if to each outcome ω is assigned the complex number

$$Z(\omega) = X(\omega) + iY(\omega)$$

so that the possible values of $Z(\omega)$ are

$$z = x + iy$$

The mathematical expectation of Z is readily found:

$$E(Z) = \int_{-\infty}^{\infty} \int_{-\infty}^{\infty} (x + iy) f_{XY}(x, y) \, dx \, dy$$

$$= \int_{-\infty}^{\infty} \int_{-\infty}^{\infty} x f_{XY}(x, y) \, dx \, dy + i \int_{-\infty}^{\infty} \int_{-\infty}^{\infty} y f_{XY}(x, y) \, dx \, dy$$

$$= E(X) + iE(Y) \quad (6.141)$$

The variance is defined as

$$\mathrm{Var}(Z) = E\big[|Z - E(Z)|^2\big] = E\{[Z - E(Z)][Z - E(Z)]^*\} \quad (6.142)$$

where the asterisk denotes the complex conjugate.

Bearing in mind that

$$|Z - E(Z)|^2 = [X - E(X)]^2 + [Y - E(Y)]^2$$

we have

$$\text{Var}(Z) = E\{[X - E(X)]^2 + [Y - E(Y)]^2\} \qquad (6.143)$$

By virtue of Eq. (6.76),

$$E\{[X - E(X)]^2 + [Y - E(Y)]^2\} = E\{[X - E(X)]^2\} + E\{[Y - E(Y)]^2\}$$

so that

$$\text{Var}(Z) = \text{Var}(X) + \text{Var}(Y) \qquad (6.144)$$

The covariance of a pair of complex random variables

$$Z_1 = X_1 + iY_1 \qquad Z_2 = X_2 + iY_2 \qquad (6.145)$$

is defined as

$$\text{Cov}(Z_1, Z_2) = E\{[Z_1 - E(Z_1)][Z_2 - E(Z_2)]^*\}$$

and equals

$$
\begin{aligned}
\text{Cov}(Z_1, Z_2) &= E\{X_1 + iY_1 - [E(X_1) + iE(Y_1)]\} \\
&\quad \times \{X_2 - iY_2 - [E(X_2) - iE(Y_2)]\} \\
&= E\{[X_1 - E(X_1)] + i[Y_1 - E(Y_1)]\} \\
&\quad \times \{[X_2 - E(X_2)] - i[Y_2 - E(Y_2)]\} \\
&= E\{[X_1 - E(X_1)][X_2 - E(X_2)]\} \\
&\quad + E\{[Y_1 - E(Y_1)][Y_2 - E(Y_2)]\} \\
&\quad + iE\{[X_2 - E(X_2)][Y_1 - E(Y_1)\} \\
&\quad - iE\{[X_1 - E(X_1)][Y_1 - E(Y_1)]\} \\
&= \text{Cov}(X_1, X_2) + \text{Cov}(Y_1, Y_2) \\
&\quad + i[\text{Cov}(X_2, Y_1) - \text{Cov}(X_1, Y_2)] \qquad (6.146)
\end{aligned}
$$

As in the case of a real pair, Z_1 and Z_2 are said to be uncorrelated if their covariance vanishes, that is, if

$$\text{Cov}(X_1, X_2) + \text{Cov}(Y_1, Y_2) = 0 \quad \text{and} \quad \text{Cov}(X_2, Y_1) + \text{Cov}(X_1, Y_2) = 0$$

$$(6.147)$$

We calculate now the second-order moment:

$$E(Z_1 Z_2^*) = E\{E(Z_1) + [Z_1 - E(Z_1)]\}\{E(Z_2^*) + [Z_2^* - E(Z_2^*)]\}$$

$$= E(Z_1)E(Z_2^*) + \text{Cov}(Z_1, Z_2) \qquad (6.148)$$

and for an uncorrelated pair,

$$E(Z_1 Z_2^*) = E(Z_1)E(Z_2^*) \qquad (6.149)$$

A pair of complex random variables is *orthogonal* if

$$E(Z_1 Z_2^*) = 0 \qquad (6.150)$$

We say that Z_1 and Z_2 are independent if

$$f_{Z_1 Z_2}(x_1, y_1; x_2, y_2) = f_{X_1 Y_1}(x_1, y_1) f_{X_2 Y_2}(x_2, y_2) \qquad (6.151)$$

n complex variables $Z_j = X_j + iY_j$, $j = 1, 2, \ldots, n$, are independent if the groups $(X_1, Y_1), (X_2, Y_2), \ldots, (X_n, Y_n)$ are independent. Their joint probability satisfies

$$f_{Z_1 Z_2 \cdots Z_n}(x_1, y_1; x_2, y_2; \ldots; x_n, y_n) = \prod_{j=1}^{n} f_{Z_j}(x_j, y_j) \qquad (6.152)$$

PROBLEMS

6.1. The random variables X and Y are said to have a uniform distribution in $x^2 + y^2 \leqslant R^2$ if

$$f_{XY}(x, y) = \begin{cases} \dfrac{1}{\pi R^2}, & x^2 + y^2 \leqslant R^2 \\ 0, & \text{otherwise} \end{cases}$$

Verify that

$$f_X(x) = \begin{cases} \dfrac{2}{\pi R}\sqrt{1 - \left(\dfrac{x}{R}\right)^2}, & |x| \leqslant R \\ 0, & \text{otherwise} \end{cases}$$

that is, that X is not uniformly distributed. Find $f_Y(y)$.

6.2. Find the conditional distribution function of X under the hypothesis that $Y \leqslant y$, $F_X(x|Y \leqslant y)$. Determine $f_X(x|Y \leqslant y)$.

6.3. Find the conditional distribution function $F_X(x|X \leqslant a,\ Y \leqslant b)$, provided that $F_{XY}(a, b) > 0$.

6.4. Find the expression for the conditional distribution function $F_{XY}(x, y|a < X \leqslant b)$.

6.5. Derive the formulas

$$f_X(x) = \int_{-\infty}^{\infty} f_X(x|y)f_Y(y)\,dy \quad \text{and} \quad f_Y(y) = \int_{-\infty}^{\infty} f_Y(y|x)f_X(x)\,dx$$

6.6. Show that the random variables defined in Prob. 6.1. are dependent.

6.7. Find $E(X)$ and $E(Y)$ for the random variables in Prob. 6.1.

6.8. X_1 and X_2 are independent standard normal random variables. Find $f_{Y_1Y_2}(y_1, y_2)$, where $Y_1 = X_1 + X_2$, $Y_2 = X_1/X_2$. Verify that Y_2 has a Cauchy density.

6.9. $X_1, X_2, \ldots, X_n$ are independent random variables. Find the correlation coefficient r_{YZ}, where $Y = \sum_{j=1}^{n} \alpha_j X_j$, and $Z = \sum_{j=1}^{n} \beta_j X_j$, and α_j and β_j are given constants.

6.10. X_1 and X_2 are jointly normal random variables with $a = b = 0$, $\sigma_1 = \sigma_2 = 1$, and correlation coefficient r. Find $E[\max(X_1, X_2)]$ and $E[\min(X_1, X_2)]$.

6.11. X_1 and X_2 are jointly random variables. Find the joint distribution of

$$Y_1 = aX_1 + bX_2$$

$$Y_2 = cX_1 + dX_2$$

for constants a, b, c, and d satisfying $ad = bc$.

6.12. $X_1, X_2, \ldots, X_n$ are independent identically distributed exponential random variables. Find $f_Y(y)$ of their sum $Y = X_1 + X_2 + \cdots + X_n$.

6.13. $X_1, X_2, \ldots, X_n$ are independent identically distributed gamma-distributed random variables. Find $f_Y(y)$ of their sum $Y = X_1 + X_2 + \cdots + X_n$.

6.14. $X_1, X_2, \ldots, X_n$ are independent and identically distributed normal random variables, $N(a, \sigma^2)$. Verify that the sample mean $\overline{X}_n = (1/n)\sum_{j=1}^{n} X_j$ is also a normal variable $N(a, \sigma^2/n)$.

RECOMMENDED FURTHER READING

Melsa, J. L., and Sage, A. P., *An Introduction to Probability and Stochastic Processes*, Prentice-Hall, Englewood Cliffs, NJ, 1973. Chap. 4: Function of Random Variables, pp. 110–188.

Papoulis, A., *Probability, Random Variables, and Stochastic Processes*, Intern. Student Ed., McGraw-Hill Kogakusha, Tokyo, 1965. Chap. 8: Sequences of Random Variables, pp. 233–278.

chapter **7**

Reliability of Structures Described by Several Random Variables

Hitherto we have been able to consider only the reliability of structures described by a single random variable. Such an analysis was, however, confined to simple cases. Now we extend the discussion to structures described by two or more random variables, beginning with the two-dimensional case.

7.1 FUNDAMENTAL CASE

We revert to the structure described in Sec. 5.1, a bar of constant cross section subjected to an axial random force resulting in random actual stresses Σ in the bar. Now, however, the resistance (the allowable stress) is also a random variable Σ_{allow}. We consider first the case when both quantities take on only positive values, so that the reliability of the structure is

$$R = P(\Sigma \leqslant \Sigma_{\text{allow}}) \tag{7.1}$$

The reliability may also be expressed as

$$R = F_Z(0) \qquad Z = \Sigma - \Sigma_{\text{allow}} \tag{7.2}$$

or

$$R = F_V(1) \qquad V = \frac{\Sigma}{\Sigma_{\text{allow}}} \tag{7.3}$$

which is illustrated in Fig. 7.1.

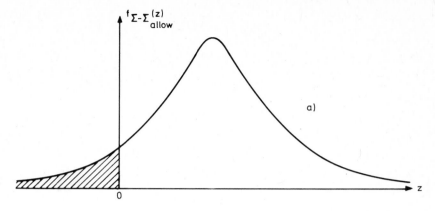

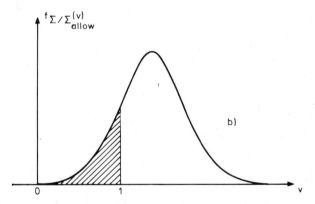

Fig. 7.1. Shaded areas represent the reliability of a structure with random applied stress Σ and random allowable stress Σ_{allow}.

Denote by $f_{\Sigma\Sigma_{\text{allow}}}(\sigma, \sigma_{\text{allow}})$ the joint probability density of Σ and Σ_{allow}. Formula (7.1) takes the form

$$R = \iint_A f_{\Sigma\Sigma_{\text{allow}}}(\sigma, \sigma_{\text{allow}}) \, d\sigma \, d\sigma_{\text{allow}} = \int_0^\infty d\sigma \int_\sigma^\infty f_{\Sigma\Sigma_{\text{allow}}}(\sigma, \sigma_{\text{allow}}) \, d\sigma_{\text{allow}}$$

or alternatively,

$$R = \iint_A f_{\Sigma\Sigma_{\text{allow}}}(\sigma, \sigma_{\text{allow}}) \, d\sigma \, d\sigma_{\text{allow}} = \int_0^\infty d\sigma_{\text{allow}} \int_0^{\sigma_{\text{allow}}} f_{\Sigma\Sigma_{\text{allow}}}(\sigma, \sigma_{\text{allow}}) \, d\sigma$$

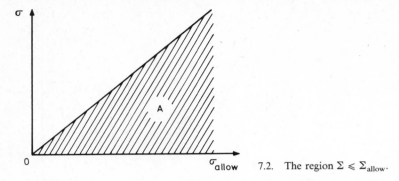

7.2. The region $\Sigma \leqslant \Sigma_{\text{allow}}$.

where the integration region is shown in Fig. 7.2. With the quite reasonable assumption of independence of Σ and Σ_{allow}, we find

$$R = \int_0^\infty \left[1 - F_{\Sigma_{\text{allow}}}(\sigma)\right] f_\Sigma(\sigma) \, d\sigma \tag{7.4}$$

or

$$R = \int_0^\infty F_\Sigma(\sigma_{\text{allow}}) f_{\Sigma_{\text{allow}}}(\sigma_{\text{allow}}) \, d\sigma_{\text{allow}} \tag{7.5}$$

where $F_\Sigma(\sigma)$ and $F_{\Sigma_{\text{allow}}}(\sigma_{\text{allow}})$ are the distribution functions of actual and allowable stresses, respectively.

It is readily shown that (7.4) and (7.5) generalize the relevant results of 5.3 and 5.21, where *either* the actual or the allowable stress was considered as the random variable. Indeed, if the allowable stress is a deterministic quantity $\bar{\sigma}_{\text{allow}}$, it has a causal density

$$f_{\Sigma_{\text{allow}}}(\sigma_{\text{allow}}) = \delta(\sigma_{\text{allow}} - \bar{\sigma}_{\text{allow}})$$

Substituting this in Eq. (7.5), we obtain

$$R = F_\Sigma(\bar{\sigma}_{\text{allow}})$$

which coincides with Eq. (5.3). If the actual stress is deterministic and the allowable stress random, we have

$$f_\Sigma(\sigma) = \delta(\sigma - \bar{\sigma})$$

where $\bar{\sigma}$ is the actual stress, and Eq. (7.4) yields

$$R = 1 - F_{\Sigma_{\text{allow}}}(\bar{\sigma})$$

which coincides with Eq. (5.21).

From here on both the actual and allowable stresses are random. Consider a few examples.

Example 7.1

The joint probability density $f_{\Sigma\Sigma_{\text{allow}}}(\sigma, \sigma_{\text{allow}})$ is known, and we seek the probability density of their difference $Z = \Sigma - \Sigma_{\text{allow}}$. Denote

$$\Sigma = X, \qquad \Sigma_{\text{allow}} = Y$$

The two-dimensional random vector $(X, -Y)$ has a joint density $f_{XY}(x, -y)$; hence by $X - Y = X + (-Y)$ we have

$$f_Z(z) = \int_{-\infty}^{\infty} f_{XY}(x, x - z)\, dx$$

For independent X and Y we have

$$f_Z(z) = \int_{-\infty}^{\infty} f_X(x) f_Y(x - z)\, dx = \int_{-\infty}^{\infty} f_X(y + z) f_Y(y)\, dy \qquad (7.6)$$

The reliability is then, according to Eq. (7.2),

$$R = \int_{-\infty}^{0} f_Z(z)\, dz \qquad (7.7)$$

Let X and Y have exponential density with parameters λ and μ, respectively, and are independent:

$$f_X(x) = \lambda e^{-\lambda x} U(x), \qquad f_Y(y) = \mu e^{-\mu y} U(y)$$

The density $f_Z(z)$ is obtained from the first of Eqs. (7.6). Note that $f_X(x) \equiv 0$ for $x < 0$ and $f_Y(x - z) \equiv 0$ for $x < z$, so we have

$$f_Z(z) = \int_{z}^{\infty} (\lambda e^{-\lambda x})\left[\mu e^{-\mu(x-z)}\right] dx = \frac{\lambda\mu}{\lambda + \mu} e^{-\lambda z}, \quad z > 0$$

$$f_Z(z) = \int_{0}^{\infty} (\lambda e^{-\lambda x})\left[\mu e^{-\mu(x-z)}\right] dx = \frac{\lambda\mu}{\lambda + \mu} e^{\mu z}, \quad z < 0$$

and $f_Z(0) = \lambda\mu/(\lambda + \mu)$. The mathematical expectation and the variance of Z are, respectively,

$$E(Z) = \frac{1}{\lambda} - \frac{1}{\mu} \qquad \text{Var}(Z) = \frac{1}{\lambda^2} + \frac{1}{\mu^2}$$

The reliability is then, in view of Eq. (7.7), $R = \lambda/(\lambda + \mu)$. Define the *central safety factor*

$$s = \frac{E(\Sigma_{\text{allow}})}{E(\Sigma)} = \frac{E(Y)}{E(X)} \tag{7.8}$$

which in our case is

$$s = \frac{\lambda}{\mu}$$

In terms of this factor, the reliability can be stated as

$$R = \frac{s}{1 + s} \tag{7.9}$$

which implies one-to-one correspondence between the reliability R and the central safety factor s. Thus, if the $R = 0.99$, the central safety factor equals 99.

If $\lambda = \mu$, then

$$f_Z(z) = \frac{\lambda}{2} e^{-\lambda |z|}$$

that is, Z has a Laplace density (see Example 3.9). The reliability is $R = 0.5$ and the central safety factor $s = 1$. Note that this one-to-one correspondence generally does not hold, as will be seen later on.

Example 7.2

Let $X = \Sigma$ have a chi-square distribution with m degrees of freedom:

$$f_X(x) = \frac{e^{-x/2} x^{m/2 - 1}}{2^{m/2} \Gamma(m/2)} U(x)$$

and let $Y = \Sigma_{\text{allow}}$ have a similar distribution with n degrees of freedom:

$$f_Y(y) = \frac{e^{-y/2} y^{n/2 - 1}}{2^{n/2} \Gamma(n/2)} U(y)$$

where both X and Y are independent variables.

Before determining the reliability of structure, we find the distribution of

$$U = \frac{X/m}{Y/n} \tag{7.10}$$

that is, of the ratio of the independent chi-square random variables divided by their respective degrees of freedom. The joint density $f_{XY}(x, y)$ is

$$f_{XY}(x, y) = \frac{x^{m/2-1}y^{n/2-1}}{2^{(m+n)/2}\Gamma(m/2)\Gamma(n/2)} e^{-(x+y)/2}U(x)U(y)$$

In order to find $f_U(u)$, we introduce the auxiliary variable $W = Y$. We then wish to find the joint density $f_{UW}(u, w)$ and obtain a marginal density $f_U(u)$ by integration. The Jacobian is

$$J = \frac{m}{n}w$$

so

$$f_{UW}(u, w) = \frac{m}{n}w\frac{1}{2^{(m+n)/2}\Gamma(m/2)\Gamma(n/2)}\left(\frac{m}{n}uw\right)^{m/2-1}$$

$$\times w^{n/2-1}\exp\left[-\frac{(m/n)uw + w}{2}\right]U(u)U(w)$$

and

$$f_U(u) = \int_0^\infty f_{UW}(u, w)\, dw$$

$$= \frac{1}{2^{(m+n)/2}\Gamma(m/2)\Gamma(n/2)}\left(\frac{m}{n}\right)^{m/2}u^{m/2-1}$$

$$\times \int_0^\infty w^{(m+n-2)/2}\exp\left[-\frac{(m/n)uw + w}{2}\right]dw$$

$$= \frac{\Gamma[(m+n)/2]}{\Gamma(m/2)\Gamma(n/2)}\left(\frac{m}{n}\right)^{m/2}\frac{u^{(m-2)/2}}{[1 + (m/n)u]^{(m+n)/2}}U(u) \quad (7.11)$$

where $U(u)$ is the unit step function. The random variable U is said to have an *F distribution with degrees of freedom m and n*. Note that an *F*-distributed random variable is often used in statistics. U is often referred to as the *variance ratio*. It is worth noting that if U has an F distribution with m and n degrees of freedom, then $1/U$ has a similar distribution with n and m degrees of freedom.

Reverting to the initial problem, we have, in accordance with Eq. (7.3),

$$R = F_V(1) \qquad V = \frac{X}{Y}$$

and by virtue of Eq. (7.10),

$$R = F_U\left(\frac{n}{m}\right) \tag{7.12}$$

where $F_U(u)$ is the (cumulative) distribution function of U.

Example 7.3

Let $X = \Sigma$ and $Y = \Sigma_{\text{allow}}$ both have uniform distributions

$$f_X(x) = \begin{cases} \dfrac{1}{x_2 - x_1}, & x_1 \leqslant x \leqslant x_2 \\ 0, & \text{otherwise} \end{cases}$$

$$f_Y(y) = \begin{cases} \dfrac{1}{y_2 - y_1}, & y_1 \leqslant y \leqslant y_2 \\ 0, & \text{otherwise} \end{cases}$$

$$F_X(x) = \begin{cases} 0, & x < x_1 \\ \dfrac{x - x_1}{x_2 - x_1}, & x_1 \leqslant x < x_2 \\ 1, & x \geqslant x_2 \end{cases}$$

$$F_Y(y) = \begin{cases} 0, & y \leqslant y_1 \\ \dfrac{y - y_1}{y_2 - y_1}, & y_1 \leqslant y < y_2 \\ 1, & y \geqslant y_2 \end{cases} \tag{7.13}$$

where $x_2 > x_1$ and $y_2 > y_1$. It is immediately seen from Eq. (7.5) that

$$R = \frac{1}{y_2 - y_1} \int_{y_1}^{y_2} F_X(y)\, dy \tag{7.14}$$

Thus, when the maximum possible value of the allowable stress Y is less than the minimum of the actual stress X (see Fig. 7.3), the reliability is zero as anticipated, indicating, under the frequency interpretation of probability, that almost every one of the large ensemble of statistically identical structures is due to fail. In the opposite case, when the minimum possible value y_1 of the allowable stress Y exceeds the maximum x_2 of the actual stress X (Fig. 7.4), the reliability is unity, since then $F_X(y)$ in (7.14) identically equals unity in the interval $\gamma \leqslant y \leqslant \delta$ and integration likewise yields unity, indicating also that almost every structure in the above ensemble is due to survive. Note that the central factor of safety

$$s = \frac{y_1 + y_2}{x_1 + x_2}$$

is always less than unity in the first case and always greater than unity in the

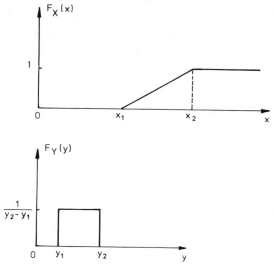

Fig. 7.3. Both X and Y have a uniform distribution ($y_2 < x_1$); the system possesses zero reliability.

second. If $y_1 < x_1$ and $y_2 > x_2$ (Fig. 7.5), we have

$$R = \frac{1}{y_2 - y_1} \int_{x_1}^{x_2} \frac{y - x_1}{x_2 - x_1} \, dy + \frac{1}{y_2 - y_1} \int_{x_2}^{y_2} dy = \frac{2y_2 - x_1 - x_2}{2(y_2 - y_1)} \quad (7.15)$$

The factor of safety s exceeds unity, but the reliability may be rather low. For example, if in some relative units $x_1 = 1$, $x_2 = 2$, $y_1 = 0.5$, $y_2 = 5.5$, we have $R = 0.8$, whereas the central safety factor is 2.

Note that, as is seen from (7.15), when

$$x_1 \to y_1 + 0 \quad \text{and} \quad x_2 \to y_2 - 0 \quad (7.16)$$

we have in the limit $R = 0.5$, corresponding to a safety factor of unity, the same situation as in Example 7.1. All other cases are covered by Problem 7.1.

The safety factor is also occasionally defined as

$$t = E\left(\frac{Y}{X}\right) = E\left(\frac{\Sigma_{\text{allow}}}{\Sigma}\right) \quad (7.17)$$

Let us find $f_U(u)$, where $U = Y/X$. Consider as an example the case where both X and Y are uniformly distributed in the interval $[0, 1]$. Then,

$$f_U(u) = \int_{-\infty}^{\infty} |x| f_Y(ux) f_X(x) \, dx$$

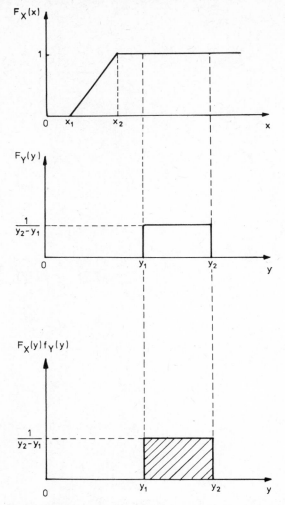

Fig. 7.4. Both X and Y have a uniform distribution ($x_2 < y_1$); the reliability (equal to the shaded area) equals unity.

Since

$$f_Y(ux) = \begin{cases} 1, & ux \leq 1 \\ 0, & \text{otherwise} \end{cases}$$

we have, for $u \leq 1$, $x \leq 1$,

$$f_U(u) = \int_0^1 x \, dx = \tfrac{1}{2}$$

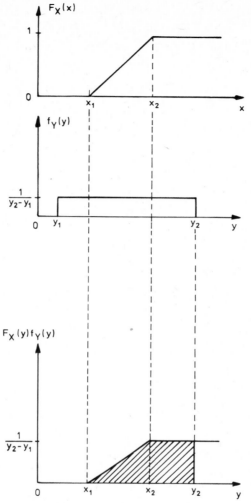

Fig. 7.5.   Both X and Y have a uniform distribution; the factor of safety is greater than unity, but the reliability (given as the shaded area) could be rather low.

and, for $u \geqslant 1$, $x \leqslant 1/u$,

$$f_U(u) = \int_0^{1/u} x \, dx = \tfrac{1}{2}\left(\frac{1}{u}\right)^2$$

The reliability is (Eq. 7.3)

$$R = 1 - P(X < Y) = 1 - F_U(1)$$

$$= 1 - \int_0^1 f_U(u) \, du = 1 - \tfrac{1}{2} = \tfrac{1}{2}$$

which checks with the case (7.16). Now,

$$t = \int_0^\infty u f_U(u) \, du = \int_0^1 \frac{u}{2} \, du + \int_1^\infty \frac{1}{2}\left(\frac{1}{u}\right) du \to \infty$$

so that the new safety factor is infinity, whereas the reliability is only 0.5!

Example 7.4

Suppose that the actual and allowable stresses are independent random variables with log-normal probability density

$$f_X(x) = \frac{1}{x\sigma_1\sqrt{2\pi}} \exp\left[-\frac{(\ln x - a)^2}{2\sigma_1^2}\right] U(x)$$

$$f_Y(y) = \frac{1}{y\sigma_2\sqrt{2\pi}} \exp\left[-\frac{(\ln y - b)^2}{2\sigma_2^2}\right] U(y)$$

where a, σ_1, b, σ_2 are the density parameters so that

$$E(X) = \exp\left(a + \tfrac{1}{2}\sigma_1^2\right)$$

$$E(Y) = \exp\left(b + \tfrac{1}{2}\sigma_2^2\right)$$

$$\mathrm{Var}(X) = \exp\left(2a + \sigma_1^2\right)\left[\exp\left(\sigma_1^2\right) - 1\right]$$

$$\mathrm{Var}(Y) = \exp\left(2b + \sigma_2^2\right)\left[\exp\left(\sigma_2^2\right) - 1\right]$$

The reliability is then

$$R = P\left(V = \frac{X}{Y} \leqslant 1\right) = F_V(1)$$

which may be rewritten as

$$R = P(\ln V \leqslant 0) = F_{\ln V}(0) \tag{7.18}$$

Note that

$$\ln V = \ln X - \ln Y$$

and since $\ln X$ and $\ln Y$ both have a normal distribution, specifically $\ln X$ is $N(a, \sigma_1^2)$ and $\ln Y$ is $N(b, \sigma_2^2)$, $\ln V$ is also normal, as a difference of normal variables, $N(a - b, \sigma_1^2 + \sigma_2^2)$, implying that V is the log-normal with the mean and variance, respectively,

$$E(V) = \exp\left[a - b + \tfrac{1}{2}(\sigma_1^2 + \sigma_2^2)\right]$$

$$\text{Var}(V) = \exp\left[2(a - b) + \sigma_1^2 + \sigma_2^2\right]\left[\exp(\sigma_1^2 + \sigma_2^2) - 1\right] \qquad (7.19)$$

Combining Eqs. (4.13) and (7.18), we get

$$R = \frac{1}{2} + \text{erf}\left(-\frac{a - b}{(\sigma_1^2 + \sigma_2^2)^{1/2}}\right) \qquad (7.20)$$

The safety factors s and t are, respectively,

$$s = \frac{\exp(a + \sigma_1^2/2)}{\exp(b + \sigma_2^2/2)}$$

$$t = \exp\left[a - b + \tfrac{1}{2}(\sigma_1^2 + \sigma_2^2)\right] \qquad (7.21)$$

For $a = b$, $\sigma_1 = \sigma_2 = \sigma$, we have $s = 1$, $t = \exp(\sigma^2)$, and $R = 0.5$.

Example 7.5

Suppose that X and Y are independent random variables with truncated normal distributions:

$$f_X(x) = \frac{A}{\sigma_1\sqrt{2\pi}} \exp\left[-\frac{(x - m_1)^2}{2\sigma_1^2}\right]\left[U(x_2 - x) - U(x_1 - x)\right]$$

$$f_Y(y) = \frac{B}{\sigma_2\sqrt{2\pi}} \exp\left[-\frac{(y - m_2)^2}{2\sigma_2^2}\right]\left[U(y_2 - y) - U(y_1 - y)\right]$$

where

$$A = \left[\text{erf}\left(\frac{x_2 - m_1}{\sigma_1}\right) - \text{erf}\left(\frac{x_1 - m_1}{\sigma_1}\right)\right]^{-1}$$

$$B = \left[\text{erf}\left(\frac{y_2 - m_2}{\sigma_2}\right) - \text{erf}\left(\frac{y_1 - m_2}{\sigma_2}\right)\right]^{-1}$$

As in Example 7.3, with uniformly distributed actual and allowable stresses, we

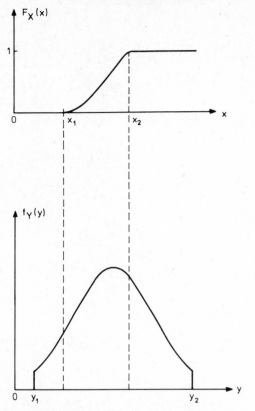

Fig. 7.6. Actual stress and allowable stress with truncated normal distribution.

conclude immediately that

$$R = 0, \quad \min \sigma = x_1 > \max \sigma_{\text{allow}} = y_2$$

$$R = 1, \quad \max \sigma = x_2 < \min \sigma_{\text{allow}} = y_1$$

Consider the case analogous to that shown in Fig. 7.6, specifically,

$$y_1 \leqslant x_1 < x_2 \leqslant y_2$$

We find then, from Eq. 7.5,

$$R = \int_{x_1}^{y_2} F_X(y) f_Y(y) \, dy$$

or, bearing in mind expression (4.28) for the truncated (cumulative) distribution function of a normal variable,

$$R = \frac{AB}{\sigma_2\sqrt{2\pi}} \int_{x_1}^{x_2} \left[\mathrm{erf}\left(\frac{y - m_1}{\sigma_1}\right) - \mathrm{erf}\left(\frac{x_1 - m_1}{\sigma_1}\right) \right] \exp\left[-\frac{(y - m_2)^2}{2\sigma_2^2} \right] dy$$

$$+ \frac{B}{\sigma_2\sqrt{2\pi}} \int_{x_2}^{y_2} \exp\left[-\frac{(y - m_2)^2}{2\sigma_2^2} \right] dy \tag{7.22}$$

For $x_2 = y_2$, the last term in Eq. (7.22) drops out. Assume now that the actual and allowable stresses are both distributed symmetrically, that is, that

$$x_1 = m_1 - k_1\sigma_1 \qquad x_2 = m_1 + k_1\sigma_1$$

$$y_1 = m_2 - k_2\sigma_2 \qquad y_2 = m_2 + k_2\sigma_2$$

where k_1 and k_2 are positive numbers. If both k_1 and k_2 are greater than 2, then according to Eq. (4.30) both A and B differ only slightly from unity and x_1 and x_2 may be replaced by $-\infty$ and $+\infty$, respectively. The second integral term can be neglected, and Eq. (7.22) replaced by

$$R = \frac{1}{\sigma_2\sqrt{2\pi}} \int_{-\infty}^{\infty} \left[\tfrac{1}{2} + \mathrm{erf}\left(\frac{y - m_1}{\sigma_1}\right) \right] \exp\left[-\frac{(y - m_2)^2}{2\sigma_2^2} \right] dy \tag{7.23}$$

which coincides formally with the expression for the reliability, provided both X and Y can be assumed in advance to have a normal distribution. In that case, rather than evaluate the above equation, we note that $Z = X - Y$ in Eq. (7.2) is likewise normal, $N(m_1 - m_2, \sigma_1^2 + \sigma_2^2)$, and the reliability, in view of Eq. (4.13), becomes

$$R = F_Z(0) = \tfrac{1}{2} + \mathrm{erf}\left(-\frac{m_1 - m_2}{\left(\sigma_1^2 + \sigma_2^2\right)^{1/2}} \right) \tag{7.24}$$

and with

$$s_Z = \frac{\left(\sigma_1^2 + \sigma_2^2\right)^{1/2}}{m_2 - m_1}$$

as the coefficient of variation of Z, we finally have

$$R = \tfrac{1}{2} + \mathrm{erf}\left(\frac{1}{s_Z} \right) \tag{7.25}$$

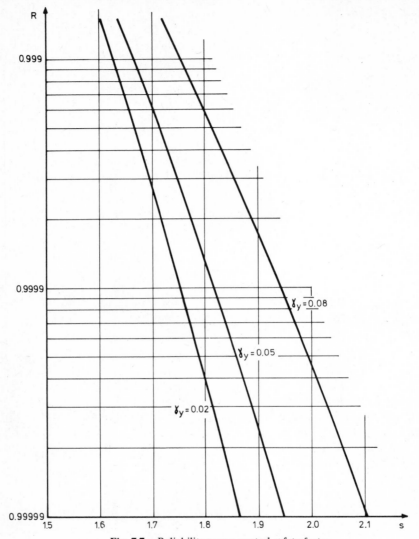

Fig. 7.7. Reliability versus central safety factor.

In spite of the formal similarity of (7.20) and (7.24), there is a basic difference: In Eq. (7.24) m_1, σ_1 and m_2, σ_2 are the mathematical expectation and standard deviation of the actual and allowable stresses, respectively, whereas in Eq. (7.20) the parameters are associated with their logarithms. Equation (7.24) may be rewritten in terms of the central safety factor and the coefficients of variation

$$\gamma_X = \frac{\sigma_X}{E(X)} = \frac{\sigma_1}{m_1} \qquad \gamma_Y = \frac{\sigma_Y}{E(Y)} = \frac{\sigma_2}{m} \qquad (7.26)$$

as follows:

$$R = \tfrac{1}{2} + \mathrm{erf}\!\left(\frac{s - 1}{\left(\gamma_X^2 + s^2 \gamma_Y^2 \right)^{1/2}} \right) \tag{7.27}$$

Figure 7.7 shows the reliability versus the central factor of safety for $\gamma_X = 0.2$ and various γ_Y. Freudenthal, Ferry Borges and Castanheta, and others investigated the influence of changes in the coefficients of variation, the factor of safety, and the shape of the distribution functions $f_X(x)$ and $f_Y(y)$ on the reliability. Their results are usually plotted as $1 - R$ versus the safety factor for different coefficients of variation.

7.2   BENDING OF BEAMS UNDER SEVERAL RANDOM CONCENTRATED FORCES

As shown in Fig. 7.8, the beam is subjected to n concentrated forces $X_1, X_2, \ldots, X_n$, with joint probability density $f_{(X)}(x_1, x_2, \ldots, x_n)$. We initially assume that the allowable stress and the beam dimensions are deterministic quantities. Since the bending moment takes on extremal values at the sections where the forces are applied, the reliability is

$$R = P\!\left\{ \bigcap_{j=1}^{n} \left[|M_j| \leqslant \sigma_{\mathrm{allow}} S \right] \right\} \tag{7.28}$$

where M_j denotes the bending moment at section x_j. These moments are

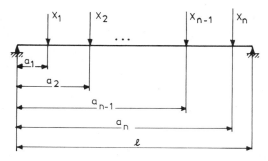

Fig. 7.8.   Beam simply supported at both edges, under n concentrated forces applied at specified cross sections.

expressed in terms of the applied forces as

$$
\begin{aligned}
M_1 &= \alpha_{11} X_1 + \alpha_{12} X_2 + \cdots + \alpha_{1n} X_n \\
M_2 &= \alpha_{21} X_1 + \alpha_{22} X_2 + \cdots + \alpha_{2n} X_n \\
&\vdots \\
M_n &= \alpha_{n1} X_1 + \alpha_{n2} X_2 + \cdots + \alpha_{nn} X_n
\end{aligned}
\tag{7.29}
$$

The joint probability density $f_{\langle M \rangle}(m_1, m_2, \ldots, m_n)$ of the bending moments is readily found via Eq. (6.127):

$$
f_{\langle M \rangle}(m_1, m_2, \ldots, m_n) = f_{\langle X \rangle}(B_{11} m_1 + B_{12} m_2 + \cdots + B_{1n} m_n, \ldots, B_{n1} m_1
$$

$$
+ B_{n2} m_2 + \cdots + B_{nn} m_n) |\det[B]|
\tag{7.30}
$$

where $[B]$ is the inverse of matrix $[A]$:

$$
[A] = \begin{bmatrix}
\alpha_{11} & \alpha_{12} & \cdots & \alpha_{1n} \\
\alpha_{21} & \alpha_{22} & \cdots & \alpha_{2n} \\
\vdots & & & \\
\alpha_{n1} & \alpha_{n2} & \cdots & \alpha_{nn}
\end{bmatrix}
$$

which was assumed to be nonsingular. When it is singular, its rank p is less than n and some (namely, $n - p$) of the moments are equal. Renumbering the unequal moments as $M_1, M_2, \ldots, M_p$, the reliability (7.28) becomes

$$
R = P \left\langle \bigcap_{j=1}^{p} \left(|M_j| \leqslant \sigma_{\text{allow}} S \right) \right\rangle
\tag{7.31}
$$

and in the general case $p \leqslant n$ can be written as

$$
R = \iint \cdots \int_A f_{\langle X \rangle}(B_{11} m_1 + B_{12} m_2 + \cdots + B_{1p} m_p, \ldots, B_{p1} m_1 + B_{p2} m_2
$$

$$
+ \cdots + B_{pp} m_p) \det[B]_{p \times p} \, dm_1 \, dm_2 \cdots dm_p
\tag{7.32}
$$

where the integration region A is defined as

$$
|m_1| \leqslant \sigma_{\text{allow}} S, \quad |m_2| \leqslant \sigma_{\text{allow}} S \quad \ldots, \quad |m_p| \leqslant \sigma_{\text{allow}} S
\tag{7.33}
$$

Example 7.6

Consider the beam, simply supported at both edges, under a pair of concentrated forces X_1 and X_2. Then,

$$[A] = l \begin{bmatrix} \xi_1(1 - \xi_1) & \xi_1(1 - \xi_2) \\ \xi_1(1 - \xi_2) & \xi_2(1 - \xi_2) \end{bmatrix}, \quad \xi_1 = \frac{a_1}{l}, \quad \xi_2 = \frac{a_2}{l}, \quad a_2 > a_1$$

with determinant

$$\det[A] = \xi_1(1 - \xi_2)(\xi_2 - \xi_1)l^2$$

so that $[A]$ is nonsingular if $\xi_1 \neq 0$, $\xi_2 \neq 1$, $\xi_1 \neq \xi_2$; that is, if none of the forces is applied at a support or applied at the same section. Then

$$[B] = [A]^{-1} = \frac{1}{l(\xi_2 - \xi_1)} \begin{bmatrix} \dfrac{\xi_2}{\xi_1} & -1 \\ -1 & \dfrac{1 - \xi_1}{1 - \xi_2} \end{bmatrix}$$

we have

$$f_{(M)}(m_1, m_2) = f_{(X)}\left[\frac{\xi_2}{\xi_1(\xi_2 - \xi_1)l} m_1 - \frac{1}{(\xi_2 - \xi_1)l} m_2, \frac{-1}{(\xi_2 - \xi_1)l} m_1 \right.$$

$$\left. + \frac{1 - \xi_1}{(1 - \xi_2)(\xi_2 - \xi_1)l} m_2 \right]$$

$$\times \frac{1}{\xi_1(1 - \xi_2)(\xi_2 - \xi_1)l^2}$$

and the reliability becomes

$$R = \frac{1}{\xi_1(1 - \xi_2)(\xi_2 - \xi_1)l^2}$$

$$\times \iint\limits_A f_{(X)}\left[\frac{\xi_2}{\xi_1(\xi_2 - \xi_1)l} m_1 - \frac{1}{(\xi_2 - \xi_1)l} m_2, \frac{-1}{(\xi_2 - \xi_1)l} m_1 \right.$$

$$\left. + \frac{1 - \xi_1}{(1 - \xi_2)(\xi_2 - \xi_1)l} m_2 \right] dm_1 \, dm_2 \qquad (7.34)$$

In the singular case, where $\xi_1 = 0$ and $\xi_2 \neq 1$, we have only one section where the extremum bending moment M acts, $x = a_2$:

$$M = l\xi_2(1 - \xi_2) X_2$$

and in view of Eq. (4.47)

$$f_{(M)}(m) = \frac{1}{l\xi_2(1 - \xi_2)} f_{X_2}\left[\frac{m}{l\xi_2(1 - \xi_2)} \right]$$

The reliability is

$$R = \frac{1}{l\xi_2(1 - \xi_2)} \int_A f_{X_2}\left[\frac{m}{l\xi_2(1 - \xi_2)} \right] dm$$

where the region A is determined by

$$|m| \leq \sigma_{\text{allow}} S$$

that is,

$$R = \frac{1}{l\xi_2(1 - \xi_2)} \int_{-\sigma_{\text{allow}}S}^{\sigma_{\text{allow}}S} f_{X_2}\left[\frac{m}{l\xi_2(1 - \xi_2)} \right] dm$$

$$= F_{X_2}\left[\frac{\sigma_{\text{allow}}S}{l\xi_2(1 - \xi_2)} \right] - F_{X_2}\left[-\frac{\sigma_{\text{allow}}S}{l\xi_2(1 - \xi_2)} \right] \qquad (7.35)$$

The case $\xi_1 \neq 0$, $\xi_2 = 1$ is treated analogously, the result being

$$R = F_{X_2}\left[\frac{\sigma_{\text{allow}}S}{l\xi_1(1 - \xi_1)} \right] - F_{X_2}\left[-\frac{\sigma_{\text{allow}}S}{l\xi_1(1 - \xi_1)} \right] \qquad (7.36)$$

The last singular case is $\xi_1 = \xi_2 \equiv \xi$, the concentrated force $Y \equiv X_1 + X_2$ is applied at section ξ, the distribution function of X is (see Example 6.4)

$$F_Y(y) = \int_{-\infty}^{\infty} dx_2 \int_{-\infty}^{y - x_2} f_{X_1 X_2}(x_1, x_2) \, dx_1$$

and the reliability is

$$R = F_Y\left[\frac{\sigma_{\text{allow}}S}{l\xi(1 - \xi)} \right] - F_Y\left[-\frac{\sigma_{\text{allow}}S}{l\xi(1 - \xi)} \right] \qquad (7.37)$$

For X_1 and X_2 being jointly normal with mathematical expectations m_1 and m_2, respectively, variances σ_1^2 and σ_2^2, respectively, and correlation coefficient

r, Y is also normal $N(m_1 + m_2, \sigma_1^2 + 2r\sigma_1\sigma_2 + \sigma_2^2)$ with Eq. (6.79) taken into account. Eq. (7.37) become

$$R = \text{erf}\left[\frac{\sigma_{\text{allow}}Sl^{-1}\xi^{-1}(1-\xi)^{-1} - (m_1 + m_2)}{\left(\sigma_1^2 + 2r\sigma_1\sigma_2 + \sigma_2^2\right)^{1/2}}\right]$$

$$- \text{erf}\left[-\frac{\sigma_{\text{allow}}Sl^{-1}\xi^{-1}(1-\xi)^{-1} + (m_1 + m_2)}{\left(\sigma_1^2 + 2r\sigma_1\sigma_2 + \sigma_2^2\right)^{1/2}}\right] \qquad (7.38)$$

Example 7.7

A circular shaft of radius c is simultaneously subjected to bending moment M and torque T, considered as random variables with given joint probability density function $f_{MT}(m, t)$; the yield stress σ_y in pure tension is a deterministic quantity. We seek the reliability of the system.

We resort to the maximum shear stress theory of failure. The maximum shear stress is

$$\tau_{\text{max}} = \left[\left(\frac{\sigma_{\text{bending}}}{2}\right)^2 + \tau_{\text{torsion}}^2\right]^{1/2}$$

or

$$\tau_{\text{max}} = \left[\left(\frac{Mc}{2I}\right)^2 + \left(\frac{Tc}{J}\right)^2\right]^{1/2}$$

where, for the circular cross section, $I = \pi c^4/4$, $J = 2I$, and

$$\tau_{\text{max}} = \frac{2}{\pi c^3}(M^2 + T^2)^{1/2} \qquad (7.39)$$

The strength requirement $\tau_{\text{max}} \leqslant \frac{1}{2}\sigma_y$ becomes

$$(M^2 + T^2)^{1/2} \leqslant \frac{\pi}{4}\sigma_y c^3$$

and the reliability reads

$$R = \text{Prob}\left([M^2 + T^2]^{1/2} \leqslant \frac{\pi}{4}\sigma_y c^3\right)$$

or with $Z = (M^2 + T^2)^{1/2}$,

$$R = F_Z\left(\frac{\pi}{4}\sigma_y c^3\right) \qquad (7.40)$$

We proceed to find the probability density function $f_Z(z)$. To do this, we first introduce the auxiliary variable

$$\Theta = \tan^{-1}\left(\frac{T}{M}\right) \tag{7.41}$$

the possible values of θ lie in the interval $[0, 2\pi]$. The inverse transformation is of the form

$$M = Z\cos\Theta \qquad T = Z\sin\Theta \tag{7.42}$$

since

$$\frac{\partial(m, t)}{\partial(z, \theta)} = \begin{vmatrix} \cos\theta & -z\sin\theta \\ \sin\theta & z\cos\theta \end{vmatrix} = z$$

we obtain

$$f_{Z\Theta} = zf_{MT}(z\cos\theta, z\sin\theta)U(z), \quad 0 \le \theta \le 2\pi$$

the marginal densities being

$$f_Z(z) = zU(z)\int_0^{2\pi} f_{MT}(z\cos\theta, z\sin\theta)\,d\theta$$

$$f_\Theta(\theta) = \int_0^\infty zf_{MT}(z\cos\theta, z\sin\theta)\,dz, \quad 0 \le \theta \le 2\pi \tag{7.43}$$

Suppose M is $N(a, \sigma^2)$ and T is $N(b, \sigma^2)$ and they are independent. Then

$$f_{Z\Theta}(z, \theta) = \frac{z}{2\pi\sigma^2}\exp\left[-\frac{(z\cos\theta - a)^2 + (z\sin\theta - b)^2}{2\sigma^2}\right] \tag{7.44}$$

From Eq. (7.43) we obtain

$$f_Z(z) = \frac{zU(z)}{2\pi\sigma^2}\int_0^{2\pi}\exp\left[-\frac{(z\cos\theta - a)^2 + (z\sin\theta - b)^2}{2\sigma^2}\right]d\theta$$

or

$$f_Z(z) = \frac{zU(z)}{2\pi\sigma^2}\exp\left[-\frac{z^2 + a^2 + b^2}{2\sigma^2}\right]\int_0^{2\pi}\exp\left[\frac{z\sqrt{a^2 + b^2}}{\sigma^2}\cos(\theta - \bar{\theta})\right]d\theta$$

where $\bar{\theta} = \tan^{-1}(b/a)$. Introducing the variables $\varphi = \theta - \bar{\theta}$, $\alpha = \sqrt{a^2 + b^2}$,

we have

$$\int_{-\bar{\theta}}^{2\pi-\bar{\theta}} \exp\left[-i\left(\frac{i\,\alpha z}{\sigma^2}\right)\cos\varphi\right]d\varphi = 2\pi I_0\left(\frac{\alpha z}{\sigma^2}\right) \tag{7.45}$$

where $I_0(z)$ is a modified Bessel function of zero order, namely,

$$I_0(z) = \frac{1}{2\pi}\int_0^{2\pi} e^{x\cos\theta}\,d\theta = \sum_{n=0}^{\infty} \frac{x^{2n}}{2^{2n}(n!)^2} \tag{7.46}$$

Hence,

$$f_Z(z) = \frac{z}{\sigma^2}\exp\left(-\frac{z^2+\alpha^2}{2\sigma^2}\right)I_0\left(\frac{\alpha z}{\sigma^2}\right)U(z) \tag{7.47}$$

The (cumulative) distribution function is then

$$F_Z(z) = \frac{1}{\sigma^2}\int_0^z z\exp\left(-\frac{z^2+\alpha^2}{2\sigma^2}\right)I_0\left(\frac{\alpha z}{\sigma^2}\right)dz \tag{7.48}$$

Applying the expression

$$\int x^n I_{n-1}(\alpha x)\,dx = \frac{x^n}{\alpha}I_n(\alpha x)$$

for positive α we find

$$F_Z(z) = \exp\left(-\frac{z^2+\alpha^2}{2\sigma^2}\right)\sum_{n=1}^{\infty}\left(\frac{z}{\alpha}\right)^n I_n\left(\frac{\alpha z}{\sigma^2}\right)U(z) \tag{7.49}$$

and combining Eqs. (7.40) and (7.49) we have the reliability

$$R = \exp\left\{-\frac{\left[(\pi/4)\sigma_y c^2\right]^2+\alpha^2}{2\sigma^2}\right\}\sum_{n=1}^{\infty}\left(\frac{\pi}{4}\frac{\sigma_y c^3}{\alpha}\right)I_n\left(\frac{\pi}{4}\frac{\sigma_y c^3\alpha}{\sigma^2}\right) \tag{7.50}$$

In the particular case where $a = b = 0$, we have

$$f_Z(z) = \frac{z}{\sigma^2}\exp\left(-z^2/2\sigma^2\right)$$

since $I_0(0) = 1$; that is, Z has a Rayleigh distribution, and the reliability reads

$$R = 1 - \exp\left(-\frac{\pi^2}{32}\frac{\sigma_y^2 c^6}{\sigma^2}\right) \tag{7.51}$$

The general case necessitates numerical integration in Eq. (7.48) or numerical evaluation of Eq. (7.49). The random variable Z with probability density as per (7.47) is said to have a *generalized Rayleigh distribution function*, for which tables are readily available (Bark et al.). For $\alpha \ll \sigma$, the series (7.46) may be reduced to two terms, and we have

$$f_Z(z) = \frac{z}{\sigma^2} \exp\left(-\frac{z^2 + \alpha^2}{2\sigma^2}\right)\left(1 + \frac{z^2\alpha^2}{4\sigma^4}\right)U(z)$$

In the opposite case, $\alpha \gg \sigma$, we may use the asymptotic representation for the modified Bessel function:

$$I_0(x) \simeq \frac{e^x}{\sqrt{2\pi x}}\left(1 + \frac{1}{8x} + \frac{9}{128x^2} + \cdots\right)$$

and

$$f_Z(z) \simeq \frac{1}{\sigma\sqrt{2\pi}} \exp\left[-\frac{(z - \alpha)^2}{2\sigma^2}\right]A(z) \tag{7.52}$$

where $A = (1 + \sigma^2/8\alpha z)(z/\alpha)^{1/2}$ is very close to unity in the vicinity of $z \sim \alpha$, and Z has a normal distribution $N(\alpha, \sigma^2)$.

Example 7.8

In this example we generalize Prob. 5.6 for the case where a single random force Q is applied at a random distance X, Σ_{allow} being a random variable as well. For simplicity we assume that all these quantities are independent and log-normal.

$$f_Q(q) = \frac{1}{q\sigma_1\sqrt{2\pi}} \exp\left[-\frac{(\ln q - m_1)^2}{2\sigma_1^2}\right]U(q)$$

$$f_X(x) = \frac{1}{x\sigma_2\sqrt{2\pi}} \exp\left[-\frac{(\ln x - m_2)^2}{2\sigma_2^2}\right]U(x)$$

$$f_{\Sigma_{\text{allow}}}(\sigma_{\text{allow}}) = \frac{1}{\sigma_{\text{allow}}\sigma_3\sqrt{2\pi}} \exp\left[-\frac{(\ln \sigma_{\text{allow}} - m_3)^2}{2\sigma_3^2}\right]U(\sigma_{\text{allow}})$$

The requirement is $Y \equiv M/\Sigma_{\text{allow}} = QX/\Sigma_{\text{allow}} \leqslant S$, where the section modulus S is a deterministic quantity. In logarithmic form, the requirement is

$$\ln Y = \ln Q + \ln X - \ln \Sigma_{\text{allow}} \leqslant \ln S$$

where $\ln Q$, $\ln X$, and $\ln \Sigma_{\text{allow}}$ are normal, hence $\ln Y$ is also normal, with mean

$$E(\ln y) = E(\ln Q) + E(\ln X) - E(\ln \Sigma_{\text{allow}})$$

$$= \exp\left(m_1 + \tfrac{1}{2}\sigma_1^2\right) + \exp\left(m_2 + \tfrac{1}{2}\sigma_2^2\right) - \exp\left(m_3 + \tfrac{1}{2}\sigma_3^2\right) \quad (7.53)$$

and variance

$$\text{Var}(\ln Y) = \text{Var}(\ln Q) + \text{Var}(\ln X) + \text{Var}(\ln \Sigma_{\text{allow}})$$

$$= \sum_{j=1}^{3} \exp\left(2m_j + \sigma_j^2\right)\left[\exp\left(\sigma_j^2\right) - 1\right] \quad (7.54)$$

Under these circumstances,

$$R = P(\ln Y \leqslant \ln S) = F_{\ln Y}(\ln S)$$

or in view of Eq. (4.13), we have

$$R = \tfrac{1}{2} + \text{erf}\left[\frac{\ln S - E(\ln Y)}{\sqrt{\text{Var}(\ln Y)}}\right] \quad (7.55)$$

7.3 BENDING OF BEAMS UNDER SEVERAL RANDOM CONCENTRATED MOMENTS

Confining ourselves to the case where all n moments have the same sign, the maximum moment M appear at the clamping support (Fig. 7.9):

$$M = M_1 + M_2 + \cdots + M_n \quad (7.56)$$

and the reliability is

$$R = P(M \leqslant \sigma_{\text{allow}} S) \quad (7.57)$$

Accordingly, the probabilistic characteristics of the maximum moment have to be known irrespective of whether the allowable stress is a random or a deterministic quantity. In some instances, the reliability calculation can be performed in closed form. Examples follow.

Example 7.9

Let $M_1, M_2, \ldots, M_n$ be independent random variables, each with a chi-square probability density with $\nu_1, \nu_2, \ldots, \nu_n$ degrees of freedom, respectively.

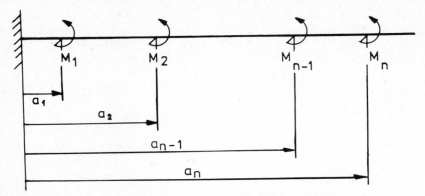

Fig. 7.9. Cantilever under n concentrated random moments applied at specified cross sections.

It can be shown (an exercise for the reader) that the maximum moment M also has a chi-square distribution, with $\nu = \nu_1 + \nu_2 + \cdots + \nu_n$ degrees of freedom, and with density

$$f_M(m) = \frac{e^{-m/2}m^{\nu/2-1}}{2^{\nu/2}\Gamma(\nu/2)}U(m) \tag{7.58}$$

The reliability R is then for deterministic σ_{allow} and S,

$$R = \int_0^{\sigma_{\text{allow}}S} f_M(m)\,dm \tag{7.59}$$

The reliability calculation may be simplified by using Table 5.1 (see Problem 5.1). An exponentially distributed random variable is a chi-square one with two degrees of freedom (compare Eqs. (4.8) and (4.9)), and as a result the sum of n independent exponentially distributed random variables likewise has a chi-square distribution (with $\nu = 2n$ degrees of freedom), and in this case the reliability is again determined by Eq. (7.59).

Example 7.10

Let $n = 2$, and let M_1 and M_2 be independent and uniformly distributed, with

$$f_{M_1}(m_1) = \begin{cases} 1, & 0 \leqslant m_1 \leqslant 1 \\ 0, & \text{otherwise} \end{cases}$$

$$f_{M_2}(m_2) = \begin{cases} 1, & 0 \leqslant m_2 \leqslant 1 \\ 0, & \text{otherwise} \end{cases}$$

By (6.51), we have for the distribution function of $M = M_1 + M_2$:

$$F_M(m) = \int_{-\infty}^{\infty} F_{M_1}(m - m_2) f_{M_2}(m_2)\, dm_2$$

with the following result:

$$F_M(m) = \begin{cases} 0, & m \leqslant 0 \\ \dfrac{m^2}{2}, & 0 \leqslant m \leqslant 1 \\ 1 - \dfrac{(2 - m)^2}{2}, & 1 \leqslant m \leqslant 2 \end{cases} \tag{7.60}$$

The reliability is then

$$R = F_M(\sigma_{\text{allow}} S) \tag{7.61}$$

The probability density of the maximum moment

$$f_M(m) = \begin{cases} 0, & m \leqslant 0 \\ m & 0 \leqslant m \leqslant 1 \\ 2 - m, & 1 \leqslant m \leqslant 2 \\ 0, & m \geqslant 2 \end{cases} \tag{7.62}$$

is often referred to as *triangular* density.

It can be shown by induction that for concentrated moments, all of them distributed uniformly in the interval (0, 1), the probability density is

$$f_M(m) = \frac{1}{(n-1)!} \sum_{k=0}^{i} \binom{n}{k} (-1)^k (m - k)^{n-1} \tag{7.63}$$

where $i < m < i + 1$; $i = 0, 1, \ldots, n - 1$.

Example 7.11

M_1 and M_2 are assumed to be independent random variables, each with an exponential distribution

$$f_{M_1}(m_1) = \lambda_1 e^{-\lambda_1 m_1} U(m_1) \qquad f_{M_2}(m_2) = \lambda_2 e^{-\lambda_2 m_2} U(m_2) \tag{7.64}$$

where λ_1 and λ_2 are positive constants. The probability density of the maximum moment will be

$$f_M(m) = \int_{-\infty}^{\infty} f_{M_1}(m_1) f_{M_2}(m - m_1)\, dm_1 = \int_0^m \lambda_1 e^{-\lambda_1 m_1} \lambda_2 e^{-\lambda_2(m - m_1)}\, dm_1$$

$$= \frac{\lambda_1 \lambda_2 (e^{-\lambda_1 m} - e^{-\lambda_2 m})}{\lambda_2 - \lambda_1} \tag{7.65}$$

The associated reliability is

$$R = P(M \leqslant \sigma_{allow}S) = \frac{1}{\lambda_2 - \lambda_1}\left[\lambda_1\exp(-\lambda_2\sigma_{allow}S) - \lambda_2\exp(-\lambda_1\sigma_{allow}S)\right]$$

For $\lambda_1 = \lambda_2$ we have, instead of Eq. (7.65),

$$f_M(m) = \lambda^2 m e^{-\lambda m} \tag{7.66}$$

that is, the maximum moment has a gamma distribution. The reliability is given by

$$R = 1 - (1 + \lambda\sigma_{allow}S)\exp(-\lambda\sigma_{allow}S) \tag{7.67}$$

For n moments applied, each with an exponential distribution with parameter λ, the maximum moment will have a gamma distribution

$$f_M(m) = \frac{\lambda(\lambda m)^{n-1}}{(n-1)!}e^{-\lambda m}$$

with the corresponding probability distribution

$$F_M(m) = 1 - e^{-\lambda m}\left[1 + \lambda m + \frac{(\lambda m)^2}{2!} + \cdots + \frac{(\lambda m)^{n-1}}{(n-1)!}\right] \tag{7.68}$$

Consequently, the reliability is given by

$$R = 1 - \exp(-\lambda\sigma_{allow}S)\left[1 + \lambda\sigma_{allow}S + \frac{(\lambda\sigma_{allow}S)^2}{2!} + \cdots + \frac{(\lambda\sigma_{allow}S)^{n-1}}{(n-1)!}\right]$$

$$\tag{7.69}$$

7.4   THE CENTRAL LIMIT THEOREM AND RELIABILITY ESTIMATE

In the preceding section we presented some closed solutions of the reliability of beams under n concentrated moments. It turns out that with a fairly large number of moments, the distribution of the moment M in a clamped edge can be approximated by a normal distribution, irrespective of the probability densities of the component moments. Actually, if the random variables $M_1, M_2, \ldots, M_n$ are independent normal variables with means $E(M_1), E(M_2), \ldots, E(M_n)$ and variances $\text{Var}(M_1), \text{Var}(M_2), \ldots, \text{Var}(M_n)$, respectively, the distribution of their sum M is also normal $N(\eta, \sigma^2)$, where

$$\eta = E(M_1) + E(M_2) + \cdots + E(M_n)$$

$$\sigma^2 = \text{Var}(M_1) + \text{Var}(M_2) + \cdots + \text{Var}(M_n) \qquad (7.70)$$

Under certain conditions (see, e.g., Feller or Uspensky), the limiting distribution of $M = M_1 + M_2 + \cdots + M_n$ is the normal distribution, irrespective of the shape of the densities $f_{M_j}(m_j)$. If all M_j have identical distributions with mean μ and variance σ^2, then the distribution of M can be approximated by the normal curve $N(n\mu, n\sigma^2)$ (central limit theorem).

The central limit theorem can be illustrated by comparing the exact distribution of M with the corresponding normal distribution. Formula (7.63) gives the probability density of the maximum moment, when the component moments had a uniform distribution in the interval $(0, 1)$:

$$E(M_i) = \tfrac{1}{2} \qquad \sigma_{M_i}^2 = \tfrac{1}{12}$$

For $n = 2$ we obtained M having the triangular density with

$$E(M) = 1 \qquad \sigma_M^2 = \tfrac{1}{6}$$

so that the approximating normal curve is

$$f_M(m) = \frac{\sqrt{3}}{\sqrt{\pi}} \exp\left[-3(m - 1)^2\right]$$

For $n = 3$ we have (an exercise for the reader)

$$E(M) = \tfrac{3}{2} \qquad \sigma_M^2 = \tfrac{1}{4} \qquad (7.71)$$

the approximating curve being

$$f_M(m) = \sqrt{\frac{2}{\pi}} \exp\left[-2\left(m - \frac{3}{2}\right)^2\right]$$

which, as seen in Fig. 7.10, is a very good approximation of the exact distribution.

The reliability of a structure, under n concentrated moments, is then approximated as

$$R \simeq \tfrac{1}{2} + \text{erf}\left(\frac{\sigma_{\text{allow}}S - \eta}{\sigma}\right)$$

where η and σ are determined by Eq. (7.70).

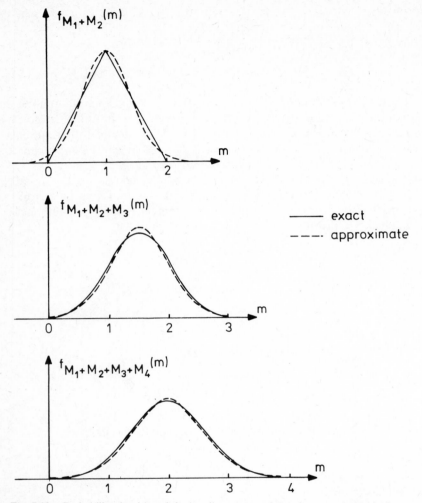

Fig. 7.10.   Probability densities of the bending moment in section at a clamped edge.

In the reliability calculation, the random variables governing the problem are often treated as though they were uncorrelated. The risk of loss of covariances was illustrated in Example 6.12. The following example illustrates the error resulting from such omissions, for the reliability calculations.

Example 7.12

Consider a cantilever subjected to a pair of random concentrated moments M_1 and $M_2 = \alpha M_1$, acting in the same cross section. M_1 and M_2 are jointly normally distributed random variables, whereas the allowable stress σ_{allow}, the section modulus S, and parameter α are deterministic quantities. We seek the error resulting from the assumption of uncorrelatedness of M_1 and M_2.

The maximum bending moment in the cantilever is $M_1 + M_2$, normally distributed with mean

$$E(M_1 + M_2) = (1 + \alpha)E(M_1)$$

and variance

$$\sigma^2_{M_1 + M_2} = \sigma^2_{M_1} + 2\,\text{Cov}(M_1, M_2) + \sigma^2_{M_2} = (1 + \alpha)^2 \sigma^2_{M_1}$$

since

$$\sigma^2_{M_2} = E\left[\alpha M_1 - E(\alpha M_1)\right]^2 = \alpha^2 \sigma^2_{M_1}$$

$$\text{Cov}(M_1, M_2) = E\{[M_1 - E(M_1)][M_2 - E(M_2)]\} = \alpha \sigma^2_{M_1}$$

The reliability of the cantilever is (without omitting $\text{Cov}(M_1 M_2)$)

$$R^{(\text{exact})} = P\left(\frac{|M_1 + M_2|}{S} \leqslant \sigma_{\text{allow}}\right) = P(-\sigma_{\text{allow}}S \leqslant |M_1 + M_2| \leqslant \sigma_{\text{allow}}S)$$

$$= \text{erf}\left[\frac{\sigma_{\text{allow}}S - (1 + \alpha)E(M_1)}{|1 + \alpha|\sigma_{M_1}}\right] - \text{erf}\left[-\frac{\sigma_{\text{allow}}S - (1 + \alpha)E(M_1)}{|1 + \alpha|\sigma_{M_1}}\right].$$

With $\text{Cov}(M_1, M_2)$ omitted, the approximate values of the variance of the maximum bending moment is

$$\tilde{\sigma}^2_{M_1 + M_2} = \sigma^2_{M_1} + \sigma^2_{M_2} = (1 + \alpha^2)\sigma^2_{M_1}$$

the corresponding reliability estimate being

$$R^{(\text{approximate})} = \text{erf}\left[\frac{\sigma_{\text{allow}}S - (1 + \alpha)E(M_1)}{(1 + \alpha^2)^{1/2}\sigma_{M_1}}\right]$$

$$- \text{erf}\left[-\frac{\sigma_{\text{allow}}S - (1 + \alpha)E(M_1)}{(1 + \alpha^2)^{1/2}\sigma_{M_1}}\right]$$

For $E(M_1)$ vanishing identically, the formulas for the reliabilities become

$$R^{(\text{exact})} = 2\,\text{erf}\left[\frac{\sigma_{\text{allow}}S}{|1 + \alpha|\sigma_{M_1}}\right] \qquad R^{(\text{approximate})} = 2\,\text{erf}\left[\frac{\sigma_{\text{allow}}S}{(1 + \alpha^2)^{1/2}\sigma_{M_1}}\right]$$

$$(7.72)$$

the associated percentage error, relative to the exact value being

$$\eta = \frac{R^{(\text{exact})} - R^{(\text{approximate})}}{R^{(\text{exact})}} \times 100\%$$

For large values of $\sigma_{\text{allow}} S / \sigma_{M_1}$, the error functions in both the exact and approximate expressions of the reliability are close to their asymptotic value 0.5 at infinity, and therefore the error resulting from omission of $\text{Cov}(M_1, M_2)$ is small. In other circumstances, however, it may be very large. Consider, for example, the case where $\alpha = -1$. Then Eq. (7.71) yields

$$R^{(\text{exact})} = 1$$

which is self-evident, since in this case $M_1 + M_2 = M_1 + (-1)M_1 = 0$, and the random event

$$\frac{|M_1 - M_2|}{S} \leq \sigma_{\text{allow}}$$

is a certain one, resulting in unity reliability.

Now, for $\sigma_{\text{allow}} S / \sigma_{M_1} = \sqrt{2}/2$,

$$R^{(\text{approximate})} = 2 \, \text{erf}(1/2) = 0.382925$$

the relative error being

$$\eta = \frac{1.0 - 0.382925}{1.0} \times 100\% = 61.7075\%$$

The relative error increases as $\sigma_{\text{allow}} S / \sigma_{M_1}$ decreases, reaching 100% for $\sigma_{\text{allow}} S / \sigma_{M_1} \to 0$.

PROBLEMS

7.1. Both $X \equiv \Sigma$ and $Y \equiv \Sigma_{\text{allow}}$ have uniform distributions as per formula (7.13). Find the reliability in the following cases:

(a) $\gamma < \alpha$ and $\alpha \leq \delta \leq \beta$

(b) $\alpha \geq \gamma, \beta \geq \delta$

7.2. Both $X \equiv \Sigma$ and $Y \equiv \Sigma_{\text{allow}}$ have truncated normal distributions, as in Example 7.5. Find the probability density $f_Z(z)$ of the difference $Z = X - Y$ for $z < 0$.

7.3. Given that the actual stress X and allowable stress Y are jointly normally distributed random variables with mathematical expectations m_1 and m_2, standard deviations σ_1 and σ_2, and correlation coefficient r.

Verify that

$$f_V(v) = \frac{\sqrt{1 - r^2}}{\pi} \left(\frac{\sigma_1 \sigma_2}{\sigma_1^2 - 2r\sigma_1\sigma_2 v + \sigma_2^2 v^2} \right)$$

$$\times \exp\left\{ -\frac{1}{2(1 - r^2)\sigma_1^2\sigma_2^2} \left[m_1^2\sigma_2^2 - 2rm_1 m_2\sigma_1\sigma_2 + m_2^2\sigma_1^2 \right] \right\}$$

$$\times \left[1 + \sqrt{\frac{\pi}{2}}\, t \exp\left(\frac{t^2}{2} \right) \mathrm{erf}(t) \right]$$

where

$$V = \frac{X}{Y} \qquad t = \frac{m_2\sigma_1^2 - rm_1\sigma_1\sigma_2 + m_1\sigma_2^2 v - rm_2\sigma_1\sigma_2 v}{\sigma_1\sigma_2\sqrt{(1 - r^2)(\sigma_2^2 v - 2r\sigma_1\sigma_2 v + \sigma_1^2)}}$$

In the particular case $m_1 = m_2 = 0$,

$$f_V(v) = \frac{\sqrt{1 - r^2}}{\pi} \left(\frac{\sigma_2/\sigma_1}{1 - 2r(\sigma_2/\sigma_1)v + (\sigma_2/\sigma_1)^2 v^2} \right)$$

With what density does the latter coincide for uncorrelated X and Y? Find the reliability of the structure.

7.4. Assume in Example 7.7

$$f_{MT}(m, t) = \frac{1}{2\pi\sigma_M\sigma_T} \exp\left[-\tfrac{1}{2}\left(\frac{m^2}{\sigma_M^2} + \frac{t^2}{\sigma_T^2} \right) \right]$$

that is, M and T are independent normal variables with zero mean and with variances σ_M^2 and σ_T^2, respectively. (a) Verify that

$$f_M(m) = \frac{m}{\sigma_M\sigma_T} \exp\left[-\frac{m^2(\sigma_M^2 + \sigma_T^2)}{4\sigma_M^2\sigma_T^2} \right] I_0\left(\frac{m^2|\sigma_M^2 - \sigma_T^2|}{4\sigma_M^2\sigma_T^2} \right)$$

and find the reliability of the shaft. (b) Repeat for the case where $\sigma_M = \sigma_T$.

7.5. Extend Example 7.7 to the case where the von Mises stress theory of failure is used instead of the maximum shear stress theory.

7.6. The clamped-clamped beam is subjected to a load as shown in the accompanying figure. Both Q and Σ_{allow} have log-normal distributions. Find the reliability of the beam.

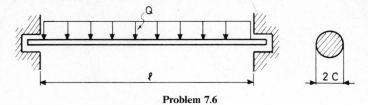

Problem 7.6

7.7. The cantilever is subjected to a pair of random independent moments with gamma distributions

$$f_{M_1}(m_1) = \frac{1}{\lambda^{\alpha+1}\Gamma(\alpha + 1)} m_1^\alpha e^{-m_1/\lambda} U(m_1)$$

$$f_{M_2}(m_2) = \frac{1}{\lambda^{\beta+1}\Gamma(\beta + 1)} m_2^\beta e^{-m_2/\lambda} U(m_2)$$

Show that the maximum moment also has a gamma distribution, and calculate the reliability.

7.8. For the system of Problem 7.7, assume that the moments have identical Rayleigh distribution

$$f_{M_1}(x) = f_{M_2}(x) = \frac{x}{a^2} \exp(-x^2/2a^2) U(x)$$

Verify that the reliability of the cantilever is

$$R = 1 - \exp(-z^2/2a^2) - \sqrt{\pi}\,\frac{z}{a}\exp(-z^2/4a^2)\text{erf}\left(\frac{z}{a\sqrt{2}}\right)$$

where $z = \sigma_{\text{allow}} S$.

7.9. A cantilever is subjected to three concentrated moments having uniform distribution in the interval $(0, 1)$. Show that

$$f_M(m) = \begin{cases} 0, & m < 0 \\ \frac{1}{2}m^2, & 0 \leqslant m < 1 \\ \frac{1}{2}m^2 - \frac{3}{2}(m-1)^2, & 1 \leqslant m < 2 \\ \frac{1}{2}m^2 - \frac{3}{2}(m-1)^2 + \frac{3}{2}(m-2)^2, & 2 \leqslant m < 3 \\ 0, & m \geqslant 3 \end{cases}$$

so that the graph consists of segments of three different parabolas, in the interval $(0, 3)$. Find the reliability of the structure, and compare with the estimate by the central limit theorem.

7.10. Prove Eq. (7.61).

7.11. The probability density (7.60) is referred to as a *generalized Erlang density of the first order*. Show by mathematical induction that the sum of n independent random variables, each of which has an exponential distribution with different parameters $\lambda_1, \lambda_2, \ldots, \lambda_n$, has a generalized Erlang distribution of $(n - 1)$th order, given by

$$f_{n-1}(m) = (-1)^{n-1} \prod_{i=1}^{n} \lambda_i \sum_{j=1}^{n} \frac{\exp(-\lambda_j m)}{\prod\limits_{k \neq j}^{n} (\lambda_j - \lambda_k)}$$

where the product $\prod_{k \neq j}^{n}$ is taken over $k = 1, 2, \ldots, j - 1, j + 1, \ldots, n$. The distribution function of M is

$$F_M(m) = (-1)^{n-1} \prod_{i=1}^{n} \lambda_i \sum_{j=1}^{n} \frac{1 - \exp(-\lambda_j m)}{\lambda_j \prod\limits_{k \neq j} (\lambda_j - \lambda_k)}.$$

7.12. Assume M and T in Example 7.7 to be uniformly distributed in the circle $m^2 + t^2 \leqslant p^2$ (see also Prob. 6.1), that is,

$$f_{MT}(m, t) = \begin{cases} \dfrac{1}{\pi p^2}, & m^2 + t^2 \leqslant p^2 \\ 0, & \text{otherwise} \end{cases}$$

Find the probability distribution and density functions of the maximum shear stress $\tau_{\max}$. Find the reliability of the circular shaft.

CITED REFERENCES

Bark, L. S., Bol'shev, M. N., Kuznetsov, P. I., and Tshernishev, A. P., *Tables of Rayleigh-Rice Distribution*, Computer Center of the Academy of Sciences of USSR, 1964.

Feller, W., *An Introduction to Probability Theory and its Applications*, 2nd ed., Vol. 1, John Wiley & Sons, New York, 1957.

Ferry Borges, J., and Castanheta, M., *Structural Safety*, 2nd ed., Nat. Civil Eng. Lab., Lisbon, Portugal, 1971.

Freudenthal, A. M., "The Safety and the Probability of Failure," *Trans. ASCE*, **121**, 1337–1397 (1956).

Freudenthal, A. B., Garrelts, J. M., and Shinozuka, M., "The Analysis of Structural Safety," *J. Structural Div.*, *Proc. ASCE*, **92** (ST1), 267–325 (1966).

Freudenthal, A. M., *Critical Appraisal of Safety Criteria and their Basic Concepts*, AIBSE, New York, 1968, pp. 13–25.

Julian, O. G., "Synopsis of First Progress Report of the Committee on Factors of Safety," *J. Struct. Div., Proc. ACSE*, **83** (ST4), 1316.1–1316.22 (1957).

Uspensky, J. V., *Introduction to Mathematical Probability*, McGraw-Hill, New York, 1937.

RECOMMENDED FURTHER READING

A large number can be found in:

"Structural Safety—A Literature Review" by the Task Committee of Structural Safety of the Administrative Committee on Analysis and Design of the Structural Division (A. H. -S. Ang., Chairman), *J. Structural Div., Proc. ASCE*, **98** (ST 4), 845–884 (1972).

chapter **8**

Elements of the Theory of Random Functions

This chapter deals with some basic aspects of the random function, which is in essence a generalization of the random variable.

8.1 DEFINITION OF A RANDOM FUNCTION

The random function was in fact already encountered in Sec. 5.7, where the initial imperfection of a beam simply supported at its ends was assumed to be codirectional with the axial half-wave $G \sin \pi\xi$. G is a random variable, which signifies that to each outcome ω of the experiment, we assign (according to a certain rule) a number $G(\omega)$, so as to have a corresponding function

$$\overline{U}(\xi, \omega) = G(\omega)\sin \pi\xi \qquad (8.1)$$

We have thus generated a set of functions (one for each ω), three members of which are shown in Fig. 8.1. The set is called a *random function* of the coordinate ξ.

An alternative approach consists in specifying the coordinate ξ_i instead of the outcome ω_i. For a specific ξ_i we obtain random variables $\overline{U}(\xi_i, \omega) = G(\omega)\sin \pi\xi_i$, namely,

$$\overline{U}(\xi_1, \omega) = G(\omega)\sin \pi\xi_1 \qquad \overline{U}(\xi_2, \omega) = G(\omega)\sin \pi\xi_2 \quad \cdots$$

$$\overline{U}(\xi_N, \omega) = G(\omega)\sin \pi\xi_N \qquad (8.2)$$

The resulting set is again a random function, denoted $\{\overline{U}(\xi), \; \xi \in [0, 1]\}$ or, more simply, $\overline{U}(\xi)$.

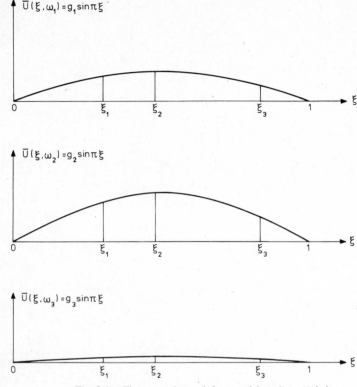

Fig. 8.1.   Three members of the set of functions $U(\xi, \zeta)$.

Similarly, the random additional displacement of a beam as per Eqs. (5.90) and (5.92) can be denoted

$$\{U(\xi, \lambda): \xi \in [0, 1], \lambda \in [0, \infty)\}$$

or, more simply, $U(\xi, \lambda)$. The maximum additional displacement, given by Eq. (5.92), is a random function of time λ only and can be denoted $E(\lambda)$. The random functions of time are often referred to as *random processes*.

With these preliminary remarks, we proceed to the formal definition of a random function:

A random function $X(t, \omega)$ is a set of random variables $\{X(t, \omega): t \in J\}$ indexed by a scalar or vector parameter t belonging to an index set J.

A random function can be visualized in four ways:

1. A genuine random function (both ω and t variable)
2. A deterministic function of t (ω specified, t variable)
3. A random number (ω variable, t specified).
4. A deterministic number (both ω and t specified)

Subsequently, we denote the random function $X(t, \omega)$ in abbreviated form $X(t)$, omitting the dependence on ω.

8.2 FIRST- AND SECOND-ORDER DISTRIBUTION FUNCTIONS

For specified t, the random function $X(t)$ is a random variable whose distribution function generally depends on t and is denoted $F_X(x; t)$:

$$F_X(x; t) = P\{X(t) \leqslant x\} \tag{8.3}$$

$F_X(x; t)$ is the probability of the event $\{X(t) \leqslant x\}$ consisting of all outcomes such that for a given t the realization of the random variable does not exceed the given number x. $F_X(x; t)$ is referred to as *a first-order distribution* of the function $X(t)$. The first-order probability density function denoted by $f_X(x; t)$ is defined as

$$f_X(x; t) = \frac{\partial F_X(x, t)}{\partial x} \tag{8.4}$$

Example 8.1

The random function $X(t)$ represents the random variable $X(t) = Y$, where Y is a continuous random variable with the probability distribution function $F_Y(y)$. Then

$$F_X(x; t) = P\{X(t) \leqslant x\} = P\{Y \leqslant x\} = F_Y(x)$$

and therefore

$$f_X(x; t) = f_Y(x)$$

where $f_Y(y)$ is the probability density of Y.

Example 8.2

The random function $X(t)$ is given by $X(t) = Y \sin \pi t$, where again $F_Y(y)$ is given and t is a deterministic parameter $t \in [\frac{1}{4}, \frac{1}{2}]$. Then

$$F_X(x; t) = P\{Y \sin \pi t \leqslant x\} = P\left\{Y \leqslant \frac{x}{\sin \pi t}\right\} = F_Y\left(\frac{x}{\sin \pi t}\right)$$

and

$$f_X(x; t) = \frac{1}{\sin \pi t} f_Y\left(\frac{x}{\sin \pi t}\right), \qquad f_Y(y) = F_Y'(y)$$

The first-order distribution and density functions represent the simplest characteristics of the random function $X(t)$. More complete characterization is obtainable by considering a pair of random variables $X(t_1)$ and $X(t_2)$. Their joint distribution function generally depends on t_1 and t_2. The *second-order distribution function*, denoted by $F_X(x_1, x_2; t_1, t_2)$ is defined by

$$F_X(x_1, x_2; t_1, t_2) = P\{X(t_1) \leqslant x_1, X(t_2) \leqslant x_2\} \tag{8.5}$$

The second-order probability density of the random function $X(t)$ is given by

$$f_X(x_1, x_2; t_1, t_2) = \frac{\partial^2 F_X(x_1, x_2; t_1, t_2)}{\partial x_1 \, \partial x_2} \tag{8.6}$$

The first-order distribution functions are obtainable from the second-order one as follows

$$F_X(x_1; t_1) = F_X(x_1, \infty; t_1, t_2)$$

$$F_X(x_2; t_2) = F_X(\infty, x_1; t_1, t_2) \tag{8.7}$$

and the corresponding relations for the densities are

$$f_X(x_1; t_1) = \int_{-\infty}^{\infty} f_X(x_1, x_2; t_1, t_2) \, dx_2$$

$$f_X(x_2; t_2) = \int_{-\infty}^{\infty} f_X(x_1, x_2; t_1, t_2) \, dx_1 \tag{8.8}$$

Example 8.3

We seek the second-order distribution function for the random function $X(t)$ in Example 8.1. We have

$$F_X(x_1, x_2; t_1, t_2) = P\{X(t_1) \leqslant x_1, X(t_2) \leqslant x_2\} = P\{Y \leqslant x_1, Y \leqslant x_2\}$$

If $x_1 < x_2$, then

$$F_X(x_1, x_2; t_1, t_2) = P\{Y \leqslant x_1\} = F_Y(x_1)$$

If, however, $x_2 \leqslant x_1$, then

$$F_X(x_1, x_2; t_1, t_2) = P\{Y \leqslant x_2\} = F_Y(x_2)$$

For the random function $X(t)$ described in Example 8.2, we have

$$F_X(x_1, x_2; t_1, t_2) = \begin{cases} F_Y\left(\dfrac{x_1}{\sin \pi t_1}\right), & \dfrac{x_1}{\sin \pi t_1} < \dfrac{x_2}{\sin \pi t_2} \\[3mm] F_Y\left(\dfrac{x_2}{\sin \pi t_2}\right), & \dfrac{x_1}{\sin \pi t_1} \geqslant \dfrac{x_2}{\sin \pi t_2} \end{cases}$$

8.3 MOMENT FUNCTIONS

The mathematical expectations $\eta(t)$ or the mean of a random function $X(t)$ is defined as the mathematical expectation of the random variable $X(t)$ for a fixed t:

$$\eta(t) = E[X(t)] = \int_{-\infty}^{\infty} x f_X(x; t) \, dx \tag{8.9}$$

and generally is a function of t.

The joint moment of the random variables $X(t_1)$ and $X(t_2)$ is called the *autocorrelation* function $R_X(t_1, t_2)$

$$R_X(t_1, t_2) = E[X(t_1)X(t_2)] = \int_{-\infty}^{\infty} \int_{-\infty}^{\infty} x_1 x_2 f_X(x_1, x_2; t_1, t_2) \, dx_1 \, dx_2 \tag{8.10}$$

and is generally a function of t_1 and t_2.

The covariance of the random variables $X(t_1)$ and $X(t_2)$ is called the *autocovariance* function $C_X(t_1, t_2)$

$$C_X(t_1, t_2) = E\{[X(t_1) - \eta(t_1)][X(t_2) - \eta(t_2)]\}$$

$$= \int_{-\infty}^{\infty} \int_{-\infty}^{\infty} [x_1 - \eta(t_1)][x_2 - \eta(t_2)] f_X(x_1, x_2; t_1, t_2) \, dx_1 \, dx_2 \tag{8.11}$$

The variance of the random variable $X(t)$, σ_X^2 is called the *variance of the random function* $X(t)$:

$$\sigma_X^2(t) = C_X(t, t) = R_X(t, t) - \eta^2(t) \tag{8.12}$$

Example 8.4

For the random function $X(t)$ in Example 8.1, we have

$$\eta(t) = E(Y)$$

$$R_X(t_1, t_2) = E(Y^2)$$

$$C_X(t_1, t_2) = \text{Var}(Y)$$

$$\sigma_X^2(t) = \text{Var}(Y)$$

For $X(t)$ in Example 8.2 we have

$$\eta(t) = E(Y)\sin \pi t$$

$$R_X(t_1, t_2) = E(Y^2)\sin \pi t_1 \sin \pi t_2$$

$$C_X(t_1, t_2) = \text{Var}(Y)\sin \pi t_1 \sin \pi t_2$$

$$\sigma_X^2(t) = \text{Var}(Y)\sin^2 \pi t$$

8.4 PROPERTIES OF THE AUTOCOVARIANCE FUNCTION

The autocovariance function has the following properties:

1. As follows from definition (8.11), it is a symmetric function of its arguments:

$$C_X(t_1, t_2) = C_X(t_2, t_1) \tag{8.13}$$

2. From property (6.72) of the covariance of a pair of random variables, it follows that

$$|C_X(t_1, t_2)| \leqslant \{\text{Var}[X(t_1)]\text{Var}[X(t_2)]\}^{1/2} = [C_X(t_1, t_1)C_X(t_2, t_2)]^{1/2} \tag{8.14}$$

This inequality enables us to introduce the normalized autocovariance function

$$c_X(t_1, t_2) = \frac{C_X(t_1, t_2)}{[C_X(t_1, t_1)C_X(t_2, t_2)]^{1/2}} = \frac{C_X(t_1, t_2)}{\sigma_X(t_1)\sigma_X(t_2)}$$

which, in view of Eq. (8.14), in turn satisfies the inequality

$$|c_X(t_1, t_2)| \leqslant 1 \tag{8.16}$$

Note that for $t_1 = t_2$

$$c_X(t_1, t_1) = c_X(t_2, t_2) = 1$$

3. The autocovariance function is nonnegative definite. For any deterministic function $\varphi(t)$, the following inequality holds:

$$\int_{-\infty}^{\infty}\int_{-\infty}^{\infty} C_X(t_1, t_2)\varphi(t_1)\varphi(t_2)\, dt_1\, dt_2 \geqslant 0 \tag{8.17}$$

This property, analogous to the nonnegative definiteness of the variance-covariance matrix as per Eq. (6.90), will be proved later on.

8.7 PROBABILITY DENSITY FUNCTION

The *probability distribution function of nth order* is defined as the probability distribution function of n random variables $X(t_1), X(t_2), \ldots, X(t_n)$:

$$F_X(x_1, x_2, \ldots, x_n; t_1, t_2, \ldots, t_n)$$

$$= P\{X(t_1) \leqslant x_1, X(t_2) \leqslant x_2, \ldots, X(t_n) \leqslant x_n\} \qquad (8.18)$$

which is a function of $2n$ variables.

If the function $F_X(x_1, x_2, \ldots, x; t_1, t_2, \ldots, t_n)$ has a derivative

$$\frac{\partial^n F_X(x_1, x_2, \ldots, x_n; t_1, t_2, \ldots, t_n)}{\partial x_1 \, \partial x_2 \, \cdots \, \partial x_n} = f_X(x_1, x_2, \ldots, x_n; t_1, t_2, \ldots, t_n)$$

$$(8.19)$$

this derivative is called the *probability density function* of nth order of $X(t)$. The set of functions $f_X(x_1; t_1), f_X(x_1, x_2; t_1, t_2), \ldots, f_X(x_1, x_2, \ldots, x_n; t_1, t_2, \ldots, t_n)$ represents the probabilistic properties of $X(t)$ in one, two, or n coordinates (including that of time, if such is the case). The probability density of nth order yields all those of lower order, except for some "pathological" cases. The random function $X(t)$ is determined probabilistically when the distribution function of any order is known. Occasionally, all of the probabilistic information is contained in the first-order probability density. The simplest example is the function whose values $X_1, X_2, \ldots, X_n$ at the respective noncoincident coordinates $t_1, t_2, \ldots, t_n$ ($t_i \neq t_j$, $i, j = 1, 2, \ldots, n$) are independent random variables. This means that the nth order probability density is a product of the first-order probability densities:

$$f_X(x_1, x_2, \ldots, x_n; t_1, t_2, \ldots, t_n) = \prod_{j=1}^{n} f_X(x_j; t_j) \qquad (8.20)$$

Consider now the processes where the second-order probability density suffices for complete characterization of a random function $X(t)$. The latter is called a *Markov random function* if the probabilistic behavior in the "future" depends solely on the "most recent past," that is, on the present state. This may be written in terms of conditional probability densities as follows:

$$f_X(x_n; t_n | x_1, \ldots, x_{n-1}; t_1, \ldots, t_{n-1})$$

$$= f_X(x_n, t_n | x_{n-1}; t_{n-1}), \quad t_1 < t_2 < \cdots < t_{n-1} < t_n \qquad (8.21)$$

or

$$f_X(x_1, x_2, \ldots, x_n; t_1, t_2, \ldots, t_n) = f_X(x_1, t_1) \prod_{j=2}^{n} f_X(x_j; t_j | x_{j-1}; t_{j-1})$$

$$(8.22)$$

Another class of random function where only the second-order probability densities are needed is that of the *random functions with independent increments*, in which for nonoverlapping $t_1 < t_2 < \cdots < t_n$, the random variables $X(t_2) - X(t_1)$, $X(t_3) - X(t_2), \ldots, X(t_n) - X(t_{n-1})$ are independent. To know the probability densities of any order for such a random function, it suffices to know those of $X(t)$ and of $X(t_j) - X(t_{j-1})$, that is, those of first and second order.

8.6 NORMAL RANDOM FUNCTION

A random function is said to be normal if the random variables $X(t_j)$, $j = 1, 2, \ldots, n$, are jointly normal, that is, if the nth-order probability density is given:

$$f_X(x_1, x_2, \ldots, x_n; t_1, t_2, \ldots, t_n)$$

$$= \frac{1}{\langle (2\pi)^n \det[V] \rangle^{1/2}} \exp\left\langle -\tfrac{1}{2}(\{X\} - \{m\})^T [V]^{-1}(\{X\} - \{m\}) \right\rangle$$

$$(8.23)$$

where

$$\{X\}^T = \begin{bmatrix} X(t_1) & X(t_2) & \cdots & X(t_n) \end{bmatrix}$$

$$\{m\}^T = \begin{bmatrix} \eta(t_1) & \eta(t_2) & \cdots & \eta(t_n) \end{bmatrix}$$

$$[V] = E\left\langle (\{X\} - \{m\})(\{X\} - \{m\})^T \right\rangle = [v_{jk}]_{n \times n} \qquad (8.24)$$

is the variance-covariance matrix with elements

$$v_{jk} = E\{X(t_j) - E[X(t_j)]\}\{X(t_k) - E[X(t_k)]\} = C_X(t_j, t_k)$$

The first-order density is given by

$$f_X(x; t) = \frac{1}{[2\pi C_X(t, t)]^{1/2}} \exp\left[-\frac{(x - E[X(t)])^2}{2C_X(t, t)} \right] \qquad (8.25)$$

It is evident that the mean function $\eta(t)$ and the autocovariance function $C_X(t_1, t_2)$ determine completely the normal random function.

8.7 JOINT DISTRIBUTION OF RANDOM FUNCTIONS

Sometimes we have to deal with two or more random functions. Consider first a pair of such functions $X(t)$ and $Y(t)$. The joint probability distribution of the random variables $X(t_1)$, $X(t_2),\ldots$, $X(t_n)$, $Y(\bar{t_1})$, $Y(\bar{t_2}),\ldots$, $Y(\bar{t_m})$ is defined as

$$F_{XY}\left(x_1, x_2,\ldots, x_n, y_1, y_2,\ldots, y_m; t_1, t_2,\ldots, t_n, \bar{t_1}, \bar{t_2},\ldots, \bar{t_m}\right)$$

$$= P\{ X(t_1) \leqslant x_1, X(t_2) \leqslant x_2,\ldots, X(t_n) \leqslant x_n, Y(\bar{t_1})$$

$$\leqslant y_1, Y(t_2) \leqslant y_2,\ldots, Y(\bar{t_m}) \leqslant y_m\} \qquad (8.26)$$

If this function has a derivative

$$\frac{\partial^{n+m} F_{XY}}{\partial x_1\, \partial x_2\, \cdots\, \partial x_n\, \partial y_1\, \partial y_2\, \cdots\, \partial y_m}$$

$$= f_{XY}\left(x_1, x_2,\ldots, x_n, y_1, y_2,\ldots, y_m; t_1, t_2,\ldots, t_n, \bar{t_1}, \bar{t_2},\ldots, \bar{t_m}\right)$$

$$(8.27)$$

the latter is called the joint, $(n + m)$th-order probability density of $X(t)$ and $Y(t)$. The latter are independent, if for arbitrary n and m

$$f_{XY}\left(x_1, x_2,\ldots, x_n, y_1, y_2,\ldots, y_m; t_1, t_2,\ldots, t_n, \bar{t_1}, \bar{t_2},\ldots, \bar{t_m}\right)$$

$$= f_X\left(x_1, x_2,\ldots, x_n; t_1, t_2,\ldots, t_n\right) f_Y\left(y_1, y_2,\ldots, y_m; \bar{t_1}, \bar{t_2},\ldots, \bar{t_m}\right)$$

$$(8.28)$$

The *cross-correlation function* of two random function $X(t)$ and $Y(t)$ is defined as

$$R_{XY}(t_1, t_2) = E\left[X(t_1) Y(t_2) \right] = \int_{-\infty}^{\infty} \int_{-\infty}^{\infty} xy f_{XY}(x, y; t_1, t_2)\, dx\, dy \qquad (8.29)$$

If the random functions are independent, then, in view of Eq. (8.28) we have

$$R_{XY}(t_1, t_2) = \int_{-\infty}^{\infty} x f_X(x; t_1)\, dx \int_{-\infty}^{\infty} y f_Y(y; t_2)\, dy$$

$$= \eta_X(t_1) \eta_Y(t_2) \qquad (8.30)$$

where $\eta_x(t)$ and $\eta_y(t)$ are the mean functions of $X(t)$ and $Y(t)$, respectively.

The *cross-covariance function* is defined as

$$C_{XY}(t_1, t_2) = E\{[X(t_1) - \eta_X(t_1)][Y(t_2) - \eta_Y(t_2)]\}$$

$$= R_{XY}(t_1, t_2) - \eta_X(t_1)\eta_Y(t_2) \qquad (8.31)$$

and the *cross-covariance* and *cross-correlation matrices*, respectively, are

$$\begin{bmatrix} C_X(t_1, t_2) & C_{XY}(t_1, t_2) \\ C_{YX}(t_1, t_2) & C_Y(t_1, t_2) \end{bmatrix}, \quad \begin{bmatrix} R_X(t_1, t_2) & R_{XY}(t_1, t_2) \\ R_{YX}(t_1, t_2) & R_Y(t_1, t_2) \end{bmatrix} \qquad (8.32)$$

Obviously,

$$C_{XY}(t_1, t_2) = C_{YX}(t_2, t_1) \qquad (8.33)$$

The cross-covariance function $C_{XY}(t_1, t_2)$ has the following property stemming from Eq. (6.72), which is valid for any correlation moment:

$$|C_{XY}(t_1, t_2)| \leqslant \sigma_X(t_1)\sigma_Y(t_2) = [C_X(t_1, t_1)C_Y(t_2, t_2)]^{1/2} \qquad (8.34)$$

A set of n random functions $X_1(t), X_2(t), \ldots, X_n(t)$ may be treated as components of the vector random function $\{X(t)\}$, with the cross-covariance matrix

$$[C_X(t_1, t_2)] = E\langle (\{X(t_1)\} - \{\eta(t_1)\})(\{X(t_2)\} - \{\eta(t_2)\})^T \rangle$$

$$= [C_{X_j X_k}(t_1, t_2)]_{n \times n}$$

where T is a transpose operation, $C_{X_j}(t_1, t_2)$ are the autocovariances, and $C_{X_j X_k}(t_1, t_2)$ the cross-covariances, defined in complete analogy with Eq. (8.31) as

$$C_{X_j X_k}(t_1, t_2) = R_{X_j X_k}(t_1, t_2) - \eta_{X_j}(t_1)\eta_{X_k}(t_2)$$

$R_{X_j X_k}(t_1, t_2)$ forms the cross-correlation matrix $[R_X(t_1, t_2)]$.

Example 8.5

We are given two random functions

$$X(t) = \sum_{i=1}^{n} G_i \sin i\pi t, \qquad Y(t) = \sum_{i=1}^{n} G_i i\pi \cos i\pi t$$

where G_i are the random variables with given probabilistic properties. The elements of the cross-covariance matrix are

$$C_X(t_1, t_2) = \sum_{i=1}^{n} \sum_{j=1}^{n} \text{Cov}(G_i, G_j) \sin i\pi t_1 \sin j\pi t_2$$

$$C_{XY}(t_1, t_2) = \sum_{i=1}^{n} \sum_{j=1}^{n} \text{Cov}(G_i, G_j) j\pi \sin i\pi t_1 \cos j\pi t_2$$

$$C_{YX}(t_1, t_2) = \sum_{i=1}^{n} \sum_{j=1}^{n} \text{Cov}(G_i, G_j) i\pi \cos i\pi t_1 \sin j\pi t_2$$

$$C_Y(t_1, t_2) = \sum_{i=1}^{n} \sum_{j=1}^{n} \text{Cov}(G_i, G_j) ij\pi^2 \cos i\pi t_1 \cos j\pi t_2$$

8.8   COMPLEX RANDOM FUNCTIONS

A *complex random function*

$$Z(t) = X(t) + iY(t) \tag{8.35}$$

is defined as a set of complex random variables. Its mean function $E[Z(t)]$ is

$$E[Z(t)] = E[X(t)] + iE[Y(t)] \tag{8.36}$$

and its autocorrelation function is

$$R_Z(t_1, t_2) = E[Z(t_1)Z^*(t_2)] \tag{8.37}$$

where the asterisk denotes the complex conjugate. By this latter definition, the mean square value of the complex random function is nonnegative,

$$R_Z(t, t) = E[Z(t)Z^*(t)] = E[|Z(t)|^2] \geqslant 0 \tag{8.38}$$

The *autocovariance* is defined as

$$C_Z(t_1, t_2) = R_Z(t_1, t_2) - \eta_Z(t_1)\eta_Z^*(t_2) \tag{8.39}$$

The autocovariance function of a complex random function has the following properties:

1.
$$C_Z(t_1, t_2) = C_Z^*(t_2, t_1) \tag{8.40}$$

In the particular case of a real function, we have Eq. (8.13).

2. For a complex random function $Z(t)$, we have

$$|\text{Re } C_z(t_1, t_2)| \leqslant [C_z(t_1, t_1) C_z(t_2, t_2)]^{1/2} \tag{8.41}$$

which is a generalization of property (8.14) of a real random function. Equation (8.41) immediately follows from the inequality

$$E\left\{\left|[C_Z^{-1}(t_1, t_1)]^{1/2}[Z(t_1) - \eta_Z(t_1)]\right.\right.$$

$$\left.\left. \pm [C_Z^{-1}(t_2, t_2)]^{1/2}[Z(t_2) - \eta_Z(t_2)]\right|^2\right\} \geqslant 0$$

since the expression under the operator of mathematical expectation is always nonnegative.

3. The analogue of property (8.17) is

$$\int_a^b \int_a^b C_Z(t_1, t_2)\varphi^*(t_1)\varphi(t_2)\, dt_1\, dt_2 \geqslant 0 \tag{8.42}$$

where $\varphi(t)$ is any deterministic complex function. This property will be proved later on.

The cross-correlation and cross-covariance functions of a pair of complex random functions $Z_1(t)$ and $Z_2(t)$ are defined, respectively, by

$$R_{Z_1 Z_2}(t_1, t_2) = E[Z_1(t_1) Z_2^*(t_2)]$$

$$C_{Z_1 Z_2}(t_1, t_2) = R_{Z_1 Z_2}(t_1, t_2) - \eta_{Z_1}(t_1)\eta_{Z_2}^*(t_2) \tag{8.43}$$

Note that

$$C_{Z_1 Z_2}(t_1, t_2) = C_{Z_2 Z_1}^*(t_2, t_1) \tag{8.44}$$

The transpose of a complex vector random function $\{Z(t)\}$ is

$$\{Z(t)\}^T = [Z_1(t) \quad Z_2(t) \quad \cdots \quad Z_n(t)]$$

and the cross-covariance matrix is

$$[C_Z(t_1, t_2)] = E\langle(\{Z(t_1)\} - \{\eta_Z(t_1)\})(\{Z(t_2)\} - \{\eta_Z(t_2)\})^{*T}\rangle$$

$$= [C_{Z_j Z_k}(t_1, t_2)]_{n \times n}$$

where $C_{Z_j Z_k}(t_1, t_2)$ are the cross-covariance functions,

$$C_{Z_j Z_k}(t_1, t_2) = R_{Z_j Z_k}(t_1, t_2) - \eta_{Z_j}(t_1)\eta_{Z_k}(t_2)$$

and $R_{Z_j Z_k}(t_1, t_2)$ forms the cross-correlation matrix $[R_Z(t_1, t_2)]_{n \times n}$.

Example 8.6

Given the complex random function

$$Z(t) = \varphi(t) + U\cos \omega t + i[\psi(t) + V\sin \omega t]$$

where $\varphi(t)$ and $\psi(t)$ are deterministic functions, ω is a real number, and U and V are random variables with

$$\text{Var}(U) = \sigma_U^2 \qquad \text{Var}(V) = \sigma_V^2 \qquad E(U) = E(V) = 0$$

The mean function is given by

$$E[Z(t)] = \varphi(t) + E(U)\cos \omega t + i[\psi(t) + E(V)\sin \omega t] = \varphi(t) + i\psi(t)$$

and the autocovariance by

$$C_Z(t_1, t_2) = E[(U\cos \omega t_1 + iV\sin \omega t_1)(U\cos \omega t_2 - iV\sin \omega t_2)]$$

$$= \sigma_U^2 \cos \omega t_1 \cos \omega t_2 + \sigma_V^2 \sin \omega t_1 \sin \omega t_2 + i\,\text{Cov}(U, V)\sin \omega(t_1 - t_2)$$

$$(8.45)$$

Example 8.7

Given two complex random functions

$$Z_1(t) = X_1(t) + iY_1(t) \qquad Z_2(t) = X_2(t) + iY_2(t)$$

with mean functions $\eta_1(t)$, $\eta_2(t)$, autocovariance functions $C_{Z_1}(t_1, t_2)$, $C_{Z_2}(t_1, t_2)$ and cross-covariance functions $C_{Z_1 Z_2}(t_1, t_2)$, $C_{Z_2 Z_1}(t_1, t_2)$. We seek the mean function $\eta(t)$ and autocovariance function $C_Z(t_1, t_2)$ of the sum of Z_1 and Z_2:

$$Z(t) = Z_1(t) + Z_2(t)$$

The mean function $\eta(t)$ is found as

$$\eta(t) = \eta_1(t) + \eta_2(t)$$

For the autocovariance function we have

$$C_Z(t_1, t_2) = E\{[Z_1(t_1) - \eta_1(t_1) + Z_2(t_1) - \eta_2(t_1)]$$

$$\times [Z_1^*(t_2) - \eta_1^*(t_2) + Z_2^*(t_2) - \eta_2^*(t_2)]\} \qquad (8.46)$$

$$= C_{Z_1}(t_1, t_2) + C_{Z_1 Z_2}(t_1, t_2) + C_{Z_2 Z_1}(t_1, t_2) + C_{Z_2}(t_1, t_2)$$

In components this equation will be written as

$$C_Z(t_1, t_2) = E\{[X_1(t_1) - EX_1(t_1) + X_2(t_1) - EX_2(t_1)]$$
$$+ i[Y_1(t_1) - EY_1(t_1) + Y_2(t_1) - EY_2(t_1)]\}$$
$$\times \{[X_1(t_2) - EX_1(t_2) + X_2(t_2) - EX_2(t_2)]$$
$$- i[Y_1(t_2) - EY_1(t_2) + Y_2(t_2) - EY_2(t_2)]\}$$
$$= C_{X_1}(t_1, t_2) + C_{X_1 X_2}(t_1, t_2) + C_{X_2 X_1}(t_1, t_2) + C_{X_2}(t_1, t_2)$$
$$+ C_{Y_1}(t_1, t_2) + C_{Y_1 Y_2}(t_1, t_2) + C_{Y_2 Y_1}(t_1, t_2) + C_{Y_2}(t_1, t_2)$$
$$+ i\big[C_{Y_1 X_1}(t_1, t_2) + C_{Y_1 X_2}(t_1, t_2) + C_{Y_2 X_1}(t_1, t_2) + C_{Y_2 X_2}(t_1, t_2)$$
$$- C_{X_1 Y_1}(t_1, t_2) - C_{X_1 Y_2}(t_1, t_2) - C_{X_2 Y_1}(t_1, t_2) - C_{X_2 Y_2}(t_1, t_2)\big]$$

$$(8.47)$$

If the functions are real, only the first line is retained in Eq. (8.47). A pair of random functions $Z_1(t)$ and $Z_2(t)$ are said to be *uncorrelated* if their cross-covariance is zero:

$$C_{Z_1 Z_2}(t_1, t_2) = 0 \qquad (8.48)$$

For an uncorrelated pair, Eq. (8.46) reads

$$C_Z(t_1, t_2) = C_{Z_1}(t_1, t_2) + C_{Z_2}(t_1, t_2) \qquad (8.49)$$

and in the case of real random functions, Eq. (8.47) yields

$$C_Z(t_1, t_2) = C_{X_1}(t_1, t_2) + C_{X_2}(t_1, t_2) \qquad (8.50)$$

That is, the autocovariance function of the sum of a pair of uncorrelated random functions equals the sum of the respective autocovariances.

8.9   STATIONARY RANDOM FUNCTIONS

The random function $X(t)$ is said to be *strictly stationary* if the probability density functions $f_{(X)}(x_1, x_2, \ldots, x_n; t_1, t_2, \ldots, t_n)$ of the random variables $X(t_1), X(t_2), \ldots, X(t_n)$, and $f_{(X)}(x_1, x_2, \ldots, x_n; t_1 + \varepsilon, t_2 + \varepsilon, \ldots, t_n + \varepsilon)$ of their counterparts $X(t_1 + \varepsilon), X(t_2 + \varepsilon), \ldots, X(t_n + \varepsilon)$ coincide for any n and $t_1, t_2, \ldots, t_n$ and are independent of ε. This means that the pair $X(t)$ and $X(t + \varepsilon)$ have identical nth-order probability densities for any t.

Analogously, we say that a pair $X(t)$ and $Y(t)$ are *jointly strictly stationary* if their $(m + n)$th-order joint probability density coincides with that of $X(t + \varepsilon)$ and $Y(t + \varepsilon)$ for any ε.

The complex function $Z(t) = X(t) + iY(t)$ is stationary if $X(t)$ and $Y(t)$ are stationary both individually and jointly.

For the first-order probability density of the stationary function, we have

$$f_X(x; t) = f_X(x; t + \varepsilon) \tag{8.51}$$

and since this must be true for any ε, it is true for $\varepsilon = -t$ and

$$f_X(x; t) = f_X(x; 0)$$

so that $f_X(x; t)$ is independent of t:

$$f_X(x; t) = f_X(x) \tag{8.52}$$

As a result, the mean function $\eta(t)$,

$$\eta(t) = \int_{-\infty}^{\infty} x f_X(x; t)\, dx = \int_{-\infty}^{\infty} x f_X(x)\, dx \equiv \eta \tag{8.53}$$

is time-invariant. The variance is also constant:

$$\mathrm{Var}[X(t)] = \sigma_X^2(t) = \int_{-\infty}^{\infty} (x - \eta)^2 f_X(x)\, dx = \mathrm{const} = \sigma_X^2 \tag{8.54}$$

The second-order probability density of a stationary random function satisfies the equality

$$f_X(x_1, x_2; t_1, t_2) = f_X(x_1, x_2; t_1 + \varepsilon, t_2 + \varepsilon)$$

Choosing $\varepsilon = -t_1$, we have

$$f_X(x_1, x_2; t_1, t_2) = f_X(x_1, x_2; 0, t_2 - t_1) \equiv f_X(x, x_2, t_2 - t_1)$$

and denoting $\tau = t_2 - t_1$,

$$f_X(x_1, x_2; t_1, t_2) = f_X(x_1, x_2, \tau) \tag{8.55}$$

so that $f_X(x_1, x_2, \tau)$ is the second-order probability density of the random variables $X(t_1)$ and $X(t_1 + \tau)$.

The autocorrelation and autocovariance functions of a stationary random function turn out to be dependent only on $t_2 - t_1$. In fact, in view of Eq.

(8.39),

$$R_X(t_1, t_2) = \int_{-\infty}^{\infty} \int_{-\infty}^{\infty} x_1 x_2^* f_X(x_1, x_2; t_1, t_2) \, dx_1 \, dx_2$$

$$= \int_{-\infty}^{\infty} \int_{-\infty}^{\infty} x_1 x_2^* f_X(x_1, x_2, \tau) \, dx_1 \, dx_2 = R_X(\tau)$$

$$C_X(t_1, t_2) = R_X(t_1, t_2) - \eta_X(t_1)\eta_X^*(t_2) = R_X(\tau) - \eta_X \eta_X^* = C_X(\tau)$$

$$(8.56)$$

The variance of $X(t)$ equals the value of the autocovariance function at $t_2 = t_1 = t$. Therefore,

$$\mathrm{Var}[X(t)] = C_X(t, t) = C_X(\tau)|_{\tau=0} = C_X(0) \qquad (8.57)$$

The autocovariance function of a stationary random function has the following properties:

1.
$$C_X(\tau) = C_X^*(-\tau) \qquad (8.58)$$

which follows immediately from Eq. (8.40). For a real function, this property becomes

$$C_X(\tau) = C_X(-\tau) \qquad (8.59)$$

That is, the autocorrelation function of a real random function is an even function.

2.
$$|\mathrm{Re}\, C_X(\tau)| \leqslant C_X(0) \qquad (8.60)$$

which follows from Eq. (8.41). For a real random function, this signifies

$$|C_X(\tau)| \leqslant C_X(0) \qquad (8.61)$$

that is, the variance is the upper bound of the autocovariance function.

3.
$$\int_a^b \int_a^b C_Z(t_2 - t_1)\varphi^*(t_1)\varphi(t_2) \, dt_1 \, dt_2 \geqslant 0 \qquad (8.62)$$

which follows from Eq. (8.42).

A random function is said to be *stationary in the wide sense (or weakly stationary)* if its mean function is constant and the autocovariance depends only on $t_2 - t_1$:

$$\eta(t) = \text{const} \qquad C_X(t_1, t_2) = C_X(\tau) \tag{8.63}$$

The random function $Z(t)$ in Example (8.6) is stationary in the wide sense if

$$\varphi(t) = \text{const} \qquad \psi(t) = \text{const}$$

$$E(U) = E(V) = 0 \qquad \sigma_U^2 = \sigma_V^2 \tag{8.64}$$

whence

$$E[Z(t)] = \text{const}$$

$$C_Z(t_1, t_2) = \sigma_U^2 \cos \omega (t_2 - t_1) - i \, \text{Cov}(U, V) \sin \omega (t_2 - t_1)$$

$$= \sigma_U^2 \cos \omega \tau - i \, \text{Cov}(U, V) \sin \omega \tau \tag{8.65}$$

While a strictly stationary random function is always stationary in the wide sense as well, the reverse is not necessarily the case. A stationary normal random function is stationary in both senses, and its higher order probability densities are uniquely determined by the mean and autocovariance functions.

A pair of random functions $X(t)$ and $Y(t)$ are *jointly stationary in the wide sense* if each of them is stationary in the wide sense. In addition, their cross-correlation function depends only on $t_2 - t_1$;

$$R_{XY}(t_1, t_2) = R_{XY}(t_2 - t_1) = R_{XY}(\tau) \tag{8.66}$$

Example 8.10

The initial imperfections $\overline{Y}_0(x)$ of an infinite beam are a stationary normal random function of the axial coordinate x with zero mean and the autocovariance function

$$C_{\overline{Y}_0}(\xi) = a^2 \exp(-b^2 \xi^2)$$

where $\xi = x_2 - x_1$, x_1 and x_2 being the axial coordinates, a^2 and b^2 positive constants. We seek the probability of $|\overline{Y}_0(x)| \leqslant \alpha$. We have $\sigma_{\overline{Y}_0}^2 = a^2$, and

$$f_{\overline{Y}_0}(\bar{y}_0) = \frac{1}{a\sqrt{2\pi}} \exp\left(-\frac{\bar{y}_0^2}{2a^2}\right)$$

The sought probability is

$$P = \int_{-\alpha}^{\alpha} f_{\overline{Y}_0}(\bar{y}_0) \, d\bar{y}_0 = \text{erf}\left(\frac{\alpha}{a}\right) - \text{erf}\left(-\frac{\alpha}{a}\right) = 2 \, \text{erf}\left(\frac{\alpha}{a}\right)$$

8.10 SPECTRAL DENSITY OF A STATIONARY RANDOM FUNCTION

The *spectral density* $S_X(\omega)$ of a stationary random function $X(t)$ is defined as the Fourier transform* of its autocorrelation function $R_X(\tau)$:

$$S_X(\omega) = \frac{1}{2\pi} \int_{-\infty}^{\infty} R_X(\tau) e^{-i\omega\tau} d\tau \qquad (8.67)$$

Resorting to Fourier inversion, we express $R_X(\tau)$ as

$$R_X(\tau) = \int_{-\infty}^{\infty} S_X(\omega) e^{i\omega\tau} d\omega \qquad (8.68)$$

Equations (8.67) and (8.68), which state that the autocorrelation function and the spectral density are interrelated through Fourier transformations, are called the *Wiener-Khintchine formulas*.

It is readily shown that the spectral density is a real function. In fact,

$$S_X(\omega) = \frac{1}{2\pi} \int_{-\infty}^{\infty} R_X(\tau) e^{-i\omega\tau} d\tau$$

$$= \frac{1}{2\pi} \int_{-\infty}^{\infty} \left[\text{Re}\, R_X(\tau)\cos\omega\tau + \text{Im}\, R_X(\tau)\sin\omega\tau \right] d\tau$$

$$+ i\frac{1}{2\pi} \int_{-\infty}^{\infty} \left[\text{Im}\, R_X(\tau)\cos\omega\tau - \text{Re}\, R_X(\tau)\sin\omega\tau \right] d\tau$$

However, $R_X(\tau) = R_X^*(-\tau)$, and hence the integrand in the second integral, is an odd function, so that the integral vanishes. Therefore,

$$S_X(\omega) = \frac{1}{2\pi} \int_{-\infty}^{\infty} \left[\text{Re}\, R_X(\tau)\cos\omega\tau + \text{Im}\, R_X(\tau)\sin\omega\tau \right] d\tau \qquad (8.69)$$

This expression indicates that $S_X(\omega)$ is a real function of ω. For a real random function $X(t)$, we have $\text{Im}\, R_X(\tau) = 0$, $\text{Re}\, R_X(\tau) = R_X(\tau)$, and

$$S_X(\omega) = \frac{1}{2\pi} \int_{-\infty}^{\infty} R_X(\tau)\cos\omega\tau \, d\tau \qquad (8.70)$$

and since $R_X(\tau)$ is then an even function of τ, $S_X(\omega)$ is even too:

$$S_X(-\omega) = S_X(\omega) \qquad (8.71)$$

For a real $X(t)$, then, the Wiener-Khintchine formula (8.68) yields

$$R_X(\tau) = \int_{-\infty}^{\infty} S_X(\omega)\cos\omega\tau \, d\omega \qquad (8.72)$$

*It should be noted that the transform may be variously defined in different literature sources.

Another basic property of the spectral density of a stationary random function is that its integral equals the mean square of $X(t)$. Indeed, since due by Eq. (8.56), $R_X(0) = E(|X|^2)$, we find, by substituting $\tau = 0$ in Eq. (8.68),

$$E(|X|^2) = \int_{-\infty}^{\infty} S_X(\omega)\, d\omega \qquad (8.73)$$

so that $S_X(\omega)$ represents the *mean square spectral density*. In other words, the mean square of the random process $X(t)$ equals the integral of the spectral density over all frequencies.

Analogously, the *cross-spectral density* of a pair of jointly stationary random functions is defined as the Fourier transform of their cross-correlation function (8.66):

$$S_{XY}(\omega) = \frac{1}{2\pi} \int_{-\infty}^{\infty} R_{XY}(\tau) e^{-i\omega\tau}\, d\tau \qquad (8.74)$$

Bearing in mind Eq. (8.44), we readily show that

$$S_{XY}(\omega) = S_{YX}^{*}(\omega) \qquad (8.75)$$

That is, $S_{XY}(\omega)$ has a Hermitian property, a generalization of the symmetry property. $S_{XY}(\omega) = S_{YX}(\omega)$ for real-valued random functions.

The relevant representation of the cross-correlation function $R_{XY}(\tau)$ is again obtained through Fourier inversion:

$$R_{XY}(\tau) = \int_{-\infty}^{\infty} S_{XY}(\omega) e^{i\omega\tau}\, d\omega \qquad (8.76)$$

Letting $\tau = 0$ in the latter, we have

$$R_{XY}(0) = E[X(t)Y^{*}(t)] = \int_{-\infty}^{\infty} S_{XY}(\omega)\, d\omega$$

The spectral densities $S_X(\omega)$ and $S_Y(\omega)$ and their cross-spectral counterparts $S_{XY}(\omega)$ and $S_{YX}(\omega)$ form the *cross-spectral density matrix*

$$[S(\omega)] = \begin{bmatrix} S_X(\omega) & S_{XY}(\omega) \\ S_{YX}(\omega) & S_Y(\omega) \end{bmatrix} \qquad (8.77)$$

For a vector random function $\{X(t)\}$ we have

$$[S_X(\omega)] = \frac{1}{2\pi} \int_{-\infty}^{\infty} [R_X(\tau)] e^{-i\omega\tau}\, d\tau \qquad (8.78)$$

where $[R_X(\tau)] = [R_{X_j X_k}(\tau)]_{n \times n}$ is the cross-correlation matrix, and $[S_X(\omega)] = [S_{X_j X_k}(\omega)]_{n \times n}$ the cross-spectral density matrix.

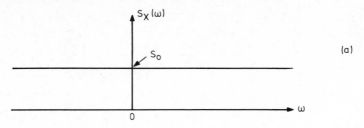

(a)

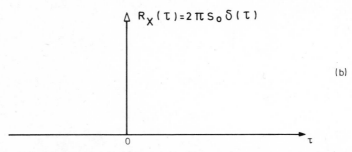

(b)

Fig. 8.2.   (a) Spectral density and (b) autocorrelation function of ideal white noise. The area under the spectral density curve is infinite.

Example 8.11

A stationary random function with constant spectral density

$$S_X(\omega) = S_0 = \text{const} \tag{8.79}$$

is called *ideal white noise** and is a useful concept because of its analytical simplicity. Its autocorrelation is given by

$$R_X(\tau) = \int_{-\infty}^{\infty} S_X(\omega) e^{i\omega\tau} \, d\omega = 2\pi S_0 \delta(\tau)$$

where $\delta(\tau)$ is the Dirac delta function. Since $R_X(\tau) = 0$ for $\tau \neq 0$, the random variables $X(t)$ and $X(t + \tau)$ for any nonzero τ are uncorrelated. Ideal white noise is physically unrealizable, since its mean square is infinite

$$E(|X|^2) = R_X(0) \to \infty \tag{8.80}$$

as is in fact the area under the spectral density curve (Fig. 8.2).

*The term was coined in analogy to "white light," which is characterized by an approximate uniform spectrum in the range of visible frequencies.

Example 8.12

A random function $X(t)$ with uniform spectral density in the interval $|\omega| < \omega_c$

$$S_X(\omega) = \begin{cases} S_0, & |\omega| < \omega_c \\ 0, & \text{otherwise} \end{cases} \tag{8.81}$$

is called *band-limited white noise*. Equation (8.72) yields

$$R_X(\tau) = S_0 \int_{-\omega_c}^{\omega_c} \cos \omega \tau \, d\omega = 2S_0 \frac{\sin \omega_c \tau}{\tau} \tag{8.82}$$

In contrast to its ideal counterpart, band-limited noise has a finite mean square:

$$E\left(|X|^2\right) = R_X(0) = 2S_0\omega_c \tag{8.83}$$

For a random process $X(t)$, ω_c is called the *cutoff frequency*. The spectral density and the autocorrelation functions of band-limited white noise are shown in Fig. 8.3.

Example 8.13

Consider a random function with the autocorrelation function

$$R_X(\tau) = d^2 e^{-\alpha|\tau|}\cos\Omega\tau \tag{8.84}$$

where α is a positive constant. The spectral density is

$$S_X(\omega) = \frac{d^2}{2\pi} \int_{-\infty}^{\infty} e^{-\alpha|\tau|}\cos\Omega\tau e^{-i\omega\tau} \, d\tau$$

$$= \frac{d^2}{4\pi} \left[\int_0^{\infty} \exp\{(-\alpha + i\Omega - i\omega)\tau\} \, d\tau \right.$$

$$+ \int_0^{\infty} \exp\{(-\alpha - i\Omega - i\omega)\tau\} \, d\tau$$

$$+ \left. \int_{-\infty}^0 \exp\{(\alpha + i\Omega - i\omega)\tau\} \, d\tau + \int_{-\infty}^0 \exp\{(\alpha - i\Omega - i\omega)\tau\} \, d\tau \right]$$

$$= \frac{d^2}{2\pi} \left[\frac{\alpha}{\alpha^2 + (\omega - \Omega)^2} + \frac{\alpha}{\alpha^2 + (\omega + \Omega)^2} \right]$$

$$= \frac{d^2\alpha}{\pi} \left[\frac{\omega^2 + \alpha^2 + \Omega^2}{\omega^4 + 2\omega^2(\alpha^2 - \Omega^2) + (\alpha^2 + \Omega^2)^2} \right] \tag{8.85}$$

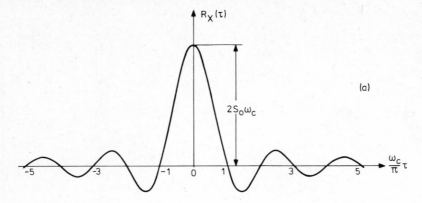

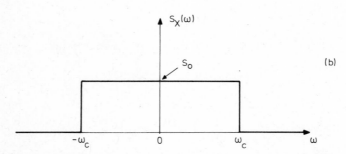

Fig. 8.3. (a) Autocorrelation function and (b) spectral density of band-limited white noise. The area under the spectral density curve (the mean square) equals $2S_0\omega_c$.

For $\alpha = 0$ we have

$$R_X(\tau) = d^2\cos\Omega\tau \tag{8.86}$$

and

$$S_X(\omega) = \frac{d^2}{2\pi}\int_{-\infty}^{\infty}\cos\Omega\tau e^{-i\omega\tau}\,d\tau$$

$$= \frac{d^2}{4\pi}\left[\int_{-\infty}^{\infty}e^{i(\Omega-\omega)\tau}\,d\tau + \int_{-\infty}^{\infty}e^{-i(\Omega+\omega)\tau}\,d\tau\right]$$

$$= \frac{d^2}{2}\left[\delta(\omega - \Omega) + \delta(\omega + \Omega)\right] \tag{8.87}$$

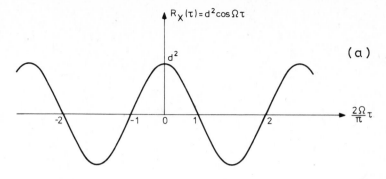

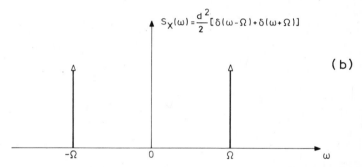

Fig. 8.4. (a) Autocorrelation function and (b) spectral density of the harmonic random function.

Note that the random function

$$X(t) = A \cos \Omega t + B \sin \Omega t \qquad (8.88)$$

where A and B are random variables such that

$$E(A) = E(B) = 0 \qquad \text{Var}(A) = \text{Var}(B) = d^2 \qquad E(AB) = 0 \quad (8.89)$$

has the autocorrelation function (8.86). For each possible pair of A and B, $X(t)$ varies harmonically; hence $X(t)$ as per Eq. (8.88) is called a *harmonic random function*. The corresponding $S_X(\omega)$ and $R_X(\tau)$ are shown in Fig. 8.4.

There are two distinct ranges of parameters, according as $S_X(\omega)$ (Eq. 8.85) has a single maximum or two symmetric maxima.

Range 1. For $\alpha^2 < 3\Omega$, $S_X(\omega)$ has two symmetric maxima at $\omega = \pm \bar{\omega}$, where

$$\bar{\omega} = (\alpha^2 + \Omega^2)^{1/4} \left[2\Omega - (\alpha^2 + \Omega^2)^{1/2} \right]^{1/2} > 0$$

The value of these maxima is given by

$$S_X(\bar{\omega}) = \frac{d^2\alpha}{4\pi\Omega\left[(d^2 + \Omega^2)^{1/2} - \Omega\right]}$$

Moreover, for $\alpha \ll 1$ and $\alpha \ll \Omega$, the spectral density is sharply peaked at the maxima. For such values of α, the random function $X(t)$ with the autocorrelation function (8.84) is close to being harmonic. We say that $X(t)$ is a *narrow-band random function*, indicating that its spectral density $S_X(\omega)$ has significant values only within a narrow interval of frequencies close to Ω (see Fig. 8.5).

Range 2.   For $\alpha^2 > 3\Omega$, $S_x(\omega)$ has a maximum at the origin $\omega = 0$ and decreases monotonically with increasing $|\omega|$. At the origin,

$$S_X(0) = \frac{d^2\alpha}{\pi(\alpha^2 + \Omega^2)}$$

A random function $X(t)$ with autocorrelation function (8.84) becomes *wide-band*, that is, its spectral density covers a wide interval of frequencies (Fig. 8.6).

For the particular case $\Omega = 0$ we have

$$R_X(\tau) = d^2 e^{-\alpha|\tau|} \qquad S_X(\omega) = \frac{d^2\alpha}{\pi}\frac{1}{\alpha^2 + \omega^2} \tag{8.90}$$

Denoting

$$d^2 = \pi\alpha S_0 \tag{8.91}$$

we have

$$S_X(\omega) = \frac{\alpha^2 S_0}{\alpha^2 + \omega^2}$$

For $\alpha \to \infty$, $S_X(\omega) \to S_0$, and according to Example 8.11, $R_X(\tau) \to 2\pi S_0\delta(\tau)$.

Realizations of wide-band and narrow-band random functions are shown in Fig. 8.7.

In different technical problems, associated with stationary random functions, use is often made of a so-called *correlation scale* defined as

$$\tau_k = \int_0^\infty |\rho(\tau)| \, d\tau$$

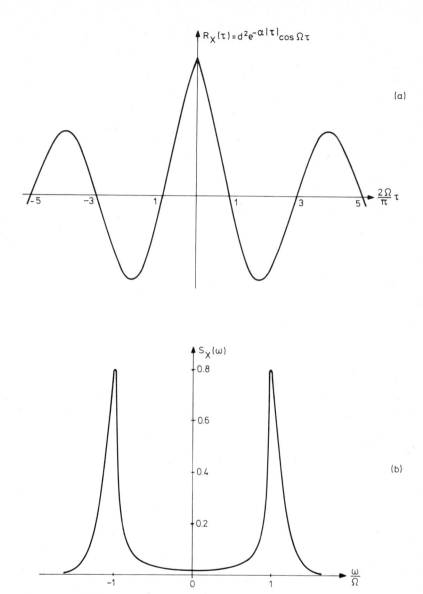

Fig. 8.5. (a) Autocorrelation function and (b) spectral density of the random function in the Example 8.13, $\alpha^2 < 3\Omega$ (range 1).

295

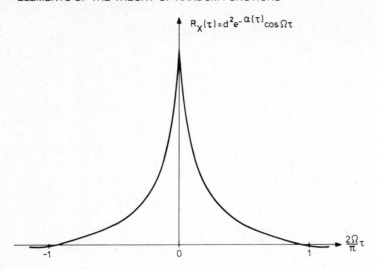

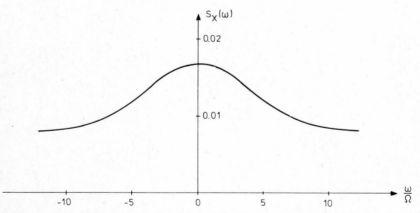

Fig. 8.6.   (a) Autocorrelation function and (b) spectral density of the random function in Example 8.13, $\alpha^2 > 3\Omega$ (range 2).

where $\rho(\tau)$ is a normalized autocorrelation function of a random function. For example, if

$$\rho(\tau) = e^{-\alpha|\tau|}, \quad \alpha > 0$$

the correlation scale is

$$\tau_k = \int_0^\infty e^{-\alpha\tau}\, d\tau = \frac{1}{\alpha}$$

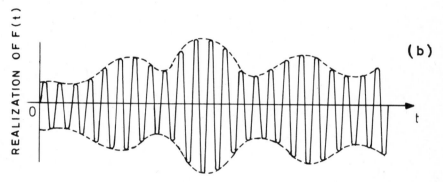

Fig. 8.7. Realization of (a) the wide-band stationary random function, (b) the narrow-band random function.

For α tending to zero, the random function reduces to a random variable with an infinite correlation scale. For α tending to infinity, the random function becomes an ideal white noise with a zero correlation scale.

8.11 DIFFERENTIATION OF A RANDOM FUNCTION

We first introduce the concept of *continuity* of a random function. As we already know, the latter may be viewed as a set of functions. If each of these is continuous at t, we can say that the random function itself is also continuous. As a less restrictive definition of continuity, we say that a random function $X(t)$ is continuous at t *in the mean square*, namely,

$$\lim_{\tau \to 0} E\{|X(t + \tau) - X(t)|^2\} = 0 \qquad (8.92)$$

A random function that is continuous at each t in some interval is called continuous in that interval. Denoting by $R_X(t_1, t_2)$ the autocorrelation function of $X(t)$, we have

$$E\{|X(t + \tau) - X(t)|^2\} = E\{[X(t + \tau) - X(t)][X(t + \tau) - X(t)]^*\}$$

$$= R_X(t + \tau, t + \tau) + R_X(t, t) - 2R_X(t, t + \tau)$$

$$(8.93)$$

so that a necessary and sufficient condition of continuity of $X(t)$ is continuity of $R_X(t_1, t_2)$ in t_1 and t_2 and $t_1 = t_2 = t$. In this case the left-hand side of Eq. (8.93) tends to zero as τ approaches zero. Consequently, a random function $X(t)$ stationary in the wide sense is continuous in the mean square if its autocorrelation function $R_X(\tau)$ is continuous at $\tau = 0$. This follows from the identity, valid for a stationary function,

$$E\{|X(t + \tau) - X(t)|^2\} = 2[R_X(0) - R_X(\tau)] \qquad (8.94)$$

We say that a random function $X(t)$ is *differentiable in the mean square* if we can find a random function $X'(t)$ such that

$$\lim_{\tau \to 0} E\left\{\left|\frac{X(t + \tau) - X(t)}{\tau} - X'(t)\right|^2\right\} = 0 \qquad (8.95)$$

Seeking the mathematical expectation, or mean, of $X'(t)$, we have

$$E[X(t + \tau) - X(t)] = \eta_X(t + \tau) - \eta_X(t)$$

so that

$$\eta_{X'}(t) = E[X'(t)] = \lim_{\tau \to 0} E\left\{\frac{X(t + \tau) - X(t)}{\tau}\right\} = \lim_{\tau \to 0} \frac{\eta_X(t + \tau) - \eta_X(t)}{\tau}$$

$$= \eta'_X(t)$$

Consequently,

$$E[X'(t)] = \frac{d\eta_X(t)}{dt} \qquad (8.96)$$

that is, the mean of the derivative of a random function equals the derivative of the mean of the latter. If, in particular, $X(t)$ is stationary in the wide sense, $\eta_X(t)$ is constant, and

$$E[X'(t)] = 0 \qquad (8.97)$$

Next we seek the autocorrelation function of $X'(t)$:

$$R_{X'}(t_1, t_2) = \lim_{\tau \to 0} E\left\{\left[\frac{X(t_1 + \tau) - X(t_1)}{\tau}\right]\left[\frac{X(t_2 + \tau) - X(t_2)}{\tau}\right]^*\right\}$$

$$= \lim_{\tau \to 0} \frac{1}{\tau^2} E\{[X(t_1 + \tau) - X(t_1)][X(t_2 + \tau) - X(t_2)]^*\}$$

$$(8.98)$$

where

$$E\{[X(t_1 + \tau) - X(t_1)][X(t_2 + \tau) - X(t_2)]^*\}$$

$$= R_X(t_1 + \tau, t_2 + \tau) - R_X(t_1, t_2 + \tau)$$

$$- R_X(t_1 + \tau, t_2) + R_X(t_1, t_2)$$

Expanding the first three terms in a Taylor series,

$$R_X(t_1 + \tau, t_2 + \tau) = R_X(t_1, t_2) + \tau\left(\frac{\partial R_X}{\partial t_1} + \frac{\partial R_X}{\partial t_2}\right)$$

$$+ \frac{\tau^2}{2!}\left(\frac{\partial^2 R_X}{\partial t_1^2} + 2\frac{\partial^2 R_X}{\partial t_1 \partial t_2} + \frac{\partial^2 R_X}{\partial t_2^2}\right) + \cdots$$

$$R_X(t_1, t_2 + \tau) = R_X(t_1, t_2) + \tau\frac{\partial R_X}{\partial t_2} + \frac{\tau^2}{2!}\frac{\partial^2 R_X}{\partial t_2^2} + \cdots$$

$$R_X(t_1 + \tau, t_2) = R_X(t_1, t_2) + \tau\frac{\partial R_X}{\partial t_1} + \frac{\tau^2}{2!}\frac{\partial^2 R_X}{\partial t_1^2} + \cdots$$

Substituting in Eq. (8.98), we find

$$R_{X'}(t_1, t_2) = \frac{\partial^2 R_X(t_1, t_2)}{\partial t_1 \partial t_2} \qquad (8.99)$$

The corresponding result for a stationary random function is

$$R_{X'}(\tau) = -\frac{d^2 R_X(\tau)}{d\tau^2} \qquad (8.100)$$

so that if the mixed derivative of the autocorrelation function at $t_1 = t_2 = t$ exists, the random function $X(t)$ is differentiable in the mean square.

The mean square value of $X'(t)$ equals

$$R_{X'}(0) = -R''_X(0) \tag{8.101}$$

or, using the Wiener-Khintchine relationship (8.68),

$$R_{X'}(0) = \int_{-\infty}^{\infty} \omega^2 S_X(\omega)\,d\omega \tag{8.102}$$

Equations (8.97) and (8.100) indicate that the derivative of a random function stationary in the wide sense is itself such a function, hence the Wiener-Khintchine relationships hold for $X'(t)$:

$$R_{X'}(\tau) = \int_{-\infty}^{\infty} S_{X'}(\omega) e^{i\omega\tau}\,d\omega$$

$$S_{X'}(\omega) = \frac{1}{2\pi} \int_{-\infty}^{\infty} R_{X'}(\tau) e^{-i\omega\tau}\,d\tau \tag{8.103}$$

Comparing the first of these equations with Eq. (8.102), we see that

$$S_{X'}(\omega) = \omega^2 S_X(\omega) \tag{8.104}$$

that is, differentiation of a random function yields its spectral density multiplied by ω^2.

Since continuity of a stationary random function implies continuity of its autocorrelation function at $\tau = 0$, the mean square of the random function, which equals the value of the autocorrelation function there, has to be finite. This leads, due to Eq. (8.102), to a condition in terms of $S_X(\omega)$:

$$\int_{-\infty}^{\infty} \omega^2 S_X(\omega)\,d\omega < \infty \tag{8.105}$$

so that $S_X(\omega)$ has to decay faster then ω^{-3}.

Example 8.14

A random function stationary in the wide sense, represented by ideal white noise, is noncontinuous in the mean square, since its autocorrelation function, given in Example 8.11, is noncontinuous at $\tau = 0$. A random function with its $R_X(\tau)$ as per Eq. (8.84) is continuous in the mean square but nondifferentiable, since, because of the term $e^{-\alpha|\tau|}$, the first derivative of $R_X(\tau)$ is noncontinuous at $\tau = 0$, and a second derivative does not exist. Band-limited white noise (Eq. 8.82) is, however, continuous and differentiable in the mean square, and so is the random function with the autocorrelation function in Example 8.10.

Consider now a random function stationary in the wide sense, with the autocorrelation function

$$R_X(\tau) = d_X^2 e^{-\alpha|\tau|}(\cos \beta\tau + \gamma \sin \beta|\tau|) \tag{8.106}$$

where $\gamma \leqslant \alpha/\beta$. [Note that noncompliance with this restriction violates a basic property of the autocorrelation function as per Eq. (8.61).] The spectral density of $X(t)$ is given by

$$S_X(\omega) = \frac{2d_X^2}{\pi} \frac{(\alpha + \gamma\beta)(\alpha^2 + \beta^2) + (\alpha - \gamma\beta)\omega^2}{(\alpha^2 + \beta^2)^2 + 2(\alpha^2 - \beta^2)\omega^2 + \omega^4} \qquad (8.107)$$

and is seen to become negative for $\gamma > \alpha/\beta$; its calculation is left to the reader as an exercise.

$R'_X(\tau)$ is given by for $\gamma = \alpha/\beta$:

$$R'_X(\tau) = -\frac{\alpha^2 + \beta^2}{\beta} d_X^2 e^{-\alpha|\tau|} \sin \beta\tau$$

The latter is continuous and differentiable at $\tau = 0$, so that $R''_X(0)$ exists. We conclude that $X(t)$ with the autocorrelation function (8.106) is continuous and differentiable in the mean square.

We next seek the cross-correlation function $R_{XX'}$ of $X(t)$ and $X'(t)$. We evaluate first the corresponding function of $X(t)$ and the difference $X(t + \tau) - X(t)$:

$$E\{X(t_1)[X(t_2 + \tau) - X(t_2)]^*\}$$
$$= R_X(t_1, t_2 + \tau) - R_X(t_1, t_2)$$

$$(8.108)$$

Expanding $R_X(t_1, t_2 + \tau)$ in a Taylor series,

$$R_X(t_1, t_2 + \tau) = R_X(t_1, t_2) + \frac{\partial R_X(t_1, t_2)}{\partial t_2} + \cdots$$

and substituting this in Eq. (8.108), we arrive at

$$R_{XX'}(t_1, t_2) = \lim_{\tau \to 0} \frac{1}{\tau} E\{X(t_1)[X(t_2 + \tau) - X(t_2)]^*\} = \frac{\partial R_X(t_1, t_2)}{\partial t_2}$$

$$(8.109)$$

whereas for a weakly stationary random function we obtain

$$R_{XX'}(t_1, t_2) = R_{XX'}(\tau) = R'_X(\tau) \qquad (8.110)$$

It is also readily shown that

$$R_{XX'}(\tau) = -R_{X'X}(\tau) \qquad (8.111)$$

It follows from Eq. (8.110) that if $X(t)$ is stationary in the wide sense, then $X(t)$ and $X'(t)$ are jointly stationary. Hence their autocorrelation function and spectral density are interrelated according to Wiener-Khintchine:

$$R_{XX'}(\tau) = \int_{-\infty}^{\infty} S_{XX'}(\omega) e^{i\omega\tau} d\omega$$

$$S_{XX'}(\omega) = \frac{1}{2\pi} \int_{-\infty}^{\infty} R_{XX'}(\tau) e^{-i\omega\tau} d\tau \qquad (8.112)$$

Substituting Eq. (8.68) into (8.110), we have

$$R_{XX'}(\tau) = \int_{-\infty}^{\infty} i\omega S_X(\omega) e^{i\omega\tau} d\omega \qquad (8.113)$$

and, with the first of Eqs. (8.112), we finally obtain

$$S_{XX'}(\omega) = i\omega S_X(\omega) \qquad (8.114)$$

so that for the real random function, $X(t)$, the cross-spectral density $S_{XX'}(\omega)$ is a purely imaginary function, and the cross-correlation function is odd:

$$R_{XX'}(\tau) = -R_{XX'}(-\tau) \qquad (8.115)$$

so that

$$R_{XX'}(0) = 0 \qquad (8.116)$$

which also follows from Eq. (8.113). This indicates that both the cross-correlation of a random function, stationary in the wide sense, and its derivative vanish. That is, $X(t)$ and $X'(t)$ are uncorrelated when calculated for the same time instances.

When the random function $X'(t)$ is differentiable in the mean square, $X''(t)$ is called the second derivative in the mean square of $X(t)$ at t. The higher-order derivatives are defined in an analogous manner. By repeated application of the above reasoning, we see that the nth derivative

$$X^{(n)}(t) = \frac{d^n X(t)}{dt^n}$$

exists if the mixed $2n$th-order derivative

$$\dot{R}_{X^{(n)}}(t_1, t_2) = \frac{\partial^{2n} R_X(t_1, t_2)}{\partial t_1^n \, \partial t_2^n} \qquad (8.117)$$

exists and is continuous. $R_{X^{(n)}}(t_1, t_2)$ is then the autocorrelation function of

$X^{(n)}(t)$. For a random function, stationary in the wide sense, Eq. (8.117) is replaced by

$$R_{X^{(n)}}(\tau) = (-1)^n R_X^{(2n)}(\tau) \tag{8.118}$$

so that $X^{(n)}(t)$ is also stationary in the wide sense, with the spectral density

$$S_{X^{(n)}}(\omega) = \omega^{2n} S_X(\omega) \tag{8.119}$$

Equation (8.118) indicates that $X^{(n)}(t)$ exists if the $2n$th-order derivative of $R_X(\tau)$ is continuous at $\tau = 0$, or, according to (8.119) and the Wiener-Khintchine relationship,

$$R_{X^{(n)}}(\tau) = \int_{-\infty}^{\infty} S_{X^{(n)}}(\omega) e^{i\omega\tau}\, d\omega$$

$$= \int_{-\infty}^{\infty} \omega^{2n} S_X(\omega) e^{i\omega\tau}\, d\omega \tag{8.120}$$

$S_X(\omega)$ decays faster than $\omega^{-(2n+1)}$ as $\omega \to \infty$.

Example 8.15

Given a normal random function $X(t)$, its first-order probability density function is (see Eq. 8.25):

$$f_X(x, t) = \frac{1}{[2\pi C_X(t, t)]^{1/2}} \exp\left\{ -\frac{[x - \eta_X(t)]^2}{2C_X(t, t)} \right\}$$

and we seek the probabilistic properties of $X'(t)$. The first-order probability density function of $X'(t)$ is

$$f_{X'}(y, t) = \frac{1}{[2\pi C_{X'}(t, t)]^{1/2}} \exp\left\{ -\frac{[y - \eta_{X'}(t)]^2}{2C_{X'}(t, t)} \right\} \tag{8.121}$$

where, in view of Eqs. (8.96) and (8.99),

$$\eta_{X'}(t) = \eta'_X(t), \quad C_{X'}(t, t) = \left. \frac{\partial^2 R_X(t_1, t_2)}{\partial t_1\, \partial t_2} \right|_{t_1 = t_2 = t} - |\eta'_X(t)|^2 \tag{8.122}$$

For a stationary random function

$$\eta_{X'}(t) = 0, \quad C_{X'}(t, t) = -R''_X(0)$$

$$f_{X'}(y, t) = \frac{1}{[-2\pi R''_X(0)]^{1/2}} \exp\left[\frac{y^2}{2R''_X(0)} \right] \tag{8.123}$$

8.12 INTEGRATION OF A RANDOM FUNCTION

We define the mean-square integral of a random function $X(t)$ as the random variable Y (a and b being given numbers):

$$Y = \int_a^b X(t)\, dt$$

such that

$$\lim_{\Delta t \to 0} E\left\{ \left| Y - \sum_{j=1}^n X(t_j)\Delta t_j \right|^2 \right\} = 0$$

$F_Y(y)$ is difficult to determine in the general case, since Y represents the *mean-square limit* of a sum of random variables $X(t_j)\Delta t_j$, although the distribution function of such a sum is determinable in closed form for some particular cases. Accordingly, we content ourselves with finding $E(y)$ and Var(y). The mathematical expectation equals

$$E(Y) = E\left\{ \int_a^b X(t)\, dt \right\} = \int_a^b E[X(t)]\, dt = \int_a^b \eta_X(t)\, dt \qquad (8.124)$$

For Var(y), as per Eq. (6.142), we write

$$\mathrm{Var}(Y) = E\left[|Y - E(Y)|^2 \right] = E\{ [Y - E(Y)][Y - E(Y)]^* \}$$

$$= E\left\{ \int_a^b [X(t_1) - \eta_X(t_1)]\, dt_1 \int_a^b [X(t_2) - \eta_X(t_2)]^* dt_2 \right\}$$

$$= \int_a^b \int_a^b E\{ [X(t_1) - \eta_X(t_1)][X(t_2) - \eta_X(t_2)]^* \}\, dt_1\, dt_2$$

which in conjunction with the definition (8.39) of the autocovariance function $C_X(t_1, t_2)$ is rewritten as

$$\mathrm{Var}(Y) = \int_a^b \int_a^b C_X(t_1, t_2)\, dt_1\, dt_2 \qquad (8.125)$$

Existence of the double integral in (8.125) is a necessary and sufficient condition for that of the integral $X(t)$ in the mean square.

We are now in a position to prove the nonnegativeness [Eq. (8.42)] of the autocovariance function $C_X(t_1, t_2)$. To do so, we consider the integral

$$Y = \int_a^b [Z(t) - \eta_Z(t)]\varphi(t)\, dt$$

where $\varphi(t)$ is a deterministic function, $Z(t)$ is a complex random function, and a and b are given numbers. Then,

$$\text{Var}(Y) = \int_a^b \int_a^b C_Z(t_1, t_2) \varphi(t_1) \varphi^*(t_2) \, dt_1 \, dt_2$$

and since the variance is a nonnegative number,

$$\int_a^b \int_a^b C_Z(t_1, t_2) \varphi(t_1) \varphi^*(t_2) \, dt_1 \, dt_2 \geqslant 0$$

This proves Eq. (8.42), from which property (8.17) for a real random function follows immediately.

According to Bochner's theorem, every nonnegative definite function has a nonnegative Fourier transform, which in turn indicates that the spectral density $S(\omega)$ is a nonnegative function:

$$S_X(\omega) \geqslant 0 \qquad (8.126)$$

Example 8.16

We maintain that a stationary random function with the autocorrelation function

$$R(\tau) = \begin{cases} R_0, & |\tau| < \tau_c \\ 0, & \text{otherwise} \end{cases} \qquad (8.127)$$

does not exist. Indeed, the spectral density of such a random function, if it existed would be

$$S(\omega) = \frac{1}{2\pi} \int_{-\infty}^{\infty} R_0 e^{-i\omega\tau} \, d\tau = \frac{R_0}{2\pi} \int_{-\tau_c}^{\tau_c} \cos \omega\tau \, d\tau = \frac{R_0}{\pi\omega} \sin \omega\tau_c$$

which can take on negative values as well. Accordingly, (8.127) cannot serve as an autocorrelation function.

Example 8.17

Given the random function

$$Y(u) = \int_a^b \varphi(u, t) X(t) \, dt \qquad (8.128)$$

where $\varphi(u, t)$ is a deterministic function of u and t, and $X(t)$ is a normal random function. Then,

$$f_Y(y, u) = \frac{1}{[2\pi C_Y(u, u)]^{1/2}} \exp\left\{\left[-\frac{[y - \eta_Y(u)]^2}{2C_Y(u, u)}\right]\right\} \qquad (8.129)$$

where

$$\eta_Y(u) = \int_a^b \varphi(u, t)\eta_X(t)\, dt$$

$$C_Y(u, u) = \int_a^b \int_a^b R_X(t_1, t_2)\varphi(u, t_1)\varphi^*(u, t_2)\, dt_1\, dt_2 - \left|\eta_Y(u)\right|^2 \quad (8.130)$$

Example 8.18

Consider a cantilever of length l, loaded by a distributed force $Q_y(x)$, a random function of the axial coordinate x, with the origin at the free end. The bending moment $M_z(x)$ is also a random function of x.

$$M_Z(x) = \int_0^x Q_y(\xi)(x - \xi)\, d\xi \quad (8.131)$$

The autocorrelation function of the bending moment is

$$R_{M_z}(x_1, x_2) = \int_0^{x_1} \int_0^{x_2} R_{Q_y}(\xi_1, \xi_2)(x_1 - \xi_1)(x_2 - \xi_2)\, d\xi_1\, d\xi_2 \quad (8.132)$$

where $R_{Q_y}(\xi_1, \xi_2)$ is the autocorrelation function of the distributed force.

For a beam simply supported at its ends

$$M_z(x) = \int_0^x Q_y(\xi)(x - \xi)\, d\xi - \frac{x}{l}\int_0^l Q_y(\xi)(l - \xi)\, d\xi \quad (8.133)$$

we have

$$R_{M_z}(x_1, x_2) = \int_0^{x_1} \int_0^{x_2} R_{Q_y}(\xi_1, \xi_2)(x_1 - \xi_1)(x_2 - \xi_2)\, d\xi_1\, d\xi_2$$

$$- \int_0^{x_1} \int_0^{l} R_{Q_y}(\xi_1, \xi_2)(x_1 - \xi_1)(l - \xi_2)\, d\xi_1\, d\xi_2$$

$$- \int_0^{l} \int_0^{x_2} R_{Q_y}(\xi_1, \xi_2)(l - \xi_1)(x_2 - \xi_2)\, d\xi_1\, d\xi_2$$

$$+ \frac{x_1 x_2}{l^2} \int_0^{l} \int_0^{l} R_{Q_y}(\xi_1, \xi_2)(l - \xi_1)(l - \xi_2)\, d\xi_1\, d\xi_2 \quad (8.134)$$

8.13 ERGODICITY OF RANDOM FUNCTIONS

Consider a random function

$$Y(t) = g[X(t)] \quad (8.135)$$

where $X(t)$ is assumed to be a strictly stationary random function and g is a given deterministic function. For determination of the mathematical expectation

$$E[Y(t)] = \int_{-\infty}^{\infty} g[x(t)] f_X(x; t) \, dx \qquad (8.136)$$

we must use ensemble averages, that is, we need an adequate population of realizations of the random function. The experimenter, however, usually has a single laboratory and a single experimental device available rather than a large number of them, say 10^3 or 10^6. For a given time interval $(0, T)$, he effects a single realization and prefers averaging in time, namely,

$$\langle Y(t) \rangle_T = \frac{1}{T} \int_0^T Y(t) \, dt \qquad (8.137)$$

The question is, therefore, how are the ensemble time averages interrelated? In other words, when is it possible to determine the probabilistic characteristics of a stationary random function from a single observation? The answer is that both averages coincide for *ergodic* random functions with a sufficiently large observation interval $(0, T)$.

We will say that a strictly stationary random function $X(t)$ is *ergodic with respect to the mean of the function $Y(t) = g[X(t)]$*, if

$$P\left\{ \lim_{T \to \infty} \langle Y(t) \rangle_T = E[Y(t)] \right\} = 1 \qquad (8.138)$$

Note that $\lim_{T \to \infty} \langle Y(t) \rangle_T$ is a random variable, whereas $E[Y(t)]$ is a constant:

$$\eta_Y = E[Y(t)] = E\{g[X(t)]\} \qquad (8.139)$$

since $X(t)$ is a stationary random function. This means that, generally,

$$\lim_{T \to \infty} \langle Y(t) \rangle_T \neq E[Y(t)]$$

in the ordinary sense. Equation (8.138) implies that (compare Problem 3.11)

$$\mathrm{Var}\left\{ \lim_{T \to \infty} \langle Y(t) \rangle_T \right\} = 0 \qquad (8.140)$$

We have,

$$E\left\{ \lim_{T \to \infty} \langle Y(t) \rangle_T \right\} = \lim_{T \to \infty} \frac{1}{T} \int_0^T E[Y(t)] \, dt$$

$$= \lim_{T \to \infty} \frac{1}{T} \int_0^T \eta_Y \, dt = \eta_Y \qquad (8.141)$$

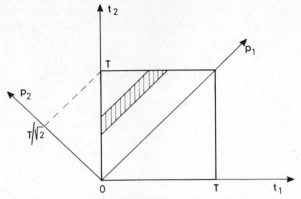

Fig. 8.8. Transformation variables in Eqs. (8.142).

and

$$\text{Var}\left\{ \lim_{T \to \infty} \langle Y(t) \rangle_T \right\}$$

$$= E\left\{ \lim_{T \to \infty} \frac{1}{T^2} \int_0^T \int_0^T [Y(t_1) - \eta_Y][Y(t_2) - \eta_Y]^* \, dt_1 \, dt_2 \right\}$$

$$= \lim_{T \to \infty} \frac{1}{T^2} \int_0^T \int_0^T C_Y(t_1, t_2) \, dt_1 \, dt_2$$

$$= \lim_{T \to \infty} \frac{1}{T^2} \int_0^T \int_0^T C_Y(t_2 - t_1) \, dt_1 \, dt_2 \qquad (8.142)$$

where

$$C_Y(t_1, t_2) = E\left\{ [Y(t_1) - \eta_Y][Y(t_2) - \eta_Y]^* \right\}$$

The double integral can be reduced to a single one through transformation of variables (see Fig. 8.8):

$$p_1 = \frac{t_1 + t_2}{\sqrt{2}} \qquad p_2 = \frac{t_2 - t_1}{\sqrt{2}}$$

Because of the evenness of $C_Y(t_2 - t_1)$, we have

$$\int_0^T \int_0^T C_Y(t_2 - t_1) \, dt_1 \, dt_2 = 2 \int_0^{T/\sqrt{2}} C_Y(\sqrt{2}\, p_2) \, dp_2 \int_{p_2}^{2T/\sqrt{2} - p_2} dp_1$$

$$= 2T \int_0^T \left(1 - \frac{\tau}{T} \right) C_Y(\tau) \, d\tau$$

so that Eq. (8.142) becomes

$$\lim_{T \to \infty} \frac{1}{T} \int_0^T \left(1 - \frac{\tau}{T}\right) C_Y(\tau) \, d\tau = 0 \qquad (8.143)$$

representing a necessary and sufficient condition for $X(t)$ to be ergodic with respect to the mean of $Y(t) = g[X(t)]$. In particular, $X(t)$ is *ergodic in the mean* if we set $g[X(t)] \equiv X(t)$:

$$P\left\{ \lim_{T \to \infty} \langle X(t) \rangle_T = E[X(t)] \right\} = 1 \qquad (8.144)$$

a necessary and sufficient condition being

$$\lim_{T \to \infty} \frac{1}{T} \int_0^T \left(1 - \frac{\tau}{T}\right) C_X(\tau) \, d\tau = 0 \qquad (8.145)$$

The random function is said to be *ergodic in the mean square* if it is ergodic with respect to the mean of the function

$$Y(t) = g[X(t)] = X^2(t) \qquad (8.146)$$

when Eq. (8.143) becomes

$$\lim_{T \to \infty} \frac{1}{T} \int_0^T \left(1 - \frac{\tau}{T}\right) C_Y(\tau) \, d\tau = \lim_{T \to \infty} \frac{1}{T} \int_0^T \left(1 - \frac{\tau}{T}\right)$$
$$\times E\{[X(t)]^2 [X(t + \tau)]^2\} \, d\tau = 0 \qquad (8.147)$$

Since the conditions for ergodicity in the mean and in the mean square are not identical, there can be random functions that are ergodic in the mean but nonergodic in the mean square and vice versa. This may be demonstrated as follows.

Example 8.19

Consider the random function

$$X(t) = U \cos \omega t + V \sin \omega t \qquad (8.148)$$

where U and V are random variables such that

$$E(U) = E(V) = 0 \qquad \sigma_U^2 = \sigma_V^2 = \sigma^2 \qquad \text{Cov}(U, V) = 0$$

It is readily checked that $X(t)$ is stationary in the wide sense. We wish to check whether $X(t)$ is ergodic in the mean and/or in the mean square.

We find

$$E[X(t)] = 0$$

$$\langle X(t) \rangle_T = \frac{1}{T} \int_0^T X(t) \, dt = \frac{1}{\omega T} [U \sin \omega T - V(\cos \omega T - 1)]$$

and

$$\lim_{T \to \infty} \langle X(t) \rangle_T = 0 = E[X(t)]$$

so that $X(t)$ is ergodic in the mean. Note also that condition (8.145) is satisfied, since

$$\lim_{T \to \infty} \frac{1}{T} \int_0^T \left(1 - \frac{\tau}{T}\right) C_X(\tau) \, d\tau = \lim_{T \to \infty} \frac{\sigma_U^2}{T} \int_0^T \left(1 - \frac{\tau}{T}\right) \cos \omega \tau \, d\tau$$

$$= \lim_{T \to \infty} \frac{\sigma_U^2}{\omega^2 T^2} (\cos \omega T - 1) = 0$$

Now, as for the mean square, we have

$$\langle [X(t)]^2 \rangle_T = \frac{1}{T} \int_0^T [X(t)]^2 \, dt$$

$$= \frac{U^2}{2} \left(1 + \frac{\sin 2\omega T}{2\omega T}\right) + \frac{V^2}{2} \left(1 - \frac{\sin 2\omega T}{2\omega T}\right)$$

$$- \frac{UV}{2\omega T} (\cos 2\omega T - 1) \qquad (8.149)$$

and

$$\lim_{T \to \infty} \frac{1}{T} \int_0^T [X(t)]^2 \, dt = \frac{U^2 + V^2}{2}$$

That is, it is dependent on the realizations of U and V, the conclusion being that in general

$$E\{[X(t)]^2\} \neq \lim_{T \to \infty} \frac{1}{T} \int_0^T [X(t)]^2 \, dt$$

that is, in general $X(t)$ in nonergodic in the mean square. In a particular case, $U^2 + V^2 = 2\sigma^2$, $X(t)$ is ergodic both in the mean and in the mean square. This happens, for example, when $U = \sqrt{2}\,\sigma \cos \varphi$ and $V = -\sqrt{2}\,\sigma \sin \varphi$, where φ is a random variable uniformly distributed on $(0, 2\pi)$. Then $X(t) = \sqrt{2}\,\sigma \cos(\omega t + \varphi)$.

Example 8.20

Consider the random function

$$X(t) = U \cos \omega t + V \sin \omega t + W \qquad (8.150)$$

with U, V, and W being random variables such that

$$E(U) = E(V) = 0 \qquad \sigma_U^2 = \sigma_V^2 = \sigma^2$$

$$E(UV) = E(UW) = E(VW) = 0 \qquad (8.151)$$

We observe that

$$E[X(t)] = E(W)$$

$$E\langle\{X(t_1) - E[X(t_1)]\}\{X(t_2) - E[X(t_2)]\}\rangle = \sigma^2 \cos \omega(t_2 - t_1) + \mathrm{Var}(W)$$

so, assuming (8.151), the random function (8.150) is stationary in the wide sense.

Calculation shows that

$$\lim_{T \to \infty} \langle X^2(t) \rangle_T = \lim_{T \to \infty} \frac{1}{T} \int_0^T X^2(t)\, dt = \tfrac{1}{2}(U^2 + V^2) + W^2$$

Now, if

$$W = \sqrt{a^2 - \tfrac{1}{2}(U^2 + V^2)} \qquad (8.152)$$

where U and V are continuously distributed on $[-c, c]$ and $[-d, d]$, respectively, and are uncorrelated. Then

$$\lim_{T \to \infty} \langle X^2(t) \rangle_T = a^2 \qquad (8.153)$$

In other words, $X(t)$ is ergodic in the mean square since

$$E[X^2(t)] = \sigma^2 + E(W^2) = \sigma^2 + (a^2 - \sigma^2) = a^2 \qquad (8.154)$$

Note that $X(t)$ is ergodic in the mean iff

$$E[X(t)] = E(W) = \lim_{T \to \infty} \langle X(t) \rangle_T = W \qquad (8.155)$$

that is, if and only if W is a deterministic constant.

This example is the reverse of the previous one, where the random function was ergodic in the mean but not in the mean square.

Example 8.21 (Scheurkogel)

Consider the differential equation

$$\xi^2 \frac{d^2 X}{dt^2} + X = (4 - aU^2 - aV^2)^{1/2} \tag{8.156}$$

with initial conditions

$$X(t)|_{t=0} = \xi U + (4 - aU^2 - aV^2)^{1/2}, \quad 0 \leqslant a \leqslant 2$$

$$\frac{dX(t)}{dt}\bigg|_{t=0} = V \tag{8.157}$$

and where

$$\xi^2 = \lim_{T \to \infty} \frac{1}{T} \int_0^T X^2(t)\, dt, \quad \xi \geqslant 0 \tag{8.158}$$

The random variables U and V are independent and identically distributed on the interval $[-1, 1]$ according to the probability density $f(u) = |u|$. The problem consists in finding the mean square $E[X^2(t)]$ of the solution $X(t)$.

We will tackle this problem in two ways: (1), assuming ergodicity of the solution, and (2) arriving at it exactly, in order to estimate the error introduced by the ergodicity assumption.

By the first approach, if we assume that $X(t)$ is ergodic in the mean square, then the mean-square value is obtainable from the initial conditions alone, without having to solve the differential equation. Thus,

$$\xi^2 = E[X^2(t)] = E[X^2(0)]$$

$$= E\left[\xi U + (4 - aU^2 - aV^2)^{1/2}\right]^2 = \tfrac{1}{2}\xi^2 + 4 - a \tag{8.159}$$

Hence,

$$\xi^2 = E[X^2(t)] = 8 - 2a$$

The problem is, however, capable of exact solution. We have

$$X(t) = U\xi \cos \frac{t}{\xi} + V\xi \sin \frac{t}{\xi} + (4 - aU^2 - aV^2)^{1/2} \tag{8.160}$$

Substituting this expression in (8.158) and solving for ξ, we obtain

$$\xi^2 = \frac{4 - a(U^2 + V^2)}{1 - (U^2 + V^2)/2} \tag{8.161}$$

Note that the value of ξ^2 is independent of the particular realizations of U and V for $a = 2$, whereas for $a \neq 2$ it is dependent. Note also that ξ is an even function of both U and V, so that the first and second terms vanish when taking the expectation of (8.160). Hence, $E[X(t)]$ is independent of t.

From (8.160) we have

$$X(t)X(t + \tau) = \tfrac{1}{2}(U^2 + V^2)\xi^2\cos\frac{\tau}{\xi} + \tfrac{1}{2}(U^2 - V^2)\xi^2\cos\frac{2t + \tau}{\xi}$$

$$+ UV\xi^2\sin\frac{2t + \tau}{\xi} + U\xi(4 - aU^2 - aV^2)^{1/2}$$

$$\times \left(\cos\frac{t}{\xi} + \cos\frac{t + \tau}{\xi}\right)$$

$$+ V\xi(4 - aU^2 - aV^2)^{1/2}$$

$$\times \left(\sin\frac{t}{\xi} + \sin\frac{t + \tau}{\xi}\right) + (4 - aU^2 - aV^2)$$

with ξ as per (8.161). Because of symmetry, only the first and last terms in the above equation contribute to the autocorrelation function of $X(t)$. Substituting

$$Z = U^2 + V^2$$

the autocorrelation function becomes

$$E[X(t)X(t + \tau)] = E\left[(4 - aZ)\left(1 + \frac{Z}{2 - Z}\cos\left(\frac{1 - Z/2}{4 - aZ}\right)^{1/2}\tau\right)\right]$$

$$\tag{8.162}$$

and is independent of t, so that $X(t)$ is stationary in the wide sense. The mean-square follows from (8.162) with $\tau = 0$:

$$E[X^2(t)] = 2a + 4(a - 2)E\left(\frac{1}{Z - 2}\right)$$

$$= 2a + 4(a - 2)\int_{-1}^{1}\int_{-1}^{1}\frac{|u||v|}{u^2 + v^2 - 2}\,du\,dv$$

$$= 2a - 8(a - 2)\ln 2 \tag{8.163}$$

For $a = 2$ we have

$$E[X^2(t)] = 4 = \xi^2$$

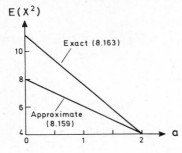

Fig. 8.9. Mean-square value $E[X^2(t)]$ as a function of parameter a.

Consequently, the solution is ergodic in the mean-square iff $a = 2$; in this case both solutions coincide.

Exact and approximate mean square values are shown in Fig. 8.9.

The conclusion from Example 8.21 is that caution should be exercised in applying the ergodicity assumption. The ergodicity or otherwise of the solution is decided by the equation and the boundary conditions themselves.

PROBLEMS

8.1. Given a pair of random functions

$$X(t) = A \sin \omega_1 t, \quad Y(t) = B \sin \omega_2 t$$

where A and B are random variables with mathematical expectations $E(A) = 1$, $E(B) = 2$, and the variance-covariance matrix

$$\begin{bmatrix} 1 & 1 \\ 1 & 9 \end{bmatrix}$$

Determine the autocorrelation functions $R_X(t_1, t_2)$, $R_Y(t_1, t_2)$ and the cross-correlation functions $R_{XY}(t_1, t_2)$, $R_{YX}(t_1, t_2)$.

8.2. Solve Problem 8.1 with the variance-covariance matrix

$$\begin{bmatrix} 1 & 3 - \varepsilon \\ 3 - \varepsilon & 9 \end{bmatrix}$$

for $\varepsilon \ll 1$. What happens when $\varepsilon \to 0$?

8.3. Check whether a random function $X(t)$ possessing the autocorrelation function

$$R_X(\tau) = d^2 e^{-\alpha|\tau|}(1 + \alpha|\tau|)$$

if differentiable or otherwise. Compare with the result obtained for $R_X(\tau)$ given in Eq. (8.106), and offer an interpretation.

8.4. Show that for a weakly stationary random function $X(t)$,

$$S_{X^{(n)}}(\omega) = (-1)^n (i\omega)^{2n} S_X(\omega)$$

8.5. Verify that the spectral density $S_{X'}(\omega)$ of the X' with $R_X(\tau)$ as in Problem 8.3 is

$$S_{X'}(\omega) = d^2 e^{-\alpha|\tau|}(1 - \alpha|\tau|)$$

8.6. Check whether $X(t)$ in Problem 8.5 has a second derivative.

8.7. The initial imperfection $Y_0(x)$ of an infinite beam is a weakly stationary, band-limited random function of the axial coordinate x. Find the spectral density of $d^2 Y_0(x)/dx^2$.

8.8. Show that the spectral density of the sum of a pair of independent random functions equals the sum of their spectral densities.

8.9. Use the nonnegativeness property of the spectral density to determine the admissible values of the parameters α and β in the autocorrelation function

$$R_X(\tau) = d^2 e^{-\alpha|\tau|}\left(\cosh \beta\tau + \frac{\alpha}{\beta} \sinh \beta|\tau|\right)$$

Check whether $X(t)$ is differentiable.

8.10. A beam, simply supported at its ends, is subjected to a distributed force Q, a random variable with given probability density function. Using the relations

$$\frac{dV_y(x)}{dx} = -Q(x) \qquad \frac{dM_z(x)}{dx} = -V_y(x)$$

show that the shear force $V_Y(x)$ and bending moment $M_z(x)$ are the random functions of x. Find

$$E[V_y(x)] \qquad E[M_z(x)] \qquad R_{V_y}(x_1, x_2) \qquad R_{M_z}(x_1, x_2)$$

and the first-order probability densities

$$f_{V_y}(v_y; x) \qquad f_{M_z}(m_z; x).$$

8.11. A cantilever is subjected to a distributed force $Q_y(x)$ with the zero mean and autocorrelation function:

$$R_{Q_y}(x_1, x_2) = e^{-\alpha|x_2 - x_1|}$$

Using Eq. (8.8), verify that

$$R_{M_z}(x_1, x_2) = \frac{1}{\alpha^4}\left[e^{-\alpha|x_1 - x_2|} - e^{-\alpha x_1} - e^{-\alpha x_2} - \alpha x_1 e^{-\alpha x_2} \right.$$

$$- \alpha x_2 e^{-\alpha x_1} + 1 + \alpha|x_1 - x_2| - \alpha x_1 x_2 + \frac{\alpha^3}{6}|x_1 - x_2|^3$$

$$\left. - \frac{\alpha^3}{6}\left(x_1^3 - 3x_1^2 x_2 - 3x_1 x_2^2 - 3x_1 x_2^2 + x_2^3 \right) \right] \quad (8.164)$$

Show that for $x \gg 1/\alpha$ the variance is

$$\mathrm{Var}[M_z(x)] = \frac{1}{\alpha^4}\left(2 - \alpha^2 x^2 + \tfrac{2}{3}\alpha^3 x^3 \right) \quad (8.165)$$

Equations (8.164) and (8.165) are due to Rzhanitsyn.

8.12. A normal random function $X(t)$ with zero mean has an autocorrelation function as per Eq. (8.106). Find the probability of $X(t) < x_0$, x_0 being a deterministic positive constant.

CITED REFERENCES

Bochner, S., *Lectures of Fourier Integrals*, Princeton, NJ, 1959.

Rzhanitsyn, A. R., "Probabilistic Calculation of Beams on a Random Load," in (B. G. Korenev and I. M. Rabinovich, Eds. *Issledovanija po Teorii Sooruzhenii* (Investigation in the Theory of Structures; in Russian), Vol. 23, "Stroizzdat" Publ. House, Moscow, 1977, pp. 158–171.

Scheurkogel, A., Private communication, Delft, 1980.

RECOMMENDED FURTHER READING

Melsa, J. L., and Sage, A. P., *An Introduction to Probability and Stochastic Processes*, Prentice-Hall, Englewood Cliffs, NJ, 1973. Chap. 5: Stochastic Processes, pp. 189–244.

Lin, Y. K., *Probabilistic Theory of Structural Dynamics*, McGraw-Hill, New York, 1967. Chap. 3: Random Processes, pp. 34–66; Chap. 4: Gaussian, Poisson, and Markov Random Processes, pp. 67–109.

Papoulis, A., *Probability, Random Variables, and Stochastic Processes*, Intern. Student Ed., McGraw-Hill Kogakusha, Tokyo, 1965; Chap. 9: General Concepts, pp. 279–335; Chap. 10: Correlation and Power Spectrum of Stationary Processes, pp. 336–384.

Parzen, E., *Stochastic Processes*, Holden-Day, San Francisco, 1962.

Scheurkogel, A., Elishakoff, I., and Kalker, J. J., "On the Error That Can Be Induced by an Ergodicity Assumption," *ASME J. Appl. Mech.*, **103**, 654–656 (1981).

chapter **9**

Random Vibration of Discrete Systems

In this chapter we turn to random vibration problems with intensive reference to the results of Chapter 8. After a review of relevant deterministic results, we proceed to cases of single- and multidegree-of-freedom systems. Finally, we demonstrate the normal mode method and the dramatic effect of the interaction between different modes, usually overlooked in the literature.

9.1 RESPONSE OF A LINEAR SYSTEM SUBJECTED TO DETERMINISTIC EXCITATION

Consider a physical system whose behavior is governed by the following differential equation with constant coefficients:

$$L_n\left(\frac{d}{dt}\right)x = f(t) \tag{9.1}$$

where $L_n(d/dt)$ is the differential operator

$$L_n\left(\frac{d}{dt}\right) = a_0\frac{d^n}{dt^n} + a_1\frac{d^{n-1}}{dt^{n-1}} + \cdots + a_{n-1}\frac{d}{dt} + a_n, \quad a_0 \neq 0 \tag{9.2}$$

and a_i are real constants.

$f(t)$ constitutes the *input*, or *excitation*, of the system, and $x(t)$ the *output*, or *response*, of the system. The linear differential equation is subject to the principle of linear superposition, namely, if $x_i(t)$ is the output of the system to input $f_i(t)$, then the output to input $\sum_{i=1}^n \alpha_i f_i(t)$ will be $\sum_{i=1}^n \alpha_i x_i(t)$, where n is some positive integer and α_i are any real numbers.

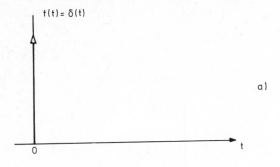

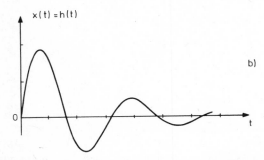

Fig. 9.1.   (a) Unit impulse applied at $t = 0$. (b) Impulse response function $h(t)$.

Let us consider an initial-value problem, with Eq. (9.1) supplemented by the initial conditions which, without loss of generality, can be assumed to be homogeneous ones.

The *impulse response function*, denoted by $h(t)$, is the response of a system with zero initial conditions to a unit impulse applied at $t = 0$ (Fig. 9.1). That is, instead of $f(t)$ we have $\delta(t)$ in Eq. (9.1)

$$a_0 \frac{d^n h}{dt^n} + a_1 \frac{d^{n-1}h}{dt^{n-1}} + \cdots + a_{n-1} \frac{dh}{dt} + a_n h = \delta(t) \qquad (9.3)$$

The general input $f(t)$ can then be viewed as a series of impulses of magnitude $f(\tau)\Delta\tau$, as shown in Fig. 9.2, where the shaded area $f(\tau)\Delta\tau$ is an impulse applied at $t = \tau$. The response to a unit impulse $\delta(t - \tau)$ applied at $t = \tau$ is the same impulse response function lagging by the time interval τ, namely, $h(t - \tau)$. Hence the increment in output Δx at time $t > \tau$ due to impulse $f(\tau)\Delta\tau$ is

$$\Delta x(t, \tau) = f(\tau)\Delta\tau h(t - \tau)$$

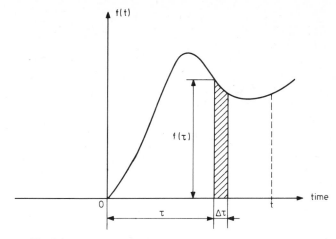

Fig. 9.2.   General input $f(t)$ considered as a series of impulses.

Summation over all impulses applied in the interval $(0, t)$ yields

$$x(t) = \int_0^t f(\tau) h(t - \tau) \, d\tau \qquad (9.4)$$

This equation is known as the *convolution* or *Duhamel integral*.
    Denoting $t - \tau = \lambda$, we have

$$x(t) = \int_0^t f(t - \lambda) h(\lambda) \, d\lambda$$

or for $\lambda = \tau$,

$$x(t) = \int_0^t f(t - \tau) h(\tau) \, d\tau \qquad (9.5)$$

Both Eqs. (9.4) and (9.5) are symmetric in terms of the input $f(t)$ and the impulse response function $h(t)$. It should be noted that $h(t - \tau)$ is identically zero for $t < \tau$ (for the time instants preceding excitation of the system). Hence, advancing the upper limit from t to infinity in Equation (9.4) would not affect the value of the integral; in these circumstances Eqs. (9.4) and (9.5) can be rewritten, for both $t > 0$ and $t < 0$, as

$$x(t) = \int_{-\infty}^{\infty} f(\tau) h(t - \tau) \, d\tau \qquad (9.6)$$

$$x(t) = \int_{-\infty}^{\infty} f(t - \tau) h(\tau) \, d\tau \qquad (9.7)$$

Consider now the response to a harmonic input. In particular, let the input be given by the real part of

$$f(t) = e^{i\omega t} \tag{9.8}$$

The solution of Eq. (9.1) consists of complementary and particular components. We will say that a system, governed by Eq. (9.1), is *asymptotically stable* if, irrespective of the initial conditions, the complementary solution (which may be identified with the response of the system, with zero input, to the initial conditions) eventually decays to zero as t becomes larger. This means that the auxiliary equation

$$a_0 r^n + a_1 r^{n-1} + \cdots + a_{n-1} r + a_n = 0 \tag{9.9}$$

has roots with negative real parts. Assume that the system under consideration is asymptotically stable; that is, that after a sufficiently long time (compared with the transient process), the *steady state response* (characterized by the presence of only a particular solution) will be reached. We seek it in a form similar to (9.8)

$$x(t) = X(\omega)e^{i\omega t} \tag{9.10}$$

Substitution of Eqs. (9.8) and (9.10) in Eq. (9.1) yields

$$x(t) = \frac{e^{i\omega t}}{L_n(i\omega)} \tag{9.11}$$

where $L_n(i\omega)$ is obtained by replacing the parameter of differentiation d/dt by $i\omega$:

$$L_n(i\omega) = a_0(i\omega)^n + a_1(i\omega)^{n-1} + \cdots + a_{n-1}(i\omega) + a_n \tag{9.12}$$

and is called the *impedance* of the system.

Denoting

$$\frac{1}{L_n(i\omega)} = H(\omega) \tag{9.13}$$

we obtain

$$x(t) = H(\omega)e^{i\omega t} \tag{9.14}$$

Here $H(\omega)$ is referred to as the *complex frequency response* or *receptance*.

If the general input $f(t)$ is representable by the Fourier integral

$$f(t) = \int_{-\infty}^{\infty} F(\omega) e^{i\omega t} \, d\omega$$

$$F(\omega) = \frac{1}{2\pi} \int_{-\infty}^{\infty} f(t) e^{-i\omega t} \, dt \qquad (9.15)$$

we may represent the output in a similar manner:

$$x(t) = \int_{-\infty}^{\infty} X(\omega) e^{i\omega t} \, d\omega$$

$$X(\omega) = \frac{1}{2\pi} \int_{-\infty}^{\infty} x(t) e^{-i\omega t} \, dt \qquad (9.16)$$

Substitution of the first of equations (9.15) and (9.16) in Eq. (9.1) leaves us with

$$X(\omega) = H(\omega) F(\omega) \qquad (9.17)$$

and $x(t)$ becomes, in view of Eq. (9.16),

$$x(t) = \int_{-\infty}^{\infty} H(\omega) F(\omega) e^{i\omega t} \, d\omega \qquad (9.18)$$

Now, for a unit impulse applied at $t = 0$,

$$f(t) = \delta(t)$$

and, due to the second of Eqs. (9.15)

$$F(\omega) = \frac{1}{2\pi} \int_{-\infty}^{\infty} \delta(t) e^{-i\omega t} \, dt = \frac{1}{2\pi} \qquad (9.19)$$

so that the response to $\delta(t)$ being the impulse-response function $h(t)$, and, bearing in mind Eq. (9.18), we have

$$h(t) = \frac{1}{2\pi} \int_{-\infty}^{\infty} H(\omega) e^{i\omega t} \, d\omega \qquad (9.20)$$

and finally

$$H(\omega) = \int_{-\infty}^{\infty} h(t) e^{-i\omega t} \, dt \qquad (9.21)$$

Here the Fourier transform equations have been replaced by the impulse-response function multiplied by the factor 2π and the complex frequency response, respectively. Had we defined the Fourier transform not as in Eq. (9.15) but as

$$f(t) = \frac{1}{2\pi} \int_{-\infty}^{\infty} F(\omega) e^{i\omega t} \, d\omega$$

$$F(\omega) = \int_{-\infty}^{\infty} f(t) e^{-i\omega t} \, dt$$

and $x(t)$ in a similar way, we would have found that $H(\omega)$ is the Fourier transform of $h(t)$ [not of $2\pi h(t)$, as in Eq. (9.20)]. In that case, however, we also would have had to modify the Wiener-Khintchine relationships derived in Chapter 8. Thus it is preferable to retain Eqs. (9.15) and (9.16) as the Fourier transform pair.

Example 9.1

Consider the mass-spring-dashpot system governed by the differential equation

$$m\ddot{x} + c\dot{x} + kx = f(t)$$

and calculate the impulse-response function $h(t)$, which satisfies

$$m\ddot{h} + c\dot{h} + kh = \delta(t) \tag{9.22}$$

with the initial conditions

$$h(0) = \dot{h}(0) = 0 \tag{9.23}$$

The external force $\delta(t)$ may be assigned a duration Δt tending to zero. Accordingly, we integrate Eq. (9.22) over the time interval Δt, to yield

$$\int_0^{\Delta t} (m\ddot{x} + c\dot{x} + kx) \, dt = m[\dot{x}(\Delta t) - \dot{x}(0)]$$

$$+ c[x(\Delta t) - x(0)] + \int_0^{\Delta t} kx \, dt = \int_0^{\Delta t} \delta(t) \, dt$$

and take the limit

$$\lim_{\Delta t \to 0} c[x(\Delta t) - x(0)] = 0$$

There are no jumps in the displacements $x(t)$, because of the too-short time interval allowed. On the other hand,

$$\lim_{\Delta t \to 0} m[\dot{x}(\Delta t) - \dot{x}(0)] = m\dot{x}(0+)$$

and since

$$\int_0^{\Delta t} \delta(t)\, dt = 1$$

we have

$$m\dot{x}(0+) = 1$$

or

$$\dot{x}(0+) = \frac{1}{m}$$

implying that the unit impulse $\delta(t)$ is actually equivalent to the instantaneous change in the velocity. We thus may view the impulse applied at $t = 0$ as the initial velocity equal to $1/m$, and instead of the nonhomogeneous equation (9.22) with homogeneous initial conditions (9.23), solve the homogeneous equation

$$m\ddot{h} + c\dot{h} + kh = 0 \qquad\qquad (9.24)$$

with nonhomogeneous initial conditions,

$$x(0) = 0, \quad \dot{x}(0) = \frac{1}{m} \qquad\qquad (9.25)$$

We next rewrite Eq. (9.24) in the form

$$\ddot{h} + 2\zeta\omega_0\dot{h} + \omega_0^2 h = 0$$

where $\omega_0 = \sqrt{k/m}$ is the natural frequency and $\zeta = c/2m\omega_0$ the viscous damping factor. Integration of the latter yields

$$h(t) = \begin{cases} A_1\exp\left[\left(-\zeta + \sqrt{\zeta^2 - 1}\,\right)\omega_0 t\right] + A_2\exp\left[\left(-\zeta - \sqrt{\zeta^2 - 1}\,\right)\omega_0 t\right], & \zeta > 1 \\ (A_1 + tA_2)\exp(-\omega_0 t), & \zeta = 1 \\ \left[A_1\exp(i\omega_d t) + A_2\exp(-i\omega_d t)\right]\exp(-\zeta\omega_0 t), & \zeta < 1 \end{cases}$$

representing an overdamped, a critically damped, and an underdamped structure, respectively, and $\omega_d = \omega_0\sqrt{1 - \zeta^2}$ is called the frequency of the damped free vibration.

Subject to the initial conditions (9.25), $h(t)$ becomes

$$h(t) = \begin{cases} \dfrac{1}{m\omega_0\sqrt{\zeta^2 - 1}} \exp(-\zeta\omega_0 t)\sinh\left(\sqrt{\zeta^2 - 1}\,\omega_0 t\right), & \zeta > 1 \\[2mm] \dfrac{t}{m} \exp(-\omega_0 t), & \zeta = 1 \\[2mm] \dfrac{1}{m\omega_d} \exp(-\zeta\omega_0 t)\sin(\omega_d t), & \zeta < 1 \end{cases} \quad (9.26)$$

In perfect analogy with this example, it can be shown that for a system described by the differential equation (9.1) the impulse-response function $h(t)$ can be found as the solution of the following homogeneous equation:

$$a_0 \frac{d^n h}{dt^n} + a_1 \frac{d^{n-1} h}{dt^{n-1}} + \cdots + a_{n-1} \frac{dh}{dt} + a_n h = 0 \qquad (9.27)$$

supplemented by the nonhomogeneous initial conditions

$$h = \frac{dh}{dt} = \cdots = \frac{d^{n-2} h}{dt^{n-2}} = 0, \quad \frac{d^{n-1} h}{dt^{n-1}} = \frac{1}{a_0}, \quad t = 0 \qquad (9.28)$$

9.2 RESPONSE OF A LINEAR SYSTEM SUBJECTED TO RANDOM EXCITATION

Let us visualize now that the excitation is represented by a random function $F(t)$ with given mean $m_F(t)$ and the autocorrelation function $R_F(t_1, t_2)$. The linear system with the impulse-response function $h(t)$ then transforms $F(t)$ into another random function $X(t)$:

$$X(t) = \int_{-\infty}^{\infty} F(\tau)h(t - \tau)\,d\tau = \int_{-\infty}^{\infty} F(t - \tau)h(\tau)\,d\tau \qquad (9.29)$$

where the integral is understood in the mean-square sense, as defined in Sec. 8.12. The mean function $m_X(t)$ and the autocorrelation function $R_X(t_1, t_2)$ are found as

$$m_X(t) = \int_{-\infty}^{\infty} m_F(t)h(t - \tau)\,d\tau = \int_{-\infty}^{\infty} m_F(t - \tau)h(\tau)\,d\tau \qquad (9.30)$$

$$R_X(t_1, t_2) = \int_{-\infty}^{\infty}\int_{-\infty}^{\infty} R_F(\tau_1, \tau_2)h(t_1 - \tau_1)h(t_2 - \tau_2)\,d\tau_1\,d\tau_2 \qquad (9.31)$$

The latter equation can also be written

$$R_X(t_1, t_2) = \int_{-\infty}^{\infty}\int_{-\infty}^{\infty} R_F(t_1 - \tau_1, t_2 - \tau_2)h(\tau_1)h(\tau_2)\,d\tau_1\,d\tau_2 \qquad (9.32)$$

When $F(t)$ replaces a stationary random function, we have instead of (9.30),

$$m_X(t) = \int_{-\infty}^{\infty} m_F h(\tau) \, d\tau = m_F \int_{-\infty}^{\infty} h(\tau) \, d\tau$$

where m_F is a constant. Finally, substituting $\omega = 0$ in Eq. (9.21) and comparing the result with the above equation, we arrive at

$$m_X(t) = m_F H(0) = \text{const} = m_X \tag{9.33}$$

The autocorrelation function $R_F(t_1, t_2)$ is a function only of the difference $t_2 - t_1$ so that $R_F(t_1 - \tau_1, t_2 - \tau_2)$ in Eq. (9.32) is a function only of $t_2 - \tau_2 - (t_1 - \tau_1) = t_2 - t_1 - \tau_2 + \tau_1$, or, with $t_2 - t_1 = \tau$,

$$R_X(t_1, t_2) = \int_{-\infty}^{\infty} \int_{-\infty}^{\infty} R_F(\tau - \tau_2 + \tau_1) h(\tau_1) h(\tau_2) \, d\tau_1 \, d\tau_2 \equiv R_X(\tau)$$

$$\tag{9.34}$$

Equations (9.33) and (9.34) imply that the output of a linear system with constant coefficients subjected to random excitation stationary in the wide sense, is itself of the same kind. To find its spectral density, we apply the Wiener-Khintchine relationship (8.67) to find

$$S_X(\omega) = \frac{1}{2\pi} \int_{-\infty}^{\infty} \int_{-\infty}^{\infty} h(\tau_1) h(\tau_2) \, d\tau_1 \, d\tau_2 \int_{-\infty}^{\infty} R_F(\tau - \tau_2 + \tau_1) e^{-i\omega\tau} \, d\tau$$

Denoting $\tau - \tau_2 + \tau_1 = \lambda$ and making use of Eq. (9.21),

$$S_X(\omega) = \int_{-\infty}^{\infty} h(\tau_1) e^{i\omega\tau_1} \, d\tau_1 \int_{-\infty}^{\infty} h(\tau_2) e^{-i\omega\tau_2} \, d\tau_2 \cdot \frac{1}{2\pi} \int_{-\infty}^{\infty} R_F(\lambda) e^{-i\omega\lambda} \, d\lambda$$

$$= H^*(\omega) H(\omega) S_F(\omega) = |H(\omega)|^2 S_F(\omega) \tag{9.35}$$

The autocorrelation function $R_X(\tau)$ can be found through (8.68):

$$R_X(\tau) = \int_{-\infty}^{\infty} S_X(\omega) e^{i\omega\tau} \, d\omega = \int_{-\infty}^{\infty} |H(\omega)|^2 S_F(\omega) e^{i\omega\tau} \, d\omega \tag{9.36}$$

while the mean square $E(X^2)$ is

$$E(X^2) = R_X(0) = \int_{-\infty}^{\infty} |H(\omega)|^2 S_F(\omega) \, d\omega$$

$$= \int_{-\infty}^{\infty} \int_{-\infty}^{\infty} R_F(\tau_1 - \tau_2) h(\tau_1) h(\tau_2) \, d\tau_1 \, d\tau_2 \tag{9.37}$$

Example 9.2

The system governed by the differential equation

$$\ddot{X} - 2a\dot{X} - 8a^2 X = F(t)$$

where a is a real constant and $F(t)$, a stationary random function, does not admit a stationary solution, since it is asymptotically unstable. One of the roots of the auxiliary equation $r^2 - 2ar - 8a^2 = 0$ is positive, irrespective of the sign of a. It is readily shown that

$$X(t) = C_1 e^{4at} + C_2 e^{-2at} + \frac{1}{6a} \int_{-\infty}^{\infty} [e^{4a(t-\tau)} - e^{-2a(t-\tau)}] U(t - \tau) F(\tau) \, d\tau$$

and the probabilistic characteristics of $X(t)$ depend on the initial conditions imposed.

Example 9.3

Consider a spring-dashpot system with negligible mass, so that is is governed by the differential equation

$$c\dot{X} + kX = F(t) \tag{9.38}$$

The complex frequency response is

$$H(\omega) = \frac{1}{ci\omega + k} \tag{9.39}$$

and Eq. (9.35) becomes

$$S_X(\omega) = \frac{S_F(\omega)}{c^2\omega^2 + k^2} \tag{9.40}$$

Assuming that $F(t)$ is a band-limited white noise,

$$S_F(\omega) = \begin{cases} S_0, & |\omega| \leqslant \omega_c \\ 0, & \text{otherwise} \end{cases} \tag{9.41}$$

Equation (9.36) then yields

$$R_X(\tau) = S_0 \int_{-\omega_c}^{\omega_c} \frac{e^{i\omega\tau} \, d\omega}{c^2\omega^2 + k^2} \tag{9.42}$$

The mean-square value equals

$$E(X^2) = S_0 \int_{-\omega_c}^{\omega_c} \frac{d\omega}{c^2\omega^2 + k^2} = \frac{2S_0}{kc} \tan^{-1}\left(\frac{c\omega_c}{k}\right) \tag{9.43}$$

For $\omega_c \to \infty$, we have at input an ideal white noise, with

$$R_X(\tau) = S_0 \lim_{\omega_c \to \infty} \int_{-\omega_c}^{\omega_c} \frac{e^{i\omega\tau} \, d\omega}{c^2\omega^2 + k^2} = \frac{S_0\pi}{ck} e^{-(k/c)|\tau|} \tag{9.44}$$

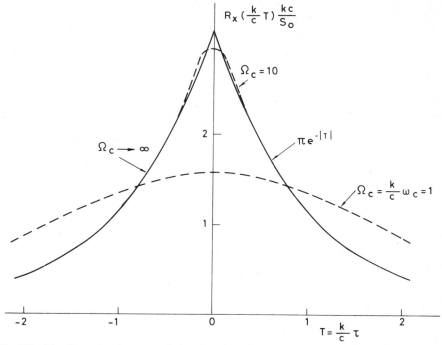

Fig. 9.3. Nondimensional autocorrelation function of displacements of spring-dashpot system with negligible mass tends to $\pi \exp(-|T|)$ at large values of Ω_c.

(see Fig. 9.3) with the mean-square

$$E(X^2) = \frac{S_0 \pi}{ck} \tag{9.45}$$

Note that in the latter case the mean square of the output turns out to be finite, in spite of the fact that the mean square of the input is infinite. Of interest is also the mean-square velocity:

$$E(\dot{X}^2) = S_0 \int_{-\omega_c}^{\omega_c} \frac{\omega^2 \, d\omega}{c^2\omega^2 + k^2} = \frac{2S_0}{c^2}\left[\omega_c - \frac{k}{c} \tan^{-1}\left(\frac{c\omega_c}{k}\right) \right] \tag{9.46}$$

for band-limited white noise. With $\omega_c \to \infty$ we obtain

$$E(\dot{X}^2) = S_0 \int_{-\infty}^{\infty} \frac{\omega^2 \, d\omega}{c^2\omega^2 + k^2} \to \infty \tag{9.47}$$

for ideal white noise. This is obvious, since $\dot{X} = [F(t) - kX]c^{-1}$, where $F(t)$ has an infinite mean square.

Example 9.4

Consider a single-degree-of-freedom system with band-limited white noise $F(t)$ as input:

$$\ddot{X} + 2\zeta\omega_0\dot{X} + \omega_0^2 X = \frac{1}{m}F(t) \tag{9.48}$$

under light damping $0 < \zeta \ll 1$. The spectral density of the displacements is readily found as

$$S_X(\omega) = \frac{1}{m^2}\frac{S_0}{\left|(i\omega)^2 + 2\zeta(i\omega)\omega_0 + \omega_0^2\right|^2}$$

$$= \frac{1}{m^2}\frac{S_0}{\left(\omega_0^2 - \omega^2\right)^2 + 4\zeta^2\omega_0^2\omega^2}, \quad |\omega| \leqslant \omega_c \tag{9.49}$$

and zero otherwise. The autocorrelation function is

$$R_X(\tau) = \frac{S_0}{m^2}\int_{-\omega_c}^{\omega_c}\frac{e^{i\omega\tau}\,d\omega}{\left(\omega_0^2 - \omega^2\right)^2 + 4\zeta^2\omega_0^2\omega^2} \tag{9.50}$$

The mean square is given by

$$E(X^2) = R_X(0) = \frac{S_0}{m^2}\int_{-\omega_c}^{\omega_c}\frac{d\omega}{\left(\omega_0^2 - \omega^2\right)^2 + 4\zeta^2\omega_0^2\omega^2} \tag{9.51}$$

It is readily verified that the denominator in Eq. (9.51) can be represented as

$$\left(\omega_0^2 - \omega^2\right)^2 + 4\zeta^2\omega_0^2\omega^2 = \left[(\omega - \omega_d)^2 + \omega_0^2\zeta^2\right]\left[(\omega + \omega_d)^2 + \omega_0^2\zeta^2\right]$$

where $\omega_d = \omega_0(1 - \zeta^2)^{1/2}$ is the frequency of damped free vibration.

Using the method of undetermined coefficients, the integrand in Eq. (9.51) is represented as

$$\frac{1}{\left(\omega_0^2 - \omega^2\right)^2 + 4\zeta^2\omega_0^2\omega^2} = \frac{M_1\omega + N_1}{(\omega - \omega_d)^2 + \omega_0^2\zeta^2} + \frac{M_2\omega + N_2}{(\omega + \omega_d)^2 + \omega_0^2\zeta^2}$$

where

$$M_1 = -M_2 = -\frac{1}{4\omega_d\omega_0^2} \qquad N_1 = N_2 = \frac{1}{2\omega_0^2}$$

and Eq. (9.51) is a sum of two integrals:

$$E(X^2) = \frac{S_0}{2m^2\omega_0^2} \int_{-\omega_c}^{\omega_c} \left[\frac{(-\omega/2\omega_d) + 1}{(\omega - \omega_d)^2 + \omega_0^2\zeta^2} + \frac{(\omega/2\omega_d) + 1}{(\omega + \omega_d)^2 + \omega_0^2\zeta^2} \right] d\omega \quad (9.52)$$

It is seen that the second integral is readily obtainable if we formally replace $-\omega_d$ by $+\omega_d$ in the result of the first integral. However,

$$\int \frac{(\mu x + \nu) \, dx}{x^2 + px + q^2} = \frac{\mu}{2} \ln(t^2 + \alpha^2) + \frac{1}{\alpha}\left(\nu - \frac{\mu p}{2}\right)\tan^{-1}\left(\frac{t}{\alpha}\right) + C \quad (9.53)$$

for $q^2 > p^2/4$, where $\alpha = \sqrt{q^2 - p^2/4}$, $t = x + p/2$, and C is the integration constant. For the first integral in (9.52) we have

$$\mu = -\frac{1}{2\omega_d} \qquad \nu = 1 \qquad p = -2\omega_d \qquad q^2 = \omega_d^2 + \omega_0^2\zeta^2 = \omega_0^2$$

$$\alpha = \sqrt{\omega_0^2 - \omega_d^2} = \omega_0\zeta \qquad t = \omega - \omega_d$$

with

$$\int_{-\omega_c}^{\omega_c} \frac{(-\omega/2\omega_d) + 1}{(\omega - \omega_d)^2 + \omega_0^2\zeta^2} d\omega = -\frac{1}{4\omega_d} \ln \frac{(\omega_c - \omega_d)^2 + \omega_0^2\zeta^2}{(\omega_c + \omega_d)^2 + \omega_0^2\zeta^2}$$

$$+ \frac{1}{2\omega_0\zeta}\left[\tan^{-1}\left(\frac{\omega_c - \omega_d}{\omega_0\zeta}\right) + \tan^{-1}\left(\frac{\omega_c + \omega_d}{\omega_0\zeta}\right)\right]$$

The final result for the mean square $E(X^2)$,

$$E(X^2) = \frac{S_0}{2m^2\omega_0^2}\left\{ \frac{1}{2\omega_d} \ln \frac{(\omega_c + \omega_d)^2 + \omega_0^2\zeta^2}{(\omega_c - \omega_d)^2 + \omega_0^2\zeta^2} \right.$$

$$\left. + \frac{1}{\omega_0\zeta}\left[\tan^{-1}\left(\frac{\omega_c - \omega_d}{\omega_0\zeta}\right) + \tan^{-1}\left(\frac{\omega_c + \omega_d}{\omega_0\zeta}\right)\right]\right\} \quad (9.54)$$

may be written down as

$$E(X^2) = \frac{S_0\pi}{2m^2\zeta\omega_0^3} I_0\left(\frac{\omega_c}{\omega_0}, \zeta\right) \quad (9.55)$$

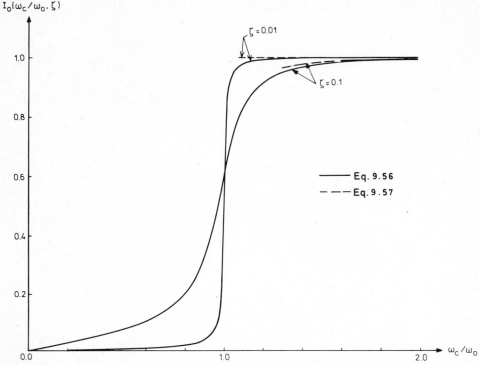

Fig. 9.4. Integral factor $I_0(\omega_c/\omega_0, \zeta)$ for mean-square displacement of single-degree-of-freedom system under band-limited white noise.

where the integral factor $I_0(\omega_c/\omega_0, \zeta)$ is given by

$$I_0\left(\frac{\omega_c}{\omega_0}, \zeta\right) = \frac{\zeta}{2\pi\sqrt{1-\zeta^2}} \ln \frac{\left(\omega_c/\omega_0 + \sqrt{1-\zeta^2}\right)^2 + \zeta^2}{\left(\omega_c/\omega_0 - \sqrt{1-\zeta^2}\right)^2 + \zeta^2}$$

$$+ \frac{1}{\pi}\left[\tan^{-1}\left(\frac{\omega_c/\omega_0 - \sqrt{1-\zeta^2}}{\zeta}\right) + \tan^{-1}\left(\frac{\omega_c/\omega_0 + \sqrt{1-\zeta^2}}{\zeta}\right)\right]$$

$$(9.56)$$

On summation of the terms in brackets, Eq. (9.56) yields the formula due to Crandall and Mark. The integral factor $I_0(\omega_c/\omega_0, \zeta)$ is shown in Fig. 9.4, as a function of ω_c/ω_0, for different values of ζ.

A simple asymptotic expression can be obtained for ω_c beyond the natural frequency ω_0. By evenness of the integrand, Eq. (9.51) may be written as

$$E(X^2) = \frac{2S_0}{\omega_0^4 m^2} \int_0^{\omega_c} \frac{d\omega}{\left(1 - \omega^2/\omega_0^2\right)^2 + \left(2\zeta\omega/\omega_0\right)^2}$$

Further,

$$E(X^2) = \frac{2S_0}{\omega_0^4 m^2}\left[\int_0^\infty \frac{d\omega}{(1 - \omega^2/\omega_0^2)^2 + (2\zeta\omega/\omega_0)^2}\right.$$

$$\left. - \int_{\omega_c}^\infty \frac{d\omega}{(1 - \omega^2/\omega_0^2)^2 + (2\zeta\omega/\omega_0)^2}\right]$$

$$= \frac{2S_0}{\omega_0^4 m^2}\left[\frac{\pi\omega_0}{4\zeta} - \int_{\omega_c}^\infty \frac{d\omega}{(1 - \omega^2/\omega_0^2)^2 + (2\zeta\omega/\omega_0)^2}\right]$$

In the last integral we change the variable $y = \omega_0/\omega$, $d\omega = -(\omega_0/y^2)\,dy$,

$$\int_{\omega_c}^\infty \frac{d\omega}{(1 - \omega^2/\omega_0^2)^2 + (2\zeta\omega/\omega_0)^2} = \omega_0\int_0^{\omega_0/\omega_c} \frac{y^2\,dy}{(y^2 - 1)^2 + (2\zeta y)^2}$$

Using asymptotic series expansion for the integrand $f(y) = y^2[(y^2 - 1)^2 + (2\zeta y)^2]^{-1}$,

$$f(y) = f(0) + \sum_{i=1}^\infty \frac{1}{i!}y^i f^{(i)}(0)$$

where $f^{(i)}(y)$ denotes the ith derivative of $f(y)$. Taking only four terms of the series, with

$$f(0) = 0 \qquad f'(0) = 0 \qquad f''(0) = 2 \qquad f'''(0) = 0$$

and

$$f^{(4)}(y) = 48(1 - 2\zeta^2)$$

we obtain

$$\int_0^{\omega_0/\omega_c} f(y)\,dy = \int_0^{\omega_0/\omega_c}\left[\frac{2y^2}{2!} + \frac{48(1 - 2\zeta^2)y^4}{4!} + \cdots\right]dy$$

$$= \frac{1}{3}\frac{1}{(\omega_c/\omega_0)^3} + \frac{2}{5}\frac{1 - 2\zeta^2}{(\omega_c/\omega_0)^5} + \cdots$$

and finally

$$E(X^2) = \frac{S_0 \pi}{2\zeta\omega_0^3 m^2}\left[1 - \frac{4}{3\pi}\frac{\zeta}{(\omega_c/\omega_0)^3} - \frac{8}{5\pi}\frac{1-2\zeta^2}{(\omega_c/\omega_0)^5} + \cdots\right] \quad (9.57)$$

This approximate formula is due to Warburton. Comparison with the exact Eqs. (9.55) and (9.56) reveals immediately that the expression in square brackets in (9.57) represents an approximation of the integral factor $I_0(\omega_c/\omega_0, \zeta)$. This approximation is also shown in Fig. 9.4, and its results are practically coincident with their exact counterparts.

Consider now the case of an undamped system ($\zeta = 0$) under band-limited white noise. Instead of Eq. (9.51), we have

$$E(X^2) = \frac{S_0}{m^2}\int_{-\omega_c}^{\omega_c}\frac{d\omega}{(\omega_0^2 - \omega^2)^2} = \frac{S_0}{2\omega_0^3 m^2}\left[\ln\frac{\omega_0 + \omega_c}{\omega_0 - \omega_c} + \frac{2\omega_c\omega_0}{\omega_0^2 - \omega_c^2}\right] \quad (9.58)$$

The mean square is finite if $\omega_c < \omega_0$, and for $\omega_c \to \omega_0$, $E(X^2) \to \infty$.

For the mean square of the velocity we have

$$E(\dot{X}^2) = \frac{S_0}{m^2}\int_{-\omega_c}^{\omega_c}\frac{\omega^2\,d\omega}{(\omega_0^2 - \omega^2)^2 + 4\zeta^2\omega_0^2\omega^2} \quad (9.59)$$

This can be rewritten as

$$E(\dot{X}^2) = \frac{S_0}{4m^2\omega_0}\int_{-\omega_c}^{\omega_c}\left[\frac{\omega}{(\omega - \omega_d)^2 + \omega_0^2\zeta^2} - \frac{\omega}{(\omega + \omega_d)^2 + \omega_0^2\zeta^2}\right]d\omega$$

which yields, with the aid of Eq. (9.53), with $\mu_1 = -\mu_2 = 1$ and $\nu_1 = \nu_2 = 0$,

$$E(\dot{X}^2) = \frac{S_0}{2m^2\omega_0}\left\{\frac{1}{2(1-\zeta^2)^{1/2}}\ln\frac{\left[\omega_c/\omega_0 - (1-\zeta^2)^{1/2}\right]^2 + \zeta^2}{\left[\omega_c/\omega_0 + (1-\zeta^2)^{1/2}\right]^2 + \zeta^2}\right.$$
$$+ \frac{1}{\zeta}\left[\tan^{-1}\left(\frac{\omega_c/\omega_0 - (1-\zeta^2)^{1/2}}{\zeta}\right)\right.$$
$$\left.\left. + \tan^{-1}\left(\frac{\omega_c/\omega_0 + (1-\zeta^2)^{1/2}}{\zeta}\right)\right]\right\} \quad (9.60)$$

This result can be put in a form analogous to Eqs. (9.55) and (9.56):

$$E(\dot{X}^2) = \frac{S_0 \pi}{2m^2 \zeta \omega_0} I_1\left(\frac{\omega_c}{\omega_0}, \zeta\right) \tag{9.61}$$

$$I_1\left(\frac{\omega_c}{\omega_0}, \zeta\right) = \frac{\zeta}{2\pi\sqrt{1-\zeta^2}} \ln \frac{\left[\omega_c/\omega_0 - (1-\zeta^2)^{1/2}\right]^2 + \zeta^2}{\left[\omega_c/\omega_0 + (1-\zeta^2)^{1/2}\right]^2 + \zeta^2}$$

$$+ \frac{1}{\pi}\left[\tan^{-1}\left(\frac{\omega_c/\omega_0 - (1-\zeta^2)^{1/2}}{\zeta}\right) + \tan^{-1}\left(\frac{\omega_c/\omega_0 + (1-\zeta^2)^{1/2}}{\zeta}\right)\right]$$

$$\tag{9.62}$$

I_1 is shown in Fig. 9.5 for $\zeta = 0.01$ and $\zeta = 0.1$. The mean-square acceleration is

$$E(\ddot{X}^2) = \frac{S_0}{m^2} \int_{-\omega_c}^{\omega_c} \frac{\omega^4 \, d\omega}{\left(\omega_0^2 - \omega^2\right)^2 + 4\zeta^2\omega_0^2\omega^2}$$

$$= \frac{S_0}{m^2} \int_{-\omega_c}^{\omega_c} \left[1 + \frac{2\omega^2\omega_0^2(1 - 2\zeta^2) - \omega_0^4}{\left(\omega_0^2 - \omega^2\right)^2 + 4\zeta^2\omega_0^2\omega^2}\right] d\omega$$

and we obtain

$$E(\ddot{X}^2) = \frac{2S_0}{m^2}\omega_c \left\{1 + \frac{1 - 4\zeta^2}{8(1-\zeta^2)^{1/2}}\left(\frac{\omega_0}{\omega_c}\right)\ln\frac{\left[\omega_c/\omega_0 - (1-\zeta^2)^{1/2}\right]^2 + \zeta^2}{\left[\omega_c/\omega_0 + (1-\zeta^2)^{1/2}\right]^2 + \zeta^2}\right.$$

$$+ \frac{1 - 4\zeta^2}{4\zeta}\left(\frac{\omega_0}{\omega_c}\right)\left[\tan^{-1}\left(\frac{\omega_c/\omega_0 - (1-\zeta^2)^{1/2}}{\zeta}\right)\right.$$

$$\left.\left. + \tan^{-1}\left(\frac{\omega_c/\omega_0 + (1-\zeta^2)^{1/2}}{\zeta}\right)\right]\right\} \tag{9.63}$$

Example 9.5

Consider a single-degree-of-freedom system as in the preceding example, but with ideal white noise $F(t)$ as input. Results for this case are obtainable from those for band-limited white noise $F(t)$ when $\omega_c \to \infty$.

$I_1(\omega_c/\omega_0, \zeta)$

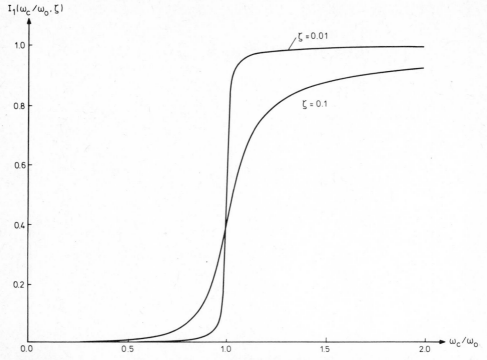

Fig. 9.5. Integral factor $I_1(\omega_c/\omega_0, \zeta)$ for mean-square velocity of a viscously damped single-degree-of-freedom system under band-limited white noise.

The autocorrelation function becomes, instead of Eq. (9.50),

$$R_X(\tau) = \frac{S_0}{m^2} \int_{-\infty}^{\infty} \frac{e^{i\omega\tau}\, d\omega}{\left(\omega_0^2 - \omega^2\right)^2 + 4\zeta^2\omega_0^2\omega^2} \qquad (9.64)$$

Introducing the notation

$$\alpha = \zeta\omega_0 \qquad \beta = \omega_0\left(1 - \zeta^2\right)^{1/2} = \omega_d \qquad \gamma = \frac{\alpha}{\beta} \qquad d^2 = \frac{S_0\pi}{2\zeta\omega_0^3 m^2}$$

$$(9.65)$$

the corresponding spectral density coincides with Eq. (8.107). Therefore,

$$R_X(\tau) = d^2 e^{-\zeta\omega_0|\tau|}\left(\cos\omega_d\tau + \frac{\zeta\omega_0}{\omega_d}\sin\omega_d|\tau|\right) \qquad (9.66)$$

and since $E(X^2) = R_X(0)$,

$$E(X^2) = d^2 = \frac{S_0\pi}{2\zeta\omega_0^3 m^2} \tag{9.67}$$

It is interesting to note that on substitution of $\zeta = c/2m\omega_0$, $\omega_0^2 = k/m$ in Eq. (9.67), we get $E(X^2) = \pi S_0/ck$, which coincides with the result (9.45) for the mean square of a massless system. The conclusion is that the mass of a single-degree-of-freedom system under ideal white noise excitation does not influence the mean-square displacement.

Comparison of the mean-square displacements found for the systems with band-limited (Eq. 9.55) and ideal (Eq. 9.67) white noise shows immediately that the former equals the latter multiplied by the integral factor I_0 in Eq. (9.56). As is seen from Fig. 9.4 for $\omega_c > \omega_0$ and a lightly damped system ($\zeta \ll 1$), the integral factor differs negligibly from unity. This means that although ideal white noise is a "mathematical fiction," it may yield a highly satisfactory result for the mean-square displacement.

As we saw in Example 8.14, the random function $X(t)$ is differentiable, so that $\dot{X}$ has a finite mean square obtainable from Eqs. (9.61) and (9.62) when $\omega_c \to \infty$:

$$E(\dot{X}^2) = \frac{S_0\pi}{2\zeta\omega_0 m^2} = \frac{S_0\pi}{mc}, \tag{9.68}$$

implying that the stiffness of a single-degree-of-freedom system under ideal white-noise excitation does not influence the mean-square velocity.

Eq. (9.68) is also obtainable directly, by the residue theorem, from the expression

$$E(\dot{X}^2) = \frac{S_0}{m^2} \int_{-\infty}^{\infty} \frac{\omega^2\,d\omega}{\left(\omega_0^2 - \omega^2\right)^2 + 4\zeta^2\omega_0^2\omega^2}$$

It is worth noting that for a system under ideal white noise excitation, comparison of Eqs. (9.67) and (9.68) yields

$$E(\dot{X}^2) = E(X^2)\omega_0^2$$

As is seen from Eq. (9.63), the mean-square acceleration tends to infinity with ω_c. This is explained by the fact that $X(t)$, the displacement of a system with ideal white noise, is not doubly differentiable, that is, $\ddot{X}(t)$ is not physically realizable since the spectral density does not satisfy the condition

$$\int_{-\infty}^{\infty} \omega^4 S_X(\omega)\,d\omega < \infty$$

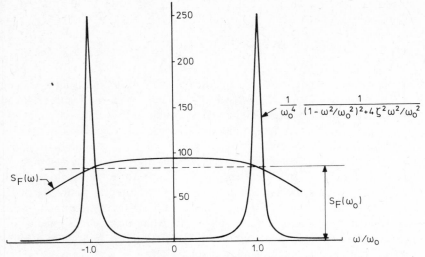

Fig. 9.6. Illustration of Laplace's asymptotic evaluation of integral (9.69).

This is also seen from Eq. (9.48) itself, since here $\ddot{X} = F(t)/m - 2\zeta\omega_0\dot{X} - \omega_0^2 X$, where X and $\dot{X}$ have finite mean squares, but $F(t)$ has an infinite one.

Equation (9.65) for a single-degree-of-freedom system with ideal white noise can be extended to a lightly damped system with nonwhite (colored) noise inputs. The term

$$\frac{1}{\left(\omega_0^2 - \omega^2\right)^2 + 4\zeta^2\omega_0^2\omega^2}$$

exhibits very sharp peaks in the vicinity of $\pm\omega_0$. In these circumstances the dominant contribution to the mean square $E(X^2)$ derives from the frequencies close to $\pm\omega_0$ (see Fig. 9.6). Accordingly, we resort to Laplace's asymptotic method for evaluation of the integral:

$$E(X^2) = \frac{1}{m^2}\int_{-\infty}^{\infty} \frac{S_F(\omega)\,d\omega}{\left(\omega_0^2 - \omega^2\right)^2 + 4\zeta^2\omega_0^2\omega^2} \tag{9.69}$$

The dominant contribution to the integral derives from the values of the integrand function close to those of ω for which the integrand is maximized. The values in question can be the minimum points $\omega = \pm\omega_0$ of the frequency function in the denominator if $S_F(\omega)$ is smooth in the neighborhood of $\pm\omega_0$:

$$E(X^2) \simeq \frac{1}{m^2} \int_{-\omega_0-\varepsilon}^{-\omega_0+\varepsilon} \frac{S_F(\omega)\, d\omega}{\left(\omega_0^2 - \omega^2\right)^2 + 4\zeta^2\omega_0^2\omega^2}$$

$$+ \frac{1}{m^2} \int_{\omega_0-\varepsilon}^{\omega_0+\varepsilon} \frac{S_F(\omega)\, d\omega}{\left(\omega_0^2 - \omega^2\right)^2 + 4\zeta^2\omega_0^2\omega^2}$$

$$\simeq \frac{S_F(-\omega_0)}{m^2} \int_{-\omega_0-\varepsilon}^{-\omega_0+\varepsilon} \frac{d\omega}{\left(\omega_0^2 - \omega^2\right)^2 + 4\zeta^2\omega_0^2\omega^2}$$

$$+ \frac{S_F(\omega_0)}{m^2} \int_{\omega_0-\varepsilon}^{\omega_0+\varepsilon} \frac{d\omega}{\left(\omega_0^2 - \omega^2\right)^2 + 4\zeta^2\omega_0^2\omega^2}$$

and since $S_F(\omega)$ is an even function, $S_F(-\omega_0) = S_F(\omega_0)$, and

$$E(X^2) \simeq \frac{S_F(\omega_0)}{m^2} \left[\int_{-\omega_0-\varepsilon}^{-\omega_0+\varepsilon} \frac{d\omega}{\left(\omega_0^2 - \omega^2\right)^2 + 4\zeta^2\omega_0^2\omega^2} \right.$$

$$\left. + \int_{\omega_0-\varepsilon}^{\omega_0+\varepsilon} \frac{d\omega}{\left(\omega_0^2 - \omega^2\right)^2 + 4\zeta^2\omega_0^2\omega^2} \right]$$

$$\simeq \frac{S_F(\omega_0)}{m^2} \int_{-\infty}^{\infty} \frac{d\omega}{\left(\omega_0^2 - \omega^2\right)^2 + 4\zeta^2\omega_0^2\omega^2} = \frac{S_F(\omega_0)\pi}{2\zeta\omega_0^3 m^2} \qquad (9.70)$$

Comparing Eqs. (9.70) and (9.57), we see that application of Laplace's asymptotic method is actually equivalent to the assumption that the input is ideal white noise with intensity $S_F(\omega_0)$. As we have shown [see discussion of Eq. (9.67)], the assumption of ideal instead of band-limited white noise as input is very satisfactory for a lightly damped system if $\omega_c > \omega_0$. This implies that for such a system Laplace's asymptotic method yields a result (9.70) very close to the exact one of Eq. (9.55). Note that the approximation given by Eq. (9.70) is perfectly exact for $S_F(\omega) = S_0(a^2\omega^2 + b^2)$, where a and b are arbitrary constants (see Appendix C, Example C2). For some cases, (9.70) can be refined. If $S_F(\omega)$ takes on significant values at $\omega = 0$ and is a decreasing function, we can also take into account the contribution of the peak in $S_F(\omega)$ at zero frequency and

$$E(X^2) \simeq \frac{S_F(\omega_0)\pi}{2\zeta\omega_0^3 m^2} + \frac{1}{m^2} \left[\frac{1}{\left(\omega_0^2 - \omega^2\right)^2 + 4\zeta^2\omega_0^2\omega^2} \right]_{\omega=0} \int_{-\infty}^{\infty} S_F(\omega)\, d\omega$$

$$= \frac{S_F(\omega_0)\pi}{2\zeta\omega_0^3 m^2} + \frac{E(F^2)}{\omega_0^4 m^2} = \frac{S_F(\omega_0)\pi}{ck} + \frac{E(F^2)}{k^2} \qquad (9.71)$$

The significance of the second term in this expression is obvious: It represents the mean-square displacement of the system under static conditions, where Eq. (9.48) reduces to $kX = F(t)$. To illustrate application of Eq. (9.71), we compare it with the exact expression (9.55) for $E(X^2)$ of the system with band-limited white noise at $\omega_c \ll \omega_0$. Then $S_F(\omega_0) = 0$, and only the second term remains in Eq. (9.71). For $\omega_c \ll \omega_0$ and $\zeta \ll 1$, we have in Eq. (9.56),

$$\frac{\omega_c/\omega_0 - (1 - \zeta^2)^{1/2}}{\zeta} < 0 \qquad \frac{\omega_c/\omega_0 + (1 - \zeta^2)^{1/2}}{\zeta} > 0$$

so that

$$\tan^{-1}\left(\frac{\omega_c/\omega_0 - (1 - \zeta^2)^{1/2}}{\zeta}\right) + \tan^{-1}\left(\frac{\omega_c/\omega_0 + (1 - \zeta^2)^{1/2}}{\zeta}\right) < \frac{\pi}{2}$$

Using the addition formula for a pair of inverse circular functions,

$$\tan^{-1}z_1 + \tan^{-1}z_2 = \tan^{-1}\left(\frac{z_1 + z_2}{1 - z_1 z_2}\right)$$

we can write Eq. (9.56) as

$$I_0\left(\frac{\omega_c}{\omega_0}, \zeta\right) = \frac{\zeta}{2\pi(1 - \zeta^2)^{1/2}} \ln \frac{\left[\omega_c/\omega_0 + (1 - \zeta^2)^{1/2}\right]^2 + \zeta^2}{\left[\omega_c/\omega_0 - (1 - \zeta^2)^{1/2}\right]^2 + \zeta^2}$$

$$+ \frac{1}{\pi} \tan^{-1}\left(\frac{2\zeta\omega_c/\omega_0}{1 - \omega_c^2/\omega_0^2}\right)$$

and then

$$\ln \frac{\left[\omega_c/\omega_0 + (1 - \zeta^2)^{1/2}\right]^2 + \zeta^2}{\left[\omega_c/\omega_0 - (1 - \zeta^2)^{1/2}\right]^2 + \zeta^2} \sim \ln \frac{1 + 2(\omega_c/\omega_0)(1 - \zeta^2)^{1/2}}{1 - 2(\omega_c/\omega_0)(1 - \zeta^2)^{1/2}}$$

$$\sim 4\frac{\omega_c}{\omega_0}(1 - \zeta^2)^{1/2}$$

Moreover,

$$\tan^{-1}\left(\frac{2\zeta\omega_c/\omega_0}{1 - \omega_c^2/\omega_0^2}\right) \sim 2\zeta\frac{\omega_c}{\omega_0}$$

so that

$$I_0\left(\frac{\omega_c}{\omega_0}, \zeta\right) \sim 4\left(\frac{\zeta}{\pi}\right)\left(\frac{\omega_c}{\omega_0}\right)$$

and with Eq. (9.55)

$$E(X^2) \simeq \frac{S_0\pi}{2m^2\zeta\omega_0^3}\left(\frac{4\zeta\omega_c}{\pi\omega_0}\right) = \frac{2S_0\omega_c}{k^2}$$

However, since

$$E(F^2) = \int_{-\omega_c}^{\omega_c} S_0\,d\omega = 2\omega_c S_0$$

we finally have $E(X^2) = E(F^2)/k^2$, which is our approximation 9.71.

Example 9.6

The system considered in Examples 9.3–9.5 involved *viscous* damping. Here we consider a single-degree-of-freedom system under so-called *structural* damping due to dissipation of energy generated through internal friction.* The relevant equation of motion is obtained formally by replacing the real stiffness coefficient k by $k(1 + i\mu)$, which is called the *complex stiffness*:

$$m\ddot{X} + k(1 + i\mu)X = F(t) \tag{9.72}$$

We now rewrite this equation as

$$\ddot{X} + \frac{K}{m}e^{i\varphi}X = \frac{1}{m}F(t) \tag{9.73}$$

where

$$K = k(1 + \mu^2)^{1/2} \qquad \varphi = \tan^{-1}\mu \tag{9.74}$$

We treat $F(t)$ as band-limited white noise as per Eq. (9.41). The complex frequence response is

$$H(\omega) = \frac{1}{(i\omega)^2 + \Omega_0^2(\cos\varphi + i\sin\varphi)} \tag{9.75}$$

*See, for example, Meirovich, pp. 55–57, or Warburton, pp. 17–19. Although some recent publications maintain that the structural damping notion may yield unsatisfactory results for transient vibration, it is widely referred to in engineering practice.

where $\Omega_0 = \sqrt{K/m}$ is a frequency parameter. The spectral density of displacements is, therefore,

$$S_X(\omega) = \begin{cases} \dfrac{1}{m^2} \left[\dfrac{S_0}{\left(\Omega_0^2\cos\varphi - \omega^2\right)^2 + \Omega_0^4\sin^2\varphi} \right], & |\omega| \leqslant \omega_c \\ 0, & \text{otherwise} \end{cases} \qquad (9.76)$$

The corresponding autocorrelation function is readily obtainable as

$$R_X(\tau) = \frac{S_0}{m^2} \int_{-\omega_c}^{\omega_c} \frac{e^{i\omega\tau}\,d\omega}{\left(\Omega_0^2\cos\varphi - \omega^2\right)^2 + \Omega_0^4\sin^2\varphi} \qquad (9.77)$$

with the mean square

$$E(X^2) = \frac{S_0}{m^2} \int_{-\omega_c}^{\omega_c} \frac{d\omega}{\left(\Omega_0^2\cos\varphi - \omega^2\right)^2 + \Omega_0^4\sin^2\varphi} \qquad (9.78)$$

For an undamped system, $\mu = 0$, $\varphi = 0$, and the result coincides with Eq. (9.58). Consider now the case where $\varphi \neq 0$. The expression for $E(X^2)$ can be put in the following form:

$$E(X^2) = \frac{S_0}{2[2(1 + \cos\varphi)]^{1/2}\Omega_0^3 m^2}$$

$$\times \left[\int_{-\omega_c}^{\omega_c} \frac{-\omega + \Omega_0[2(1 + \cos\varphi)]^{1/2}}{\omega^2 + \Omega_0^2 - \omega\Omega_0[2(1 + \cos\varphi)]^{1/2}}\,d\omega \right.$$

$$\left. + \int_{-\omega_c}^{\omega_c} \frac{\omega + \Omega_0[2(1 + \cos\varphi)]^{1/2}}{\omega^2 + \Omega_0^2 + \omega\Omega_0[2(1 + \cos\varphi)]^{1/2}}\,d\omega \right]$$

which, making use of Eq. (9.53), reduces to

$$E(X^2) = \frac{S_0}{2m^2\Omega_0^3[2(1 + \cos\varphi)]^{1/2}}$$

$$\times \left\{ \ln\left[\frac{1 + (\omega_c/\Omega_0)^2 + (\omega_c/\Omega_0)[2(1 + \cos\varphi)]^{1/2}}{1 + (\omega_c/\Omega_0)^2 - (\omega_c/\Omega_0)[2(1 + \cos\varphi)]^{1/2}} \right] \right.$$

$$+ 2\left(\frac{1 + \cos\varphi}{1 - \cos\varphi} \right)^{1/2} \left[\tan^{-1}\left(\frac{2\omega_c/\Omega_0 + [2(1 + \cos\varphi)]^{1/2}}{[2(1 - \cos\varphi)]^{1/2}} \right) \right.$$

$$\left. \left. + \tan^{-1}\left(\frac{2\omega_c/\Omega_0 - [2(1 + \cos\varphi)]^{1/2}}{[2(1 - \cos\varphi)]^{1/2}} \right) \right] \right\} \qquad (9.79)$$

When $\omega_c \rightarrow \infty$, the expression for the autocorrelation function becomes

$$R_X(\tau) = \frac{S_0}{m^2} \int_{-\infty}^{\infty} \frac{e^{i\omega\tau}\, d\omega}{\left(\Omega_0^2 \cos\varphi - \omega^2\right)^2 + \Omega_0^4 \sin^2\varphi}$$

Introducing the notation

$$\alpha = \frac{1}{2}\left[2(1 - \cos\varphi)\right]^{1/2} \qquad \beta = \Omega_0(1 - \alpha^2)^{1/2} = \Omega_d \qquad \gamma = \frac{\alpha\Omega_0}{\beta}$$

$$d^2 = \frac{S_0\pi}{2m^2\Omega_0^3\alpha} \tag{9.80}$$

the corresponding spectral density coincides with Eq. (8.107), and consequently,

$$R_X(\tau) = d^2 e^{-\alpha\Omega_0|\tau|}\left(\cos\Omega_d\tau + \frac{\alpha\Omega_0}{\Omega_d}\sin\Omega_d|\tau|\right)$$

with the mean-square displacement

$$E(X^2) = d^2 = \frac{S_0\pi}{2m^2\Omega_0^3\alpha} \tag{9.81}$$

Note that the expressions in this example can be written in terms of the frequency $\omega_0 = \sqrt{k/m}$ of natural vibrations of an undamped system and of the damping parameter μ, as defined in Eq. (9.74). We have

$$\Omega_0 = \omega_0(1 + \mu^2)^{1/4} \qquad \cos\varphi = \frac{1}{(1 + \tan^2\varphi)^{1/2}} = \frac{1}{(1 + \mu^2)^{1/2}} \tag{9.82}$$

and Eq. (9.81) becomes

$$E(X^2) = \frac{S_0\pi}{\omega_0^3 m^2} \frac{\left[1 + (1 + \mu^2)^{1/2}\right]^{1/2}}{\mu\left[2(1 + \mu^2)\right]^{1/2}} \tag{9.83}$$

For $\mu \ll 1$, Eq. (9.83) becomes

$$E(X^2) = \frac{S_0\pi}{\mu\omega_0^3 m^2} \tag{9.84}$$

For a system with band-limited white noise, we have, instead of Eq. (9.79),

$$E(X^2) = \frac{S_0\pi}{m^2\Omega_0^3\left[2(1 - \cos\varphi)\right]^{1/2}} I_0'\left(\frac{\omega_c}{\Omega_0}, \cos\varphi\right)$$

where

$$I_0'\left(\frac{\omega_c}{\Omega_0}, \cos\varphi\right) = \frac{1}{2\pi}\left(\frac{1 - \cos\varphi}{1 + \cos\varphi}\right)^{1/2}$$

$$\times \ln\left[\frac{1 + (\omega_c/\Omega_0)^2 + (\omega_c/\Omega_0)[2(1 + \cos\varphi)]^{1/2}}{1 + (\omega_c/\Omega_0)^2 - (\omega_c/\Omega_0)[2(1 + \cos\varphi)]^{1/2}}\right]$$

$$+ \frac{1}{\pi}\left[\tan^{-1}\left(\frac{\sqrt{2}\,\omega_c/\Omega_0 + (1 + \cos\varphi)^{1/2}}{(1 - \cos\varphi)^{1/2}}\right)\right.$$

$$\left. + \tan^{-1}\left(\frac{\sqrt{2}\,\omega_c/\Omega_0 - (1 + \cos\varphi)^{1/2}}{(1 - \cos\varphi)^{1/2}}\right)\right] \qquad (9.85)$$

or, in terms of ω_0 and μ

$$E(X^2) = \frac{S_0\pi}{m^2\omega_0^3}\frac{\left[1 + (1 + \mu^2)^{1/2}\right]^{1/2}}{\mu[2(1 + \mu^2)]^{1/2}}J_0\left(\frac{\omega_c}{\omega_0}, \mu\right) \qquad (9.86)$$

where

$$J_0\left(\frac{\omega_c}{\omega_0}, \mu\right) = \frac{1}{2\pi}\frac{\left[(1 + \mu^2)^{1/2} - 1\right]^{1/2}}{\left[(1 + \mu^2)^{1/2} + 1\right]^{1/2}}$$

$$\times \ln\left\{\frac{(\omega_c/\omega_0)^2 + (\omega_c/\omega_0)\sqrt{2}\left[(1 + \mu^2)^{1/2} + 1\right]^{1/2} + (1 + \mu^2)^{1/2}}{(\omega_c/\omega_0)^2 - (\omega_c/\omega_0)\sqrt{2}\left[(1 + \mu^2)^{1/2} + 1\right]^{1/2} + (1 + \mu^2)^{1/2}}\right\}$$

$$+ \frac{1}{\pi}\left[\tan^{-1}\left(\frac{\sqrt{2}\,\omega_c/\omega_0 + \left[(1 + \mu^2)^{1/2} + 1\right]^{1/2}}{\left[(1 + \mu^2)^{1/2} - 1\right]^{1/2}}\right)\right.$$

$$\left. + \tan^{-1}\left(\frac{\sqrt{2}\,\omega_c/\omega_0 - \left[(1 + \mu^2)^{1/2} + 1\right]^{1/2}}{\left[(1 + \mu^2)^{1/2} - 1\right]^{1/2}}\right)\right] \qquad (9.87)$$

The integral factor $J_0(\omega_c/\omega_0, \mu)$ is shown in Fig. 9.7, as a function of ω_c/ω_0 for different values of μ. It is worth emphasizing that the coefficient preceding the integral factor in Eq. (9.86) for a system with band-limited white noise is no

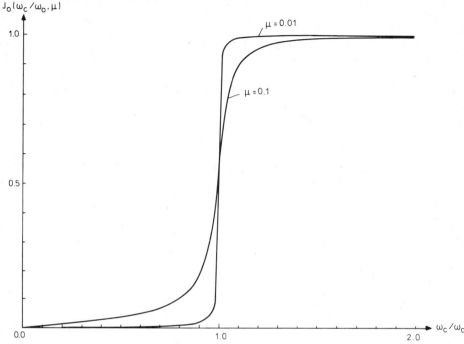

Fig. 9.7. Integral factor $J_0(\omega_c/\omega_0, \mu)$ for mean-square displacement of a structurally damped single-degree-of-freedom system under band-limited white noise.

other than the mean-square displacement for that with ideal white noise, as per Eq. (9.83). As is seen from Fig. 9.7, the integral factor J_0 approaches unity for $\omega_c/\omega_0 > 1$. This again implies, as in the case of the viscously damped system, that for rapid calculation, the ideal white noise assumption may be very handy.

Example 9.7

Let $F(t)$ in Eq. (9.48) have an autocorrelation function $R_F(\tau) = d^2 e^{-\alpha^2\tau^2}$ so that the mean square of F is

$$E(F^2) = R_F(0) = d^2$$

The spectral density $S_F(\omega)$ is

$$S_F(\omega) = \frac{1}{2\pi}\int_{-\infty}^{\infty} d^2 e^{-\alpha^2\tau^2} e^{-i\omega\tau}\, d\tau = \frac{d^2}{2\alpha\sqrt{\pi}}\exp\left(-\frac{\omega^2}{4\alpha^2}\right)$$

making use of integral (A.2) in Appendix A. In accordance with Eq. (9.71),

$$E(X^2) \simeq \frac{d^2}{m^2\omega_0^3} \left[\frac{\sqrt{\pi}}{4\alpha\zeta} \exp\left(-\frac{\omega_0^2}{4\alpha^2} \right) + \frac{1}{\omega_0} \right]$$

It is readily seen that for certain combinations of parameters, the second term in this equation may contribute significantly to the total result.

For a normal random function it suffices to know the mean and autocorrelation functions, since the first-order probability density depends only on the mean and on the variance; these are the quantities of interest to us in this case. If the system obeys Eq. (9.2), we have for the mean $m_X(t)$, in terms of the mean of the input $m_F(t)$,

$$L_n\left(\frac{d}{dt} \right) m_X(t) = m_F(t)$$

For a stable system and stationary $F(t)$, the solution reads:

$$m_X(t) = \frac{m_F}{a_n} = \text{const} = m_X \tag{9.88}$$

The mean square $E(X^2)$ can be found as the integral over the spectral density $S_X(\omega)$ (see 9.13 and 9.37):

$$E(X^2) = \int_{-\infty}^{\infty} \frac{S_F(\omega)\, d\omega}{L_n(i\omega) L_n(-i\omega)} \tag{9.89}$$

and the variance is found as

$$\text{Var}(X) = E(X^2) - m_X^2 \tag{9.90}$$

When $S_F(\omega)$ is a polynomial function, the integral (9.89) can be evaluated in closed form (see Appendix C).

The nonstationary response of simple mechanical systems is discussed below.

Example 9.8

Consider the transient response of a single-degree-of-freedom system, with equation of motion as per Eq. (9.48). We assume the initial conditions

$$X(0) = A \qquad \dot{X}(0) = B \tag{9.91}$$

where A and B are random variables and $F(t)$ is a stationary random function with spectral density $S_F(\omega)$:

$$X(t) = Ae^{-\zeta\omega_0 t}\left(\cos\omega_d t + \frac{\zeta\omega_0}{\omega_d}\sin\omega_d t\right) + \frac{B}{\omega_d}e^{-\zeta\omega_0 t}\sin\omega_d t$$

$$+ \int_0^t h(t-\tau)F(\tau)\,d\tau \tag{9.92}$$

where $h(t-\tau)$ is given in Eq. (9.26) for $\zeta < 1$

$$h(t-\tau) = \frac{1}{m\omega_d}e^{-\omega_0\zeta(t-\tau)}\sin\omega_d(t-\tau)$$

The mean $m_X(t)$ of $X(t)$ is then

$$m_X(t) = E[X(t)] = E(A)e^{-\zeta\omega_0 t}\left(\cos\omega_d t + \frac{\zeta\omega_0}{\omega_d}\sin\omega_d t\right)$$

$$+ \frac{E(B)}{\omega_d}e^{-\zeta\omega_0 t}\sin\omega_d t + \int_0^t h(t-\tau)E[F(\tau)]\,d\tau$$

Assume further for simplicity that with probability unity, A and B are deterministic constants and $E[F(\tau)] = 0$. Therefore,

$$m_X(t) = Ae^{-\zeta\omega_0 t}\left(\cos\omega_d t + \frac{\zeta\omega_0}{\omega_d}\sin\omega_d t\right) + \frac{B}{\omega_d}e^{-\zeta\omega_0 t}\sin\omega_d t \tag{9.93}$$

so that the mean function of the output depends only on the initial conditions.
    For the variance Var[$X(t)$] we obtain

$$\text{Var}[X(t)] = \int_0^t\int_0^t h(t-\tau_1)h(t-\tau_2)R_F(\tau_1, \tau_2)\,d\tau_1\,d\tau_2$$

Bearing in mind Eq. (8.72), we have

$$\text{Var}[X(t)] = \int_0^t\int_0^t\int_{-\infty}^{\infty} S_F(\omega)\cos\omega(\tau_1 - \tau_2)h(t-\tau_1)h(t-\tau_2)\,d\omega\,d\tau_1\,d\tau_2$$

Thanks to convergence of the integrals, we can change the order of integration and obtain

$$\text{Var}[X(t)] = \int_{-\infty}^{\infty}\frac{S_F(\omega)\,d\omega}{m^2\omega_d^2}\int_0^t\int_0^t\exp[-\zeta\omega_0(2t - \tau_1 - \tau_2)]\sin\omega_d(t-\tau_1)$$

$$\times \sin\omega_d(t-\tau_2)\cos\omega(\tau_1 - \tau_2)\,d\tau_1\,d\tau_2 \tag{9.94}$$

Evaluation of the double integral yields

$$\mathrm{Var}[X(t)] = \frac{1}{m^2} \int_{-\infty}^{\infty} \frac{S_F(\omega)}{\left(\omega_0^2 - \omega^2\right)^2 + 4\zeta^2\omega_0^2\omega^2}$$

$$\times \left\{ 1 + e^{-2\omega_0\zeta t}\left[1 + \frac{2\omega_0}{\omega_d}\zeta \sin \omega_d t \cos \omega_d t \right.\right.$$

$$- 2e^{\omega_0\zeta t}\left(\cos \omega_d t + \frac{\omega_0\zeta}{\omega_d} \sin \omega_d t \right) \cos \omega t$$

$$\left.\left. - 2e^{\omega_0\zeta t}\frac{\omega}{\omega_d} \sin \omega_d t \sin \omega t + \frac{(\omega_0\zeta)^2 - \omega_d^2 + \omega^2}{\omega_d^2} \sin^2 \omega_d t \right]\right\} d\omega$$

$$(9.95)$$

This result is due to Caughey and Stumpf. It is seen that as $t \to \infty$

$$\mathrm{Var}[X(t)] = \frac{1}{m^2} \int_{-\infty}^{\infty} \frac{S_F(\omega)\, d\omega}{\left(\omega_0^2 - \omega^2\right)^2 + 4\zeta^2\omega_0^2\omega^2} \qquad (9.96)$$

and coincides with Eq. (9.51), that is, with the solution of the associated stationary problem. For $F(t)$ represented by ideal white noise $S_F(\omega) = S_0$, we have Uhlenbeck's result from Eq. (9.95):

$$\mathrm{Var}[X(t)] = \frac{S_0\pi}{2\zeta\omega_0^3 m^2}\left[1 - \frac{e^{-2\omega_0\zeta t}}{\omega_d^2}\left(\omega_d^2 + \omega_0\omega_d\zeta \sin 2\omega_d t + 2\omega_0^2\zeta^2\sin^2\omega_d t \right)\right]$$

$$(9.97)$$

For excitation with slowly varying spectral density in the vicinity of the natural frequency ω_0, and for a lightly damped structure, Laplace's method may again be used, to yield

$$\mathrm{Var}[X(t)] \simeq \frac{S_F(\omega_0)\pi}{2\zeta\omega_0^3 m^2}\left[1 - \frac{e^{-2\omega_0\zeta t}}{\omega_d^2}\left(\omega_d^2 + \omega_0\omega_d\zeta \sin 2\omega_d t + 2\omega_0^2\zeta^2\sin^2\omega_d t \right)\right]$$

$$(9.98)$$

Plots of Eq. (9.97) are shown in Fig. 9.8 for $\zeta = 0$, 0.025, 0.05, and 0.10. It is seen that the response variance approaches the stationary value as time

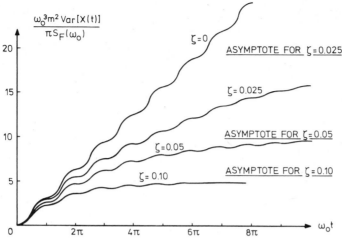

Fig. 9.8. Transient response of a single-degree-of-freedom system under random excitation with ideal white noise. (Reproduced from Caughey and Stumpf).

increases, so that only a small error is involved in treating the output process as though it were stationary, provided the input is applied for a sufficiently long time. The larger damping values result in lower stationary values and allow the response to become stationary in a shorter time.

Another interesting feature of Fig. 9.8 is that the time-varying variance does not overshoot the stationary variance level. This is, however, not a universal property of the transient response. Figure 9.9 shows the nondimensional variance of a single-degree-of-freedom system, with the input autocorrelation function

$$R_F(\tau) = R_0 e^{-\alpha|\tau|} \cos \rho \tau \tag{9.99}$$

In this case, as is seen from the figure, the mean-square value of the response does overshoot the stationary value.

Example 9.9

Consider the response to nonstationary excitation of a linear, time-invariant system which obeys the equation

$$c\dot{X} + kX = c(\beta + \gamma t)U(t)F(t) \tag{9.100}$$

c, k, β, and γ being deterministic constants and $F(t)$ a random function stationary in the wide sense, in particular, ideal white noise with $E(F) = 0$, $S_F(\omega) = S_0$. It is seen that with $\beta = 1/c$, $\gamma = 0$, Eq. (9.100) reduces to Eq. (9.38) in Example 9.3. The initial condition is assumed to be a zero one, and

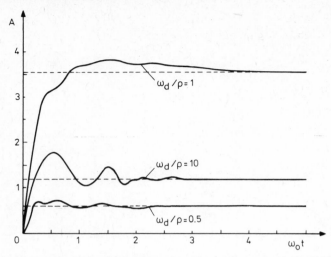

Fig. 9.9. Transient response of a single-degree-of-freedom system under random excitation; $F(t)$ has an autocorrelation function as per Eq. (9.99), $A = \omega_d m\{E[X^2(t)]/R_0\}$. (Reproduced from Barnoski and Maurer).

we seek the mean square of the particular solution. With $k/c = a$, we have

$$\dot{X} + ax = (\beta + \gamma t)U(t)F(t) \qquad (9.101)$$

The response is anticipated to be nonstationary, as is the excitation. The autocorrelation function of the excitation $Y(t) = (\beta + \gamma t)U(t)F(t)$ is

$$R_Y(t_1, t_2) = 2\pi S_0(\beta + \gamma t_1)(\beta + \gamma t_2)\delta(t_2 - t_1)U(t_1)U(t_2) \qquad (9.102)$$

$X(t)$ is readily found from Eq. (9.100):

$$X(t) = \int_0^t e^{-a(t-t_1)}Y(t_1)\, dt_1 \qquad (9.103)$$

The mean-square response is

$$E[X^2(t)] = R_X(t, t) = \int_0^t \int_0^t R_Y(t_1, t_2)e^{-a(t-t_1)}e^{-a(t-t_2)}\, dt_1\, dt_2$$

Integration yields

$$E[X^2(t)] = 2\pi S_0 \int_0^t \int_0^t (\beta + \gamma t_1)(\beta + \gamma t_2)\delta(t_2 - t_1)$$

$$\times \exp\{-a(2t - t_1 - t_2)\}\, dt_1\, dt_2$$

$$= 2\pi S_0 e^{-2at}\left\{ \frac{\beta^2}{2a}(e^{2at} - 1) + 2\beta\gamma\left[\frac{e^{2at}}{4a^2}(2ae^{2at} - 1) - \frac{1}{4a^2}(2a - 1)\right]\right.$$

$$\left. + \frac{\gamma^2}{4a^3}\left[e^{2at}(2a^2t^2 - 2at + 1) - 1\right]\right\} \qquad (9.104)$$

Note that for $\beta = 1/c$, $\gamma = 0$ we arrive at

$$E\left[X^2(t)\right] = \frac{S_0\pi}{ck}\left(1 - e^{-2kt/c}\right) \qquad (9.105)$$

and for $t \to \infty$, Eq. (9.105) coincides with the stationary solution as per Eq. (9.45).

Examples 9.8 and 9.9 are illustrations of determination of a nonstationary random response. Further examples for analysis of linear discrete systems can be found in the papers by Barnoski and Maurer, and Holman and Hart. Spectral analysis of nonstationary random functions is discussed by Priestley and by Bendat and Piersol.

9.3 RANDOM VIBRATION OF A MULTIDEGREE-OF-FREEDOM SYSTEM

The equation of motion of a system having several degrees of freedom is given by

$$[m]\{\ddot{X}\} + [c]\{\dot{X}\} + [k]\{X\} = \{F(t)\} \qquad (9.106)$$

where $\{X\}$ and $\{F\}$ are the vectors of the *generalized displacements* and *generalized forces*, respectively; $[m]$, $[c]$, and $[k]$ are the $n \times n$ mass, damping, and stiffness matrices, respectively. $\{\dot{X}(t)\}$ is a velocity vector, $\{\ddot{X}(t)\}$ is an acceleration vector.

We are given the mean vector function $\{m_F(t)\}$ and the cross-covariance matrix

$$\left[C_F(t_1, t_2)\right] = \left[C_{F_i F_j}(t_1, t_2)\right]_{n \times n}$$

of the generalized forces, where

$$C_{F_i F_j}(t_1, t_2) = E\left\{\left[F_i(t_1) - m_{F_i}(t_1)\right]\left[F_j(t_2) - m_{F_j}(t_2)\right]\right\} \qquad (9.107)$$

We wish to find the mean vector $\{m_X(t)\}$ and the cross covariance matrix

$$\left[C_X(t_1, t_2)\right] = \left[C_{X_i X_j}(t_1, t_2)\right]_{n \times n}$$

of the generalized displacements, where

$$C_{X_i X_j}(t_1, t_2) = E\left\{\left[X_i(t_1) - m_{X_i}(t_1)\right]\left[X_j(t_2) - m_{X_j}(t_2)\right]\right\} \qquad (9.108)$$

We first confine ourselves to the deterministic problem of free undamped vibration.

9.3.1   Free Undamped Vibration.   This is obtained by putting $[c] = [0]$, $\{F(t)\}$ $= \{0\}$ in Eq. (9.106). Converting to lowercase notation for the deterministic displacement function, we have

$$[m]\{\ddot{x}\} + [k]\{x\} = \{0\} \tag{9.109}$$

For free vibration we express the solution of Eq. (9.109) in the form

$$\{x\} = \{y\}\sin(\omega t + \alpha) \tag{9.110}$$

where ω is a natural frequency and y is the vibration amplitude. Substitution of (9.110) in (9.109) yields

$$[k - \omega^2 m]\{y\}\sin(\omega t + \alpha) = \{0\}$$

Since this equation has to hold for any t, we are left with

$$[k - \omega^2 m]\{y\} = \{0\}$$

This equation has nontrivial solutions if the determinant of the matrix $[k - \omega^2 m]$ vanishes, that is, if

$$\Delta(\omega^2) = \det[k - \omega^2 m] = 0 \tag{9.111}$$

This equation, called the characteristic, generally yields n positive roots $\omega_1^2, \omega_2^2, \ldots, \omega_n^2$, where $\omega_1^2 \leqslant \omega_2^2 \leqslant \cdots \leqslant \omega_n^2$ and $\omega_1, \omega_2, \ldots, \omega_n$ are the system's natural frequencies. With each natural frequency ω_i, we can associate the corresponding natural mode $\{y^{(i)}\}$, satisfying the homogeneous set of equations

$$[k - \omega_i^2 m]\{y^{(i)}\} = \{0\} \tag{9.112}$$

which has (by Eq. 9.111) a nonunique solution, so that

$$\{y^{(i)}\} = \lambda_i\{u^{(i)}\}$$

λ_i being arbitrary nonzero constants. We normalize the natural modes by setting

$$\{v^{(i)}\} = \frac{1}{\mu_i}\{y^{(i)}\} = \frac{\lambda_i}{\mu_i}\{u^{(i)}\} \tag{9.113}$$

so that

$$\{v^{(i)}\}^T[m]\{v^{(i)}\} = 1$$

The normalized natural modes are referred to as the *normal modes*. Note that the normal modes associated with different natural frequencies are orthogonal, that is,

$$\{v^{(i)}\}^T [m]\{v^{(j)}\} = 0, \quad i \neq j$$

Consequently, the two latter equations can be put in condensed form as

$$\{v^{(i)}\}^T [m]\{v^{(j)}\} = \delta_{ij} \tag{9.114}$$

where δ_{ij} is Kronecker's delta

$$\delta_{ij} = \begin{cases} 1, & i = j \\ 0, & i \neq j \end{cases} \tag{9.115}$$

If the natural frequencies happen to coincide, the corresponding natural modes are not necessarily orthogonal to each other, but can be orthogonalized through their linear combinations (see Example 9.11).

Premultiplication of the equation obtained by substitution of Eq. (9.113) in (9.112) by $\{v^{(i)}\}^T$, yields

$$\{v^{(i)}\}^T [k]\{v^{(i)}\} = \omega_i^2 \tag{9.116}$$

We now define a modal matrix $[v]$, whose ith column is the vector $\{v^{(i)}\}$:

$$[v] = \begin{bmatrix} v^{(1)} & v^{(2)} & \cdots & v^{(n)} \end{bmatrix}$$

Equations (9.114) and (9.116) then become, respectively,

$$[v]^T [m][v] = [I] \qquad [v]^T [k][v] = \lceil \omega^2 \rfloor \tag{9.117}$$

where $[I]$ is an identity matrix with elements δ_{ij} (diagonal matrix with unities on the diagonal), and $\lceil \omega^2 \rfloor$ a diagonal matrix of the natural frequencies squared.

Consider now the linear transformation

$$\{x\} = [v]\{q\} \tag{9.118}$$

Since $[v]$ is a constant matrix, we have also

$$\{\ddot{x}\} = [v]\{\ddot{q}\}$$

and Eq. (9.109) becomes

$$[m][v]\{\ddot{q}\} + [k][v]\{q\} = \{0\}$$

Premultiplying this equation by $[v]^T$ and with Eq. (9.117) in mind, we have

$$\{\ddot{q}\} + \lceil \omega^2 \rfloor \{q_i\} = \{0\} \tag{9.119}$$

For fixed i, Eq. (9.119) coincides with that of an undamped single-degree-of-freedom structure. The $q_i(t)$ are called the *principal coordinates*, and transformation of the set of coupled equations of motion into an uncoupled one is referred to as the *normal mode method*.

9.3.2   Deterministic Response via the Normal Mode Method.

We shall now deal with the deterministic version of Eq. (9.106). We assume, for the sake of simplicity, that the damping matrix $[c]$ is representable in the form

$$[c] = \alpha[m] + \beta[k] \tag{9.120}$$

where α and β are some nonnegative constants. For a treatment of nonproportional damping, consult, for example, Hurty and Rubinstein.

Then, through the transformation (9.118), Eq. (9.106) becomes

$$[m][v]\{\ddot{q}\} + [c][v]\{\dot{q}\} + [k][v]\{q\} = \{f(t)\} \tag{9.121}$$

or, premultiplying by $[v]^T$ and in view of Eq. (9.117), it takes the form

$$\{\ddot{q}\} + \lceil \alpha + \beta\omega^2 \rfloor \{\dot{q}\} + \lceil \omega^2 \rfloor \{q\} = [v]^T\{f(t)\}$$

where $\lceil \alpha + \beta\omega^2 \rfloor = \alpha[I] + \beta\lceil \omega^2 \rfloor$.

Denoting $\alpha + \beta\omega_j^2 = 2\zeta_j\omega_j$, we obtain

$$\{\ddot{q}\} + \lceil 2\zeta\omega \rfloor \{\dot{q}\} + \lceil \omega^2 \rfloor \{q\} = [v]^T\{f(t)\} \tag{9.122}$$

which represents n uncoupled equations of motion

$$\ddot{q}_j + 2\zeta_j\omega_j\dot{q}_j + \omega_j^2 q_j = \sum_{k=1}^{n} v_k^{(j)} f_k(t) \tag{9.123}$$

or, again denoting

$$\varphi_j(t) = \sum_{k=1}^{n} v_k^{(j)} f_k(t) \tag{9.124}$$

the equations become

$$\ddot{q}_j + 2\zeta_j\omega_j\dot{q}_j + \omega_j^2 q_j = \varphi_j(t) \tag{9.125}$$

This equation is analogous to Eq. (9.48) for a damped single-degree-of-freedom structure. For other particular cases in which the damping matrix becomes diagonal, the reader is referred to the paper by Caughey (1960).

Now we refer to Eqs. (9.22) and (9.26), where we replace $m \to 1$, $\omega_0 \to \omega_j$, $\zeta \to \zeta_j$, $\omega_d \to \omega_j(1 - \zeta_j^2)^{1/2}$, $h(t) \to h_j(t)$ to yield

$$
h_j(t) =
\begin{cases}
\dfrac{1}{\omega_j(\zeta_j^2 - 1)^{1/2}} \exp(-\zeta_j\omega_j t)\sinh\left[(\zeta_j^2 - 1)^{1/2}\omega_j t\right], & \zeta_j > 1 \\[2ex]
t\exp(-\omega_j t), & \zeta_j = 1 \\[2ex]
\dfrac{1}{\omega_{dj}} \exp(-\zeta_j\omega_j t)\sin(\omega_{dj} t), & \zeta_j < 1
\end{cases}
$$

$$(9.126)$$

and, in accordance with Eq. (9.29),

$$q_j(t) = \int_{-\infty}^{t} \varphi_j(\tau)h_j(t - \tau)\, d\tau \tag{9.127}$$

The output functions $x_l(t)$, $l = 1, 2, \ldots, n$, are found then by making use of Eqs. (9.118) and (9.127):

$$x_l(t) = \sum_{j=1}^{n} v_l^{(j)} q_j = \sum_{j=1}^{n} v_l^{(j)} \int_{-\infty}^{t} \varphi_j(\tau)h_j(t - \tau)\, d\tau \tag{9.128}$$

For inputs representable by Fourier integrals,

$$f_j(t) = \int_{-\infty}^{\infty} F_j(\omega)e^{i\omega t}\, d\omega$$

$$F_j(\omega) = \frac{1}{2\pi}\int_{-\infty}^{\infty} f_j(t)e^{-i\omega t}\, dt \tag{9.129}$$

we have

$$\varphi_j(t) = \int_{-\infty}^{\infty} \Phi_j(\omega)e^{i\omega t}\, d\omega$$

$$\Phi_j(\omega) = \frac{1}{2\pi}\int_{-\infty}^{\infty} \varphi_j(t)e^{-i\omega t}\, dt \tag{9.130}$$

where, following Eq. (9.124),

$$\Phi_j(\omega) = \sum_{k=1}^{n} v_k^{(j)} F_k(\omega) \tag{9.131}$$

The steady-state outputs of Eq. (9.123) are representable in the familiar form

$$q_j(t) = \int_{-\infty}^{\infty} Q_j(\omega) e^{i\omega t}\, d\omega$$

$$Q_j(\omega) = \frac{1}{2\pi} \int_{-\infty}^{\infty} q_j(t) e^{-i\omega t}\, dt \tag{9.132}$$

and substitution of Eqs. (9.129) and (9.132) in Eq. (9.123) yields

$$Q_j(\omega) = H_j(\omega)\Phi_j(\omega) = \sum_{k=1}^{n} v_k^{(j)} H_j(\omega) F_k(\omega) \tag{9.133}$$

where

$$H_j(\omega) = \frac{1}{(i\omega)^2 + 2\zeta_j \omega_j (i\omega) + \omega_j^2} \tag{9.134}$$

and

$$q_j(t) = \int_{-\infty}^{\infty} H_j(\omega)\Phi_j(\omega) e^{i\omega t}\, d\omega$$

$$= \sum_{k=1}^{n} v_k^{(j)} \int_{-\infty}^{\infty} H_j(\omega) F_k(\omega) e^{i\omega t}\, d\omega \tag{9.135}$$

The original generalized coordinates are found, as before, by applying Eq. (9.118):

$$x_l(t) = \sum_{j=1}^{n} v_l^{(j)} q_j(t) = \sum_{j=1}^{n} v_l^{(j)} \int_{-\infty}^{\infty} H_j(\omega)\Phi_j(\omega) e^{i\omega t}\, d\omega$$

$$= \sum_{j=1}^{n} v_l^{(j)} \left[\sum_{k=1}^{n} v_k^{(j)} \int_{-\infty}^{\infty} H_j(\omega) F_k(\omega) e^{i\omega t}\, d\omega \right] \tag{9.136}$$

which is another form of Eq. (9.128).

Equations (9.128) and (9.136) can also be put in a different form. For this purpose we assume that in Eq. (9.128)

$$f_k(t) = \delta(t)\, \delta_{km}, \qquad k = 1, 2, \ldots, n \tag{9.137}$$

that is, that unit impulse excitation is applied in the direction of the mth generalized coordinate

$$f_m(t) = \delta(t)$$

$$f_1(t) = f_2(t) = \cdots = f_{m-1}(t) = f_{m+1}(t) = \cdots = f_n(t) = 0 \tag{9.138}$$

and denote the response, represented by the (l, m)th impulse response function, by $q_{lm}(t)$. For $\varphi_j(t)$, we have, from Eqs. (9.137) and (9.124),

$$\varphi_j(t) = \sum_{k=1}^{n} v_k^{(j)} \delta(t) \delta_{km} = v_m^{(j)} \delta(t)$$

Now we substitute this expression into Eq. (9.128) and obtain

$$q_{lm}(t) = \sum_{j=1}^{n} v_l^{(j)} \int_{-\infty}^{t} \varphi_j(\tau) h_j(t-\tau) \, d\tau$$

$$= \sum_{j=1}^{n} v_l^{(j)} \int_{-\infty}^{t} v_m^{(j)} \delta(\tau) h_j(t-\tau) \, d\tau$$

$$= \sum_{j=1}^{n} v_l^{(j)} v_m^{(j)} h_j(t) \tag{9.139}$$

In terms of the (l, m)th impulse response function, Eq. (9.128), the response due to the general excitation, reads

$$x_l(t) = \sum_{k=1}^{n} \int_{-\infty}^{t} g_{lk}(t-\tau) f_k(\tau) \, d\tau \tag{9.140}$$

On the other hand, as in Eq. (9.19), for $f_k(t)$ as per Eq. (9.137) we have

$$F_k(\omega) = \frac{1}{2\pi} \delta_{km}, \quad k = 1, 2, \ldots, n \tag{9.141}$$

and by Eq. (9.136) and the definition of the (l, m)th impulse response function

$$g_{lm}(t) = \frac{1}{2\pi} \sum_{j=1}^{n} v_l^{(j)} v_m^{(j)} \int_{-\infty}^{\infty} H_j(\omega) e^{i\omega t} \, d\omega \tag{9.142}$$

Comparison of Eqs. (9.139) and (9.142) shows that

$$h_j(t) = \frac{1}{2\pi} \int_{-\infty}^{\infty} H_j(\omega) e^{i\omega t} \, d\omega$$

$$H_j(\omega) = \int_{-\infty}^{\infty} h_j(t) e^{-i\omega t} \, dt \tag{9.143}$$

which is in perfect analogy with Eqs. (9.20) and (9.21). Denoting now the (l, m)th complex frequency response $G_{lm}(\omega)$ as

$$G_{lm}(\omega) = \sum_{j=1}^{n} v_l^{(j)} v_m^{(j)} H_j(\omega) \tag{9.144}$$

Eq. (9.136) becomes

$$x_l(t) = \sum_{k=1}^{n} \int_{-\infty}^{\infty} G_{lk}(\omega) F_k(\omega) e^{i\omega t} \, d\omega \qquad (9.145)$$

whereas the relationship between $g_{lm}(t)$ and $G_{lm}(\omega)$ is readily obtained by comparing Eqs. (9.142) and (9.144):

$$g_{lm}(t) = \frac{1}{2\pi} \int_{-\infty}^{\infty} G_{lm}(\omega) e^{i\omega t} \, d\omega$$

$$G_{lm}(\omega) = \int_{-\infty}^{\infty} g_{lm}(t) e^{-i\omega t} \, dt \qquad (9.146)$$

Now we define the frequency response matrix $[g(t)]$ as follows:

$$[g(t)] = [g_{lm}(t)]_{n \times n}$$

and the matrix of the complex frequency responses $[G(\omega)]$

$$[G(\omega)] = [G_{lm}(\omega)]_{n \times n}$$

Equations (9.146) then can be put as

$$[g(t)] = \frac{1}{2\pi} \int_{-\infty}^{\infty} [G(\omega)] e^{i\omega t} \, d\omega$$

$$[G(\omega)] = \int_{-\infty}^{\infty} [g(t)] e^{-i\omega t} \, dt \qquad (9.147)$$

that is, the Fourier transform pair is replaced, respectively, by the impulse response function matrix $[g(t)]$ multiplied by 2π and the complex frequency response matrix $[G(\omega)]$. Now, instead of Eqs. (9.140) and (9.145) we can operate with

$$\{x(t)\} = \int_{-\infty}^{\infty} [g(t-\tau)]\{f(\tau)\} \, d\tau$$

$$\{x(t)\} = \int_{-\infty}^{\infty} [G(\omega)]\{F(\omega)\} e^{i\omega t} \, d\omega \qquad (9.148)$$

In Eq. (9.148) the fact that $[g(t-\tau)] = [0]$ for $t < \tau$ is taken into account in analogy with the single-degree-of-freedom system.

9.3.3 Random Response via the Normal Mode Method.

We now recapitulate the original problem of the random response of a multidegree-of-freedom system, Eq. (9.106). This linear system transforms the random vector function $\{F(t)\}$ into another such function $\{X(t)\}$. In accordance with Eq. (9.148) for the realizations $\{f(t)\}$ and $\{x(t)\}$ of $\{F(t)\}$ and $\{X(t)\}$, respectively, $\{X(t)\}$ is representable either as

$$\{X(t)\} = \int_{-\infty}^{\infty} [g(t - \tau)]\{F(\tau)\} \, d\tau \qquad (9.149)$$

or as

$$\{X(t)\} = \int_{-\infty}^{\infty} [g(t)]\{F(t - \tau)\} \, d\tau$$

The mean vector function $\{m_X(t)\} = E[\{X(t)\}]$ is readily obtained as

$$\{m_X(t)\} = \int_{-\infty}^{\infty} [g(t - \tau)]\{m_F(\tau)\} \, d\tau$$

For a nonstationary random vector function $\{F(t)\}$, this equation yields $\{m_X(t)\}$ if $\{m_F(t)\}$ is given. For a stationary function, $\{m_F(\tau)\} = \{c\}$, where $\{c\}$ is a vector of constants, and

$$\{m_X(t)\} = \left[\int_{-\infty}^{\infty} g(t - \tau) \, d\tau\right]\{c\} \qquad (9.150)$$

so that $\{m_X(t)\}$ is also a vector of constants. Without loss of generality, we put $\{m_F(t)\} = \{0\}$, so that $\{m_X(t)\} = \{0\}$. The cross-covariance matrix then coincides with the cross-correlation matrix. $[R_X(t_1, t_2)]$ is then

$$[R_X(t_1, t_2)] = E\{X(t_1)\}\{X(t_2)\}^{T*}$$

$$= \int_{-\infty}^{\infty} \int_{-\infty}^{\infty} [g(t_1 - \tau_1)][R_F(\tau_1, \tau_2)][g(t_2 - \tau_2)]^{T*} \, d\tau_1 \, d\tau_2$$

$$(9.151)$$

where T denotes the transpose operation. $[R_X(t_1, t_2)]$ can also be written as follows, in conjunction with the second of Eqs. (9.149):

$$[R_X(t_1, t_2)] = \int_{-\infty}^{\infty} \int_{-\infty}^{\infty} [g(\tau_1)][R_F(t_1 - \tau_1, t_2 - \tau_2)][g(\tau_2)]^{T*} \, d\tau_1 \, d\tau_2$$

$$(9.152)$$

Seeking the mean-square values of random responses $X_j(t)$, we denote the jth row of the matrix $[g(t)]$ as

$$[g_j(t)] = [g_j^{(1)} \quad g_j^{(2)} \quad \cdots \quad g_j^{(n)}] \qquad (9.153)$$

and Eq. (9.152) yields

$$E\left[X_j^2(t)\right] = \int_{-\infty}^{\infty} \int_{-\infty}^{\infty} \left[g_j(\tau_1)\right]\left[R_F(t - \tau_1, t - \tau_2)\right]\left[g_j(\tau_2)\right]^{T^*} d\tau_1 \, d\tau_2$$

(9.154)

If $\{F(t)\}$ is a stationary vector random function, the cross correlations $R_{F_j F_k}(t_1 - \tau_1, t_2 - \tau_2)$ in Eq. (9.152) are functions of the difference of arguments. Hence,

$$\left[R_F(t_1 - \tau_1, t_2 - \tau_2)\right] = \left[R_F(t_2 - t_1 - \tau_2 + \tau_1)\right]$$

Denoting $t_2 - t_1 = \tau$, we have

$$\left[R_X(t_1, t_2)\right] = \int_{-\infty}^{\infty} \int_{-\infty}^{\infty} \left[g(\tau_1)\right]\left[R_F(\tau - \tau_2 + \tau_1)\right]\left[g(\tau_2)\right]^{T^*} d\tau_1 \, d\tau_2$$

$$\equiv \left[R_X(\tau)\right]$$

(9.155)

Again, in perfect analogy with the single-degree-of-freedom system, the output of a stable linear multidegree-of-freedom system is also stationary in the wide sense. Recalling the Wiener-Khintchine relationships (8.67), (8.74), and (8.78),

$$\left[R_X(\tau)\right] = \int_{-\infty}^{\infty} \left[S_X(\omega)\right] e^{i\omega\tau} \, d\omega$$

$$\left[S_X(\omega)\right] = \frac{1}{2\pi} \int_{-\infty}^{\infty} \left[R_X(\tau)\right] e^{-i\omega\tau} \, d\tau$$

(9.156)

where $[S_X(\omega)]$ is the cross spectral density matrix of displacements, we have

$$\left[S_X(\omega)\right] = \left[\int_{-\infty}^{\infty} g(\tau_1) e^{i\omega\tau_1} \, d\tau_1\right]\left[\frac{1}{2\pi} \int_{-\infty}^{\infty} R_F(\lambda) e^{-i\omega\lambda} \, d\lambda\right]\left[\int_{-\infty}^{\infty} g(\tau_2) e^{-i\omega\tau_2}\right]^T$$

$$= \left[G(\omega)\right]\left[S_F(\omega)\right]\left[G(\omega)\right]^{T^*}$$

(9.157)

which generalizes Eq. (9.35). Bearing in mind Eq. (9.144), which reads in matrix notation as

$$\left[G(\omega)\right] = \left[v\right]\left[H\right]\left[v\right]^T$$

(9.158)

we obtain

$$\left[S_X(\omega)\right] = \left[v\right]\lceil H(\omega) \rfloor\left[v\right]^T\left[S_F(\omega)\right]\left[v\right]\lceil H \rfloor^*\left[v\right]^T$$

(9.159)

This expression can be condensed if we note that by Eq. (9.131) the following relationship is obtainable between the spectral densities of $\{\Phi\}$ and $\{F\}$:

$$[S_\Phi(\omega)] = [v]^T[S_F(\omega)][v] \tag{9.160}$$

and finally

$$[S_X(\omega)] = [v][\![H(\omega)]\!][S_\Phi(\omega)][\![H(\omega)]\!]^*[v]^T \tag{9.161}$$

The cross-correlation matrix becomes, in view of the first of Eqs. (9.156),

$$[R_X(\tau)] = [v]\int_{-\infty}^{\infty} [\![H(\omega)]\!][S_\Phi(\omega)][\![H(\omega)]\!]^* e^{i\omega\tau}\, d\omega\, [v]^T \tag{9.162}$$

At $\tau = 0$ we have

$$[R_X(0)] = [v]\int_{-\infty}^{\infty} [\![H(\omega)]\!][S_\Phi(\omega)][\![H(\omega)]\!]^*\, d\omega\, [v]^T \tag{9.163}$$

For componentwise representation, we denote the jth and kth row matrices of $[v]$, respectively, as

$$[v_j] = \begin{bmatrix} v_j^{(1)} & v_j^{(2)} & \cdots & v_j^{(n)} \end{bmatrix}$$

$$[v_k] = \begin{bmatrix} v_k^{(1)} & v_k^{(2)} & \cdots & v_k^{(n)} \end{bmatrix} \tag{9.164}$$

Then

$$E(X_j X_k^*) = R_{X_j X_k}(0) = [v_j]\int_{-\infty}^{\infty} [\![H(\omega)]\!][S_\Phi(\omega)][\![H(\omega)]\!]^*\, d\omega\, [v_k]^T \tag{9.165}$$

or

$$E(X_j X_k^*) = \sum_{\alpha=1}^{n}\sum_{\beta=1}^{n} v_j^{(\alpha)} v_k^{(\beta)} E_{\alpha\beta} \tag{9.166}$$

where

$$E_{\alpha\beta} = \int_{-\infty}^{\infty} S_{\Phi_\alpha \Phi_\beta}(\omega) H_\alpha(\omega) H_\beta^*(\omega)\, d\omega \tag{9.167}$$

Equation (9.166) can be rewritten as follows:

$$E(X_j X_k^*) = \sum_{\alpha=1}^{n} v_j^{(\alpha)} v_k^{(\alpha)} E_{\alpha\alpha} + \sum_{\substack{\alpha=1 \\ \alpha\neq\beta}}^{n}\sum_{\beta=1}^{n} v_j^{(\alpha)} v_k^{(\beta)} E_{\alpha\beta} \tag{9.168}$$

Since

$$S_{\Phi_\alpha \Phi_\beta}(\omega) = S_{\Phi_\beta \Phi_\alpha}^*(\omega) \qquad (9.169)$$

also

$$E_{\alpha\beta} = E_{\beta\alpha}^* \qquad (9.170)$$

For $j = k$, we have the mean-square displacements as

$$E\left(|X_j|^2\right) = \sum_{\alpha=1}^{n} v_j^{(\alpha)} v_j^{(\alpha)} E_{\alpha\alpha} + \sum_{\substack{\alpha=1 \\ \alpha \neq \beta}}^{n} \sum_{\beta=1}^{n} v_j^{(\alpha)} v_j^{(\beta)} E_{\alpha\beta}$$

For fixed α and β, we can begin by summing the terms containing $E_{\alpha\beta}$ and $E_{\beta\alpha}$ in Eq. (9.168), to give

$$v_j^{(\alpha)} v_j^{(\beta)} E_{\alpha\beta} + v_j^{(\beta)} v_j^{(\alpha)} E_{\beta\alpha} = 2 v_j^{(\alpha)} v_j^{(\beta)} \mathrm{Re}\left(E_{\alpha\beta}\right)$$

where $\mathrm{Re}(E_{\alpha\beta})$ denotes the real part of $E_{\alpha\beta}$. Consequently, Eq. (9.168) can be rewritten as

$$E\left(|X_j|^2\right) = \sum_{\alpha=1}^{n} \left[v_j^{(\alpha)}\right]^2 E_{\alpha\alpha} + 2 \sum_{\substack{\alpha=1 \\ \alpha < \beta}}^{n} \sum_{\beta=1}^{n} v_j^{(\alpha)} v_j^{(\beta)} \mathrm{Re}\left(E_{\alpha\beta}\right) \qquad (9.171)$$

and the mean-square displacement turns out to be a real quantity, as it should be.

In formulating the mean-square response, the term $[v_j^{(\alpha)}]^2 E_{\alpha\alpha}$ in Eq. (9.171) can be interpreted as the contribution of interaction of the αth normal mode with itself, and is referred to as *modal autocorrelation*. The first sum is that of the modal autocorrelations and therefore represents the result of interaction of like modes. The term $2 v_j^{(\alpha)} v_j^{(\beta)} \mathrm{Re}(E_{\alpha\beta})$, where $\alpha \neq \beta$, can be interpreted as the contribution of interaction of the αth and βth normal modes and is referred to as modal *cross correlation*; the second sum is that of the modal cross correlations and represents the result of interaction of unlike normal modes.

In literature, the contribution of the modal cross correlations is often disregarded, and consequently $E(|X_j|^2)$ is given by the approximation

$$E\left(|X_j|^2\right)_{\text{approximate}} = \sum_{\alpha=1}^{n} \left[v_j^{(\alpha)}\right]^2 E_{\alpha\alpha} \qquad (9.172)$$

At the same time, we are in a position to determine weighted "averages" unaffected by the cross correlations. Thus let us construct the *mass average*,

$$A_m = \frac{1}{M} \sum_{p=1}^{n} \sum_{q=1}^{n} m_{pq} E\left(X_p X_q^*\right), \quad M = \sum_{p=1}^{n} \sum_{q=1}^{n} m_{pq} \qquad (9.173)$$

where m_{pq} are the elements of matrix $[m]$. Substituting Eq. (9.166) for $E(X_p X_q^*)$ in the latter, we obtain

$$A_m M = \sum_{p=1}^{n} \sum_{q=1}^{n} m_{pq} \left(\sum_{\alpha=1}^{n} \sum_{\beta=1}^{n} v_p^{(\alpha)} v_q^{(\beta)} E_{\alpha\beta} \right)$$

and changing the order of summations,

$$A_m M = \sum_{\alpha=1}^{n} \sum_{\beta=1}^{n} E_{\alpha\beta} \left(\sum_{p=1}^{n} \sum_{q=1}^{n} m_{pq} v_p^{(\alpha)} v_q^{(\beta)} \right) \qquad (9.174)$$

By virtue of the orthogonality property as per (9.114), we equate the expression in parentheses to zero for $\alpha \neq \beta$ and to unity for $\alpha = \beta$, so that

$$A_m = \frac{1}{M} \sum_{\alpha=1}^{n} E_{\alpha\alpha} \qquad (9.175)$$

that is, without cross correlation terms.

Analogously, we can construct the *stiffness average*:

$$A_k K = \sum_{p=1}^{n} \sum_{q=1}^{n} k_{pq} E(X_p X_q^*) = \sum_{p=1}^{n} \sum_{q=1}^{n} k_{pq} \left(\sum_{\alpha=1}^{n} \sum_{\beta=1}^{n} v_p^{(\alpha)} v_q^{(\beta)} E_{\alpha\beta} \right)$$

$$= \sum_{\alpha=1}^{n} \sum_{\beta=1}^{n} E_{\alpha\beta} \left(\sum_{p=1}^{n} \sum_{q=1}^{n} k_{pq} v_p^{(\alpha)} v_q^{(\beta)} \right), \qquad K = \sum_{p=1}^{n} \sum_{q=1}^{n} k_{pq}$$

By the second of Eqs. (9.117), we again have

$$\sum_{p=1}^{n} \sum_{q=1}^{n} k_{pq} v_p^{(\alpha)} v_q^{(\beta)} = \omega_{\alpha\beta}^2 \delta_{\alpha\beta}$$

and

$$A_k = \frac{1}{K} \sum_{\alpha=1}^{n} \omega_{\alpha\alpha}^2 E_{\alpha\alpha} \qquad (9.176)$$

again without cross correlation terms.

9.4   ILLUSTRATION OF THE ROLE OF MODAL CROSS CORRELATIONS

When, instead of averages we are concerned with the exact distribution of the mean-square responses $E(|X_j|^2)$, we have to use the exact general equation 9.171; that is, the modal cross correlations must be taken into account. The question then is whether it is justifiable to disregard the modal cross correlations and use approximation (9.172) instead. Rather than attempt to formulate general guidelines in this matter, let us illustrate the role of the modal cross correlations on an example. Consider the two-degrees-of-freedom system shown in Fig. 9.10, where $F_1(t)$ is a random excitation with zero mean and ideal white noise spectral density $S_{F_1}(\omega) = S_0$. The problem consists in finding the mean-square values of the displacements $X_1(t)$ and $X_2(t)$ and velocities $\dot{X}_1(t)$ and $\dot{X}_2(t)$. The parameter ε characterizes the degree of coupling of the two masses. We immediately note that at $\varepsilon \rightarrow 0$ we should find the result (9.67) for $E(|X_1|^2)$, whereas for the mean-square value of the displacement $X_2(t)$ we should find zero, as the response of an unexcited system.

The desired mean-square values will be obtained by the normal mode method. The differential equations governing the motion of the system are (see the free-body diagrams in Fig. 9.10b)

$$m\ddot{X}_1 + (1 + \varepsilon)c\dot{X}_1 - \varepsilon c\dot{X}_2 + (1 + \varepsilon)kX_1 - \varepsilon kX_2 = F_1(t)$$

$$m\ddot{X}_2 - \varepsilon c\dot{X}_1 + (1 + \varepsilon)c\dot{X}_2 - \varepsilon kX_1 + (1 + \varepsilon)kX_2 = 0 \qquad (9.177)$$

The mass, stiffness, and damping matrices are, respectively,

$$[m] = \begin{bmatrix} m & 0 \\ 0 & m \end{bmatrix}$$

$$[k] = \begin{bmatrix} k(1 + \varepsilon) & -k\varepsilon \\ -k\varepsilon & k(1 + \varepsilon) \end{bmatrix}$$

$$[c] = \begin{bmatrix} c(1 + \varepsilon) & -c\varepsilon \\ -c\varepsilon & c(1 + \varepsilon) \end{bmatrix} \qquad (9.178)$$

so that the coefficients α and β in Eq. (9.120) are

$$\alpha = 0 \qquad \beta = \frac{c}{k} \qquad (9.179)$$

Substituting matrices $[m]$ and $[k]$ into Eq. (9.111) we arrive at the characteristic equation

$$\Delta(\omega^2) = \begin{vmatrix} k(1 + \varepsilon) - m\omega^2 & -k\varepsilon \\ -k\varepsilon & k(1 + \varepsilon) - m\omega^2 \end{vmatrix}$$

$$= (m\omega^2)^2 - 2(1 + \varepsilon)km\omega^2 + \left[(1 + \varepsilon)^2 - \varepsilon^2\right]k^2 = 0 \qquad (9.180)$$

The natural frequencies are

$$\omega_1 = \left(\frac{k}{m}\right)^{1/2}, \qquad \omega_2 = \left[\frac{k}{m}(1 + 2\varepsilon)\right]^{1/2} \qquad (9.181)$$

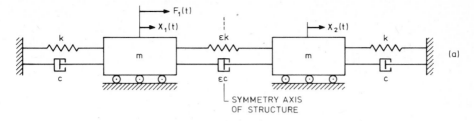

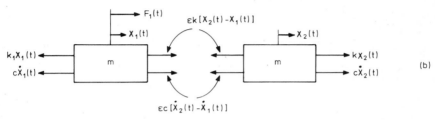

Fig. 9.10. Two-degree-of-freedom symmetric system, illustrating the influence of the cross correlation terms. (a) When ε tends to zero, the natural frequencies of a system shift together toward a natural frequency of a single-degree-of-freedom system. (b) The derivation of Eqs. (9.177).

For the first natural mode, we obtain from Eq. (9.112), on substituting ω_1^2,

$$\begin{bmatrix} k(1+\varepsilon) - m\omega_1^2 & -k\varepsilon \\ -k\varepsilon & k(1+\varepsilon) - m\omega_1^2 \end{bmatrix} \begin{Bmatrix} y_1^{(1)} \\ y_2^{(1)} \end{Bmatrix} = \begin{Bmatrix} 0 \\ 0 \end{Bmatrix}$$

which in turn yields

$$\{y^{(1)}\} = \lambda_1 \begin{Bmatrix} 1 \\ 1 \end{Bmatrix}$$

where λ_1 is an arbitrary nonzero real number.

For the second natural mode, we obtain from Eq. (9.112), on substituting ω_2^2,

$$\begin{bmatrix} k(1+\varepsilon) - m\omega_2^2 & -k\varepsilon \\ -k\varepsilon & k(1+\varepsilon) - m\omega_2^2 \end{bmatrix} \begin{Bmatrix} y_1^{(2)} \\ y_2^{(2)} \end{Bmatrix} = \begin{Bmatrix} 0 \\ 0 \end{Bmatrix}$$

so that

$$\{y^{(2)}\} = \lambda_2 \begin{Bmatrix} 1 \\ -1 \end{Bmatrix}$$

We now normalize the natural modes by letting, as in Eq. (9.113),

$$\{v^{(1)}\} = \frac{\lambda_1}{\mu_1}\begin{Bmatrix} 1 \\ 1 \end{Bmatrix} \qquad \{v^{(2)}\} = \frac{\lambda_2}{\mu_2}\begin{Bmatrix} 1 \\ -1 \end{Bmatrix}$$

and construct the modal matrix

$$[v] = \begin{bmatrix} \dfrac{\lambda_1}{\mu_1} & \dfrac{\lambda_2}{\mu_2} \\ \dfrac{\lambda_1}{\mu_1} & -\dfrac{\lambda_2}{\mu_2} \end{bmatrix}$$

The first of Eqs. (9.117) then becomes

$$\begin{bmatrix} \dfrac{\lambda_1}{\mu_1} & \dfrac{\lambda_1}{\mu_1} \\ \dfrac{\lambda_2}{\mu_2} & -\dfrac{\lambda_2}{\mu_2} \end{bmatrix}\begin{bmatrix} m & 0 \\ 0 & m \end{bmatrix}\begin{bmatrix} \dfrac{\lambda_1}{\mu_1} & \dfrac{\lambda_2}{\mu_2} \\ \dfrac{\lambda_1}{\mu_1} & -\dfrac{\lambda_2}{\mu_2} \end{bmatrix} = \begin{bmatrix} 1 & 0 \\ 0 & 1 \end{bmatrix}$$

which in turns yields

$$\frac{\lambda_1}{\mu_1} = \frac{\lambda_2}{\mu_2} = \frac{1}{\sqrt{2m}}$$

The modal matrix $[v] = [v^{(1)} \quad v^{(2)}]$ becomes

$$[v] = \frac{1}{\sqrt{2m}}\begin{bmatrix} 1 & 1 \\ 1 & -1 \end{bmatrix} \tag{9.182}$$

and the matrix $\lceil \omega^2 \rfloor$ is

$$\lceil \omega^2 \rfloor = \frac{k}{m}\begin{bmatrix} 1 & 0 \\ 0 & 1 + 2\varepsilon \end{bmatrix} \tag{9.183}$$

The spectral matrix of the excitations in the original coordinates $X_1(t)$ and $X_2(t)$ reads

$$[S_F(\omega)] = \begin{bmatrix} S_0 & 0 \\ 0 & 0 \end{bmatrix} \tag{9.184}$$

and in the normal coordinates, as in Eq. (9.160),

$$[S_\Phi(\omega)] = [v]^T[S_F(\omega)][v] = \frac{S_0}{2m}\begin{bmatrix} 1 & 1 \\ 1 & 1 \end{bmatrix} \tag{9.185}$$

Before proceeding to determine $[S_X(\omega)]$, we first consider the expression

$$\lceil H(\omega) \rfloor [S_\Phi(\omega)] \lceil H(\omega) \rfloor^* = \frac{S_0}{2m} \begin{bmatrix} |H_1(\omega)|^{(2)} & H_1(\omega)H_2^*(\omega) \\ H_1^*(\omega)H_2(\omega) & |H_2(\omega)|^2 \end{bmatrix}$$

(9.186)

For $[v_1]$ and $[v_2]$ we have

$$[v_1] = \frac{1}{\sqrt{2m}}[1 \quad 1] \qquad [v_2] = \frac{1}{\sqrt{2m}}[1 \quad -1]$$ (9.187)

and from Eq. (9.161) we obtain

$$S_{X_1}(\omega) = [v_1]\lceil H(\omega) \rfloor [S_\Phi(\omega)]\lceil H(\omega) \rfloor^*[v_1]^T$$

$$= \frac{S_0}{4m^2}\{|H_1(\omega)|^2 + |H_2(\omega)|^2 + [H_1(\omega)H_2^*(\omega) + H_1^*(\omega)H_2(\omega)]\}$$

(9.188)

$$S_{X_2}(\omega) = [v_2]\lceil H(\omega) \rfloor [S_\Phi(\omega)]\lceil H(\omega) \rfloor^*[v_2]^T$$

$$= \frac{S_0}{4m^2}\{|H_1(\omega)|^2 + |H_2(\omega)|^2 - [H_1(\omega)H_2^*(\omega) + H_1^*(\omega)H_2(\omega)]\}$$

(9.189)

Using the identity $H_1(\omega)H_2^*(\omega) + H_1^*(\omega)H_2(\omega) = 2 \operatorname{Re} H_1(\omega)H_2^*(\omega)$, the expressions for $S_{X_1}(\omega)$ and $S_{X_2}(\omega)$ can be written as

$$S_{X_1}(\omega) = \frac{S_0}{4m^2}\left[|H_1(\omega)|^2 + |H_2(\omega)|^2 + 2 \operatorname{Re} H_1(\omega)H_2^*(\omega)\right]$$

$$S_{X_2}(\omega) = \frac{S_0}{4m^2}\left[|H_1(\omega)|^2 + |H_2(\omega)|^2 - 2 \operatorname{Re} H_1(\omega)H_2^*(\omega)\right] \quad (9.190)$$

The spectral densities of the velocities are

$$S_{\dot{X}_1}(\omega) = \omega^2 S_{X_1}(\omega) \qquad S_{\dot{X}_2}(\omega) = \omega^2 S_{X_2}(\omega)$$ (9.191)

For the mean-square values, we obtain, using the corresponding spectral densities,

$$E(|X_1|^2) = \frac{S_0}{4m^2}\left\{\int_{-\infty}^{\infty}\left[|H_1(\omega)|^2 + |H_2(\omega)|^2\right]d\omega\right.$$

$$\left. +2\int_{-\infty}^{\infty} \text{Re }H_1(\omega)H_2^*(\omega)\,d\omega\right\} \qquad (9.192)$$

$$E(|\dot{X}_1|^2) = \frac{S_0}{4m^2}\left\{\int_{-\infty}^{\infty}\omega^2\left[|H_1(\omega)|^2 + |H_2(\omega)|^2\right]d\omega\right.$$

$$\left. +2\int_{-\infty}^{\infty}\omega^2 \text{Re }H_1(\omega)H_2^*(\omega)\,d\omega\right\} \qquad (9.193)$$

$$E(|X_2|^2) = \frac{S_0}{4m^2}\left\{\int_{-\infty}^{\infty}\left[|H_1(\omega)|^2 + |H_2(\omega)|^2\right]d\omega\right.$$

$$\left. -2\int_{-\infty}^{\infty} \text{Re }H_1(\omega)H_2^*(\omega)\,d\omega\right\} \qquad (9.194)$$

$$E(|\dot{X}_2|^2) = \frac{S_0}{4m^2}\left\{\int_{-\infty}^{\infty}\omega^2\left[|H_1(\omega)|^2 + |H_2(\omega)|^2\right]d\omega\right.$$

$$\left. -2\int_{-\infty}^{\infty}\omega^2 \text{Re }H_1(\omega)H_2^*(\omega)\,d\omega\right\} \qquad (9.195)$$

From Eq. (9.67) we have, with $\omega_0 \rightarrow \omega_j$, $\zeta \rightarrow \zeta_j$, and $m \rightarrow 1$,

$$\int_{-\infty}^{\infty}|H_j(\omega)|^2\,d\omega = \int_{-\infty}^{\infty}\frac{d\omega}{\left(\omega_j^2 - \omega^2\right)^2 + 4\zeta_j^2\omega_j^2\omega^2} = \frac{\pi}{2\zeta_j\omega_j^3} \qquad (9.196)$$

Now, to evaluate the integral

$$2\int_{-\infty}^{\infty} \text{Re }H_1(\omega)H_2^*(\omega)\,d\omega$$

$$= 2\int_{-\infty}^{\infty} \text{Re }\frac{d\omega}{\left[(i\omega)^2 + 2\zeta_1\omega_1(i\omega) + \omega_1^2\right]\left[(-i\omega)^2 + 2\zeta_2\omega_2(-i\omega) + \omega_2^2\right]}$$

$$= 2\int_{-\infty}^{\infty}\frac{\left\{\left[(-i\omega)^2 + \omega_1^2\right]\left[(i\omega)^2 + \omega_2^2\right] + 4\zeta_1\zeta_2\omega_1\omega_2(-i\omega)(i\omega)\right\}\,d\omega}{|(i\omega)^2 + 2\zeta_1\omega_1(i\omega) + \omega_1^2|^2|(-i\omega)^2 + 2\zeta_2\omega_2(-i\omega) + \omega_2^2|^2}$$

$$(9.197)$$

we resort to Eq. (C.1) in Appendix C. We first note that the roots of equation

$$(i\omega)^2 + 2\zeta_j\omega_j(i\omega) + \omega_j^2 = 0$$

have positive imaginary parts, that is, they lie in the upper half of the ω plane. Hence we denote

$$L_4(i\omega) = \left[(i\omega)^2 + 2\zeta_1\omega_1(i\omega) + \omega_1^2\right]\left[(i\omega)^2 + 2\zeta_2\omega_2(i\omega) + \omega_2^2\right]$$

$$= (i\omega)^4 + 2(\zeta_1\omega_1 + \zeta_2\omega_2)(i\omega)^3 + \left(\omega_1^2 + \omega_2^2 + 4\zeta_1\zeta_2\omega_1\omega_2\right)(i\omega)^2$$

$$+ 2(\zeta_1\omega_1\omega_2^2 + \zeta_2\omega_2\omega_1^2)(i\omega) + \omega_1^2\omega_2^2$$

so that the a_j's in Eq. (9.2) have the form

$$a_0 = 1 \qquad a_1 = 2(\zeta_1\omega_1 + \zeta_2\omega_2) \qquad a_2 = \omega_1^2 + \omega_2^2 + 4\zeta_1\zeta_2\omega_1\omega_2$$

$$a_3 = 2\omega_1\omega_2(\zeta_2\omega_1 + \zeta_1\omega_2) \qquad a_4 = \omega_1^2\omega_2^2$$

The expression $S_F(\omega)$ in Eq. (C.2) reads

$$S_F(\omega) = 2\left\{\left[(-i\omega)^2 + \omega_1^2\right]\left[(i\omega)^2 + \omega_2^2\right] + 4\zeta_1\zeta_2\omega_1\omega_2(-i\omega)(i\omega)\right\}$$

and

$$b_0 = 0 \qquad b_1 = 2 \qquad b_2 = 2(\omega_1^2 + \omega_2^2 - 4\zeta_1\zeta_2\omega_1\omega_2) \qquad b_3 = 2\omega_1^2\omega_2^2$$

Substitution in Eq. (C.6) yields

$$2\int_{-\infty}^{\infty} \operatorname{Re} H_1(\omega)H_2^*(\omega)\, d\omega$$

$$= \frac{8\pi(\zeta_1\omega_1 + \zeta_2\omega_2)}{\left(\omega_1^2 - \omega_2^2\right)^2 + 4\left[\zeta_1\zeta_2\omega_1\omega_2(\omega_1^2 + \omega_2^2) + (\zeta_1^2 + \zeta_2^2)\omega_1^2\omega_2^2\right]} \qquad (9.198)$$

The mean-square displacements are

$$E(|X_1|^2) = \frac{\pi S_0}{4m^2}\left\{ \frac{1}{2\zeta_1\omega_1^3} + \frac{1}{2\zeta_2\omega_2^3}\right.$$

$$+ \frac{8(\zeta_1\omega_1 + \zeta_2\omega_2)}{\left(\omega_1^2 - \omega_2^2\right)^2 + 4\left[\zeta_1\zeta_2\omega_1\omega_2(\omega_1^2 + \omega_2^2) + (\zeta_1^2 + \zeta_2^2)\omega_1^2\omega_2^2\right]}\right\}$$

$$(9.199)$$

$$E\left(|X_2|^2\right) = \frac{\pi S_0}{4m^2}\left\{\frac{1}{2\zeta_1\omega_1^3} + \frac{1}{2\zeta_2\omega_2^3}\right.$$

$$\left. - \frac{8(\zeta_1\omega_1 + \zeta_2\omega_2)}{\left(\omega_1^2 - \omega_2^2\right)^2 + 4\left[\zeta_1\zeta_2\omega_1\omega_2\left(\omega_1^2 + \omega_2^2\right) + \left(\zeta_1^2 + \zeta_2^2\right)\omega_1^2\omega_2^2\right]}\right\}$$

$$(9.200)$$

Note that the weighted averages A_m and A_k are not affected by the cross correlations.

Analogously, from Eq. (9.68), we have, with $\omega_0 \to \omega_j$, $\zeta \to \zeta_j$, and $m \to 1$,

$$\int_{-\infty}^{\infty} \omega^2 |H_j(\omega)|^2 \, d\omega = \frac{\pi}{2\zeta_j\omega_j} \tag{9.201}$$

Using again Eq. (C.6), we arrive at the following expression for the cross correlation term:

$$\int_{-\infty}^{\infty} \omega^2 \operatorname{Re} H_1(\omega) H_2^*(\omega) \, d\omega$$

$$= \frac{4\pi(\zeta_1\omega_2 + \zeta_2\omega_1)\omega_1\omega_2}{\left(\omega_1^2 - \omega_2^2\right)^2 + 4\left[\zeta_1\zeta_2\omega_1\omega_2\left(\omega_1^2 + \omega_2^2\right) + \left(\zeta_1^2 + \zeta_2^2\right)\omega_1^2\omega_2^2\right]}$$

$$(9.202)$$

The mean-square velocities are

$$E\left(|\dot{X}_1|^2\right) = \frac{\pi S_0}{4m^2}\left\{\frac{1}{2\zeta_1\omega_1} + \frac{1}{2\zeta_2\omega_2}\right.$$

$$\left. + \frac{8(\zeta_1\omega_2 + \zeta_2\omega_1)\omega_1\omega_2}{\left(\omega_1^2 - \omega_2^2\right)^2 + 4\left[\zeta_1\zeta_2\omega_1\omega_2\left(\omega_1^2 + \omega_2^2\right) + \left(\zeta_1^2 + \zeta_2^2\right)\omega_1^2\omega_2^2\right]}\right\}$$

$$(9.203)$$

$$E\left(|\dot{X}_2|^2\right) = \frac{\pi S_0}{4m^2}\left\{\frac{1}{2\zeta_1\omega_1} + \frac{1}{2\zeta_2\omega_2}\right.$$

$$\left. - \frac{8(\zeta_1\omega_2 + \zeta_2\omega_1)\omega_1\omega_2}{\left(\omega_1^2 - \omega_2^2\right)^2 + 4\left[\zeta_1\zeta_2\omega_1\omega_2\left(\omega_1^2 + \omega_2^2\right) + \left(\zeta_1^2 + \zeta_2^2\right)\omega_1^2\omega_2^2\right]}\right\}$$

$$(9.204)$$

Substituting the explicit expressions of the natural frequencies (9.181) and the damping coefficients $\zeta_j = c\omega_j/2k$, we arrive at the following equations for the mean-square values in terms of the original parameters:

$$E\left(|X_1|^2\right) = \frac{\pi S_0}{4kc}\left(1 + \frac{1}{(1 + 2\varepsilon)^2} + \frac{2(c^2/km)}{\varepsilon^2/(1 + \varepsilon) + (1 + 2\varepsilon)(c^2/km)}\right)$$

$$(9.205a)$$

$$E\left(|X_2|^2\right) = \frac{\pi S_0}{4kc}\left(1 + \frac{1}{(1 + 2\varepsilon)^2} - \frac{2(c^2/km)}{\varepsilon^2/(1 + \varepsilon) + (1 + 2\varepsilon)(c^2/km)}\right)$$

$$(9.205b)$$

$$E\left(|\dot{X}_1|^2\right) = \frac{\pi S_0}{4mc}\left(1 + \frac{1}{1 + 2\varepsilon} + \frac{2(c^2/km)}{\varepsilon^2/(1 + \varepsilon) + (1 + \varepsilon)(c^2/km)}\right) \quad (9.205c)$$

$$E\left(|\dot{X}_2|^2\right) = \frac{\pi S_0}{4mc}\left(1 + \frac{1}{1 + 2\varepsilon} - \frac{2(c^2/km)}{\varepsilon^2/(1 + \varepsilon) + (1 + \varepsilon)(c^2/km)}\right)$$

$$(9.205d)$$

For ε tending to zero, $E(|X_1|^2) \to \pi S_0/kc$, and $E(|\dot{X}_1|^2) \to \pi S_0/mc$. That is, we again have the results (9.67) and (9.68) for the single-degree-of-freedom system. Moreover, $E(|X_2|^2) \to 0$ and $E(|\dot{X}_2|^2) \to 0$, since the second mass becomes a separate unexcited system. These results are expected, as noted above.

The first two terms in Eqs. (9.205) are associated with the modal autocorrelations, and the third term with the modal cross correlation. Significantly, the contributions of the two correlations for $\varepsilon = 1$, $c^2/km = 0.01$ are 96.72% and 3.28%, respectively. However, for ε tending to zero, they tend to contribute equally, so that the error incurred by disregarding the cross correlations is 50%. With the cross correlations omitted we can no longer arrive at the results (9.67) and (9.68) associated with the single-degree-of-freedom system. The percentage error η_1 in evaluating $E(|X_1|^2)$, defined as

$$\eta_1 = \frac{E\left(|X_1|^2\right)_{\text{exact}} - E\left(|X_1|^2\right)_{\text{approximate}}}{E\left(|X_1|^2\right)_{\text{exact}}} \times 100\% \qquad (9.206)$$

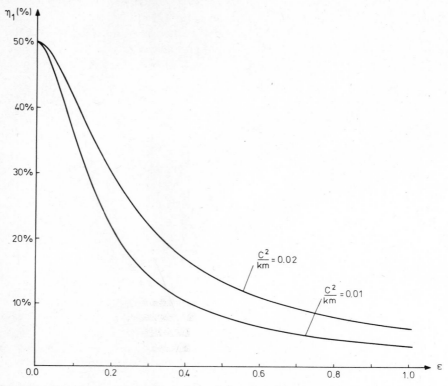

Fig. 9.11.   Percentage error associated with omission of modal cross correlations in evaluation of $E(|X_1|^2)$ for different values of c^2/km.

where $(E|X_1|^2)_{\text{exact}}$ is given by the first of Eq. (9.205) and $E(|X_1|^2)_{\text{approximate}}$ by (9.172)

$$E(|X_1|^2)_{\text{approximate}} = \frac{\pi S_0}{4kc}\left(1 + \frac{1}{(1 + 2\varepsilon)^2}\right)$$

obtained by omitting the cross-correlation term, is plotted in Fig. 9.11.

The error incurred by omission of the cross correlation terms is even larger for the second mass. The corresponding value of the percentage error η_2, defined in analogy with Eq. (9.206), tends to infinity as ε approaches zero. Indeed, when $\varepsilon \to 0$, $E(|X_2|^2)_{\text{exact}} \to 0$, whereas

$$E(|X_2|^2)_{\text{approximate}} = \frac{\pi S_0}{4kc}\left(1 + \frac{1}{(1 + 2\varepsilon)^2}\right) \to \frac{\pi S_0}{2kc} \tag{9.207}$$

and consequently $\eta_2 \to \infty$. η_2 is plotted in Fig. 9.12. (Note analogy with Example 6.12.)

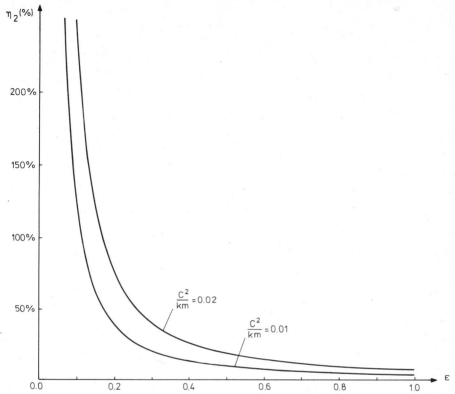

Fig. 9.12.   Percentage error associated with omission of the modal cross correlations in evaluation of $\left(E|X_2|^2 \right)$ for different values of c^2/km.

The reason for this rather dramatic contribution is revealed by Eq. 9.198 for the cross-correlation term: At $\varepsilon \to 0$, the natural frequencies crowd together, as is seen from Eq. (9.181), and in these circumstances the contribution of the cross-correlation term is of the same order of magnitude as that of its autocorrelation counterpart. By contrast, when the natural frequencies are far apart, the cross-correlation term is of an order of magnitude of $\zeta^2/(\omega_1 - \omega_2)^2$ times the autocorrelation term and may be omitted. In other words, the cross-correlation term can be omitted if the following strong inequality holds:

$$\zeta_1^2 \ll \left| 1 - \frac{\omega_1^2}{\omega_2^2} \right| \qquad \zeta_2^2 \ll \left| 1 - \frac{\omega_1^2}{\omega_2^2} \right| \tag{9.208}$$

This qualitative statement is due to Bolotin.

Let us now consider what happens when the coupling coefficient ε tends to infinity. The two-degrees-of-freedom system then degenerates into the single-degree-of-freedom system (Fig. 9.13), and the relevant differential equation

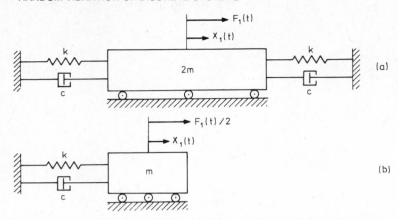

Fig. 9.13. When ε tends to infinity in the two-degrees-of-freedom system shown in Fig. 9.9, two masses act in concert as a single one (a) under half the original force, and the system behaves as in (b).

reads

$$m\ddot{X}_1 + c\dot{X}_1 + kX_1 = \tfrac{1}{2}F_1(t) \qquad (9.209)$$

that is, the resulting system behaves as a single-degree-of-freedom system with half the original force $F_1(t)$. Since the spectral density of $F_1(t)$ is S_0, so that of $\tfrac{1}{2}F_1(t)$ is $\tfrac{1}{4}S_0$ and Eqs. (9.67) and (9.68) yield, respectively, the mean-square displacement and velocity are

$$E\left(|X_1|^2\right) = \frac{S_0\pi}{4kc} \qquad E\left(|\dot{X}_1|^2\right) = \frac{S_0\pi}{4mc} \qquad (9.210)$$

These results are deducible from the first and third of Eqs. (9.205) when $\varepsilon \to \infty$. In this limiting case the cross correlation term tends to zero.

For a relatively large value of ε, when the natural frequencies are far apart and the cross correlation term may be disregarded, the following interesting conclusions follow from Eqs. (9.205):

$$E\left(|X_1|^2\right) \simeq E\left(|X_2|^2\right) \qquad E\left(|\dot{X}_1|^2\right) \simeq E\left(|\dot{X}_2|^2\right) \qquad (9.211)$$

implying that the mean-square displacements of the two masses are equal, although the excitation force is applied to one of them! The same is the case with the mean-square velocities. An analogous, and equally unexpected at first glance, finding was arrived at by Crandall and Wittig for some continuous systems, and will be discussed in Chapter 10.

The symmetric distribution of the mean-square displacements and velocities when the modal cross correlations may be disregarded, can be explained by the symmetry of the structure about the axis through the midpoint of the coupling spring. In fact, as is seen from the modal matrix $[v]$ of (Eq. 9.182), the first normal mode is associated with movement of both masses in the *same* direction with equal amplitudes $v_1^{(1)} = v_2^{(1)}$ (symmetric normal mode). The coupling spring then behaves as though it were free, while the masses behave as though they were uncoupled. The second normal mode is associated with movement of masses in *opposite* directions with equal amplitudes, so that $v_1^{(2)} = -v_2^{(2)}$ (asymmetric normal mode). Equation (9.172) contains, however, only *squares* of the amplitudes, corresponding to the symmetric and asymmetric modes, and therefore

$$\left[v_1^{(1)}\right]^2 = \left[v_2^{(1)}\right]^2 \qquad \left[v_1^{(2)}\right]^2 = \left[v_2^{(2)}\right]^2$$

As a consequence, we arrive at Eq. (9.211) directly from Eq. (9.172).

It is worth emphasizing that such a symmetric response, despite the lack of symmetry in the excitation, is an approximate result. However small (but nonzero) the cross correlation,

$$E\left(|X_1|^2\right) > E\left(|X_2|^2\right) \qquad E\left(|\dot{X}_1|^2\right) > E\left(|\dot{X}_2|^2\right) \tag{9.212}$$

Only at $\varepsilon \to \infty$ do we have $E(|X_1|^2) \to E(|X_2|^2)$. However, already for $\varepsilon = 2$ and $c^2/km \ll 1$, $E(|X_1|^2) \simeq E(|X_2|^2)$, and they differ from the $\varepsilon \to \infty$ result only by about 4%. $E(|\dot{X}_1|^2)$ also approximately equals $E(|\dot{X}_2|^2)$, the difference from the corresponding $\varepsilon \to \infty$ result being about 20%.

Note that approximate results like (9.211) are characteristic of a symmetric system; they are no longer valid for a nonsymmetric system (see Prob. 9.4).

Example 9.10

Having solved the problem of determining a probabilistic response by the normal mode approach, we will demonstrate how it can be evaluated *directly* without recourse to this approach, for the example of the two-degrees-of-freedom system considered in Sec. 9.4. We first put in Eq. (9.177):

$$F_1(t) = e^{i\omega t} \tag{9.213}$$

with a view to determining (as in Sec. 9.1) the complex frequency responses, or receptances, $H_1(\omega)$ and $H_2(\omega)$, so that

$$X_1(t) = H_1(\omega)e^{i\omega t} \qquad X_2(t) = H_2(\omega)e^{i\omega t} \tag{9.214}$$

Substitution of Eqs. (9.213) and (9.214) in (9.177) yields

$$\left[-m\omega^2 + i(1 + \varepsilon)c\omega + (1 + \varepsilon)k\right]H_1(\omega) + \left[-i\varepsilon c\omega - \varepsilon k\right]H_2(\omega) = 1$$

$$\left[-i\varepsilon c\omega - \varepsilon k\right]H_1(\omega) + \left[-m\omega^2 + i(1 + \varepsilon)c\omega + (1 + \varepsilon)k\right] = 0$$

with

$$H_1(\omega) = \frac{1}{\Delta(\omega)} \left[-m\omega^2 + i\varepsilon(1 + \varepsilon)c + (1 + \varepsilon)k \right]$$

$$H_2(\omega) = \frac{1}{\Delta(\omega)} (ic\omega + k)\varepsilon$$

where

$$\Delta(\omega) = m^2\omega^4 - 2i\omega^3(1 + \varepsilon)cm - \omega^2 \left[(1 + 2\varepsilon)c^2 + 2(1 + \varepsilon)km \right]$$

$$+ 2i\omega(1 + 2\varepsilon)ck + (1 + 2\varepsilon)k^2$$

The second pair of frequency responses is obtained as follows:

$$H_{\dot{X}_1}(\omega) = i\omega H_1(\omega) \qquad H_{\dot{X}_2}(\omega) = i\omega H_2(\omega)$$

Now,

$$E\left(|X_j|^2\right) = \int_{-\infty}^{\infty} |H_j(\omega)|^2 S_F(\omega) \, d\omega$$

$$E\left(|\dot{X}_j|^2\right) = \int_{-\infty}^{\infty} |i\omega H_j(\omega)|^2 S_F(\omega) \, d\omega$$

Since these quantities were already obtained in Sec. 9.4 by the normal mode approach, we shall deal here only with $E(|X_1|^2)$, for $S_F(\omega) = S_0$,

$$E\left(|X_1|^2\right) = S_0 \int_{-\infty}^{\infty} |H_1(\omega)|^2 \, d\omega$$

$$= S_0 \int_{-\infty}^{\infty} \frac{1}{|\Delta(\omega)|^2} \left\{ \left[-m\omega^2 + (1 + \varepsilon)k \right]^2 + c^2\omega^2(1 + \varepsilon)^2 \right\} \, d\omega$$

$$(9.215)$$

This integral is evaluated according to Appendix C and yields a result coincident with Eq. (9.205a). For further examples of direct evaluation, see book by Crandall and Mark. This direct evaluation method accounts automatically for both modal auto- and cross correlations. For many-degrees-of-freedom systems, however, its realization becomes cumbersome, and use of normal mode method is then advisable.

Example 9.11

Consider now a system with coincident natural frequencies (see Fig. 9.14). White-noise excitation is applied to the bottom mass. We are interested in the

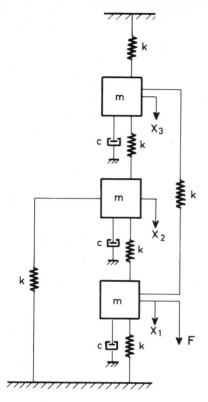

Fig. 9.14. Three-degrees-of-freedom system with two coincident natural frequencies.

mean-square value $R_{11} = E(|X_1'|^2)$ and the error incurred by omission of the cross-correlation term.

The equations of motion are readily obtainable as Eq. (9.106) with

$$[m] = \begin{bmatrix} m & 0 & 0 \\ 0 & m & 0 \\ 0 & 0 & m \end{bmatrix} \qquad [c] = \begin{bmatrix} c & 0 & 0 \\ 0 & c & 0 \\ 0 & 0 & c \end{bmatrix}$$

$$[k] = \begin{bmatrix} 3k & -k & -k \\ -k & 3k & -k \\ -k & -k & 3k \end{bmatrix} \tag{9.216}$$

The natural frequencies are found from

$$\begin{vmatrix} 3k - m\omega^2 & -k & -k \\ -k & 3k - m\omega^2 & -k \\ -k & -k & 3k - m\omega^2 \end{vmatrix} = 0$$

which reduces to

$$(4k - m\omega^2)\big[(3k - m\omega^2)(2k - m\omega^2) - 2k^2\big] = 0$$

The natural frequencies are

$$\omega_1^2 = \frac{k}{m}, \quad \omega_2^2 = \omega_3^2 = 4\frac{k}{m} \tag{9.217}$$

The natural mode associated with the first natural frequency is

$$\{y^{(1)}\} = \lambda_1[1 \quad 1 \quad 1]^T$$

whereas that with the second is

$$\{y^{(2)}\} = \lambda_2[1 \quad 1 \quad -2]^T$$

The third mode can be found from the requirement that it be orthogonal to the first two. Denoting

$$\{y^{(3)}\} = \lambda_3[1 \quad a \quad b]^T$$

we have

$$\{y^{(3)}\}^T[m]\{y^{(1)}\} = m\lambda_1\lambda_3(1 + a + b) = 0$$

$$\{y^{(3)}\}^T[m]\{y^{(2)}\} = m\lambda_2\lambda_3(1 + a - 2b) = 0$$

so that $b = 0$ and $a = -1$. Therefore,

$$\{y^{(3)}\} = \lambda_3[1 \quad -1 \quad 0]^T$$

The modal matrix becomes

$$[v] = \frac{1}{\sqrt{6m}}\begin{bmatrix} \sqrt{2} & 1 & \sqrt{3} \\ \sqrt{2} & 1 & -\sqrt{3} \\ \sqrt{2} & -2 & 0 \end{bmatrix}$$

Substituting in Eq. (9.106) $\{X\} = [v]\{Q\}$ and premultiplying the result by $[v]^T$, we obtain

$$[v]^T[m][v]\{\ddot{Q}\} + [v]^T[c][v]\{\dot{Q}\} + [v]^T[k][v]\{Q\} = [v]^T\{F\} \tag{9.218}$$

After manipulations, we find

$$[v]^T[m][v] = [I]$$

$$[v]^T[c][v] = \frac{c}{m}[I]$$

$$[v]^T[k][v] = \frac{k}{m}\begin{bmatrix} 1 & 0 & 0 \\ 0 & 4 & 0 \\ 0 & 0 & 4 \end{bmatrix}$$

Equations (9.218) finally become

$$\ddot{Q}_1 + \frac{c}{m}\dot{Q}_1 + \frac{k}{m}Q_1 = \frac{1}{\sqrt{3m}}F_1$$

$$\ddot{Q}_2 + \frac{c}{m}\dot{Q}_2 + 4\frac{k}{m}Q_2 = \frac{1}{\sqrt{6m}}F_1$$

$$\ddot{Q}_3 + \frac{c}{m}\dot{Q}_3 + 4\frac{k}{m}Q_3 = \frac{1}{\sqrt{2m}}F_1 \qquad (9.219)$$

The correlation matrix of generalized velocities $\dot{Q}_1(t)$, $\dot{Q}_2(t)$, and $\dot{Q}_3(t)$, $[r] = E\{\dot{Q}\}\{\dot{Q}\}^T$, is readily obtainable as

$$r_{11} = \frac{S_0\pi}{3c} \qquad r_{22} = \frac{S_0\pi}{6c} \qquad r_{33} = \frac{S_0\pi}{2c} \qquad (9.220)$$

$$r_{jk} = r_{kj} = \frac{4\pi S_0\left(\zeta_j\omega_k + \zeta_k\omega_j\right)\omega_j\omega_k A_{jk}}{\left(\omega_j^2 - \omega_k^2\right)^2 + 4\left[\zeta_j\zeta_k\omega_j\omega_k\left(\omega_j^2 + \omega_k^2\right) + \left(\zeta_j^2 + \zeta_k^2\right)\omega_j^2\omega_k^2\right]}$$

$$\text{for } j \neq k$$

$$A_{12} = \frac{1}{3\sqrt{2}\,m} \qquad A_{13} = \frac{1}{\sqrt{6}\,m} \qquad A_{23} = \frac{1}{2\sqrt{3}\,m}$$

$$r_{23} = \sqrt{3}\,r_{22}$$

Now we return to the sought quantities

$$[R] = E\{\dot{X}\}\{\dot{X}\}^T = E[v]\{\dot{Q}\}([v]\{\dot{Q}\})^T$$

$$= E\left([v]\{\dot{Q}\}\{\dot{Q}\}^T[v]^T\right) = [v]E\left(\{\dot{Q}\}\{\dot{Q}\}^T\right)[v]^T$$

$$= [v][r][v]^T$$

the expression for R_{11} being

$$R_{11} = \frac{1}{m}\left(\tfrac{1}{3}r_{11} + \tfrac{1}{6}r_{22} + \tfrac{1}{2}r_{33} + \frac{2}{\sqrt{18}}r_{12} + \frac{2}{\sqrt{6}}r_{13} + \frac{1}{\sqrt{3}}r_{23}\right)$$

Omission of the cross-correlation term $r_{23}/\sqrt{3}$ entails the error

$$\eta = \frac{r_{23}/\sqrt{3}}{R_{11}} \times 100\% \simeq \frac{r_{23}/\sqrt{3}}{\tfrac{1}{3}r_{11} + \tfrac{1}{6}r_{22} + \tfrac{1}{2}r_{33} + r_{23}/\sqrt{3}} \times 100\% = 30\%$$

PROBLEMS

9.1. A single-degree-of-freedom system (Eq. 9.48) is subjected to a random load (see figure) with exponential autocorrelation function $R_F(\tau) = \pi \alpha S_0 e^{-\alpha|\tau|}$.

    (a) Find the mean-square value of the displacement, velocity, and acceleration.

    (b) Check that for $\alpha \to \infty$, the values obtained in (a) coincide with those of a system under ideal white-noise excitation.

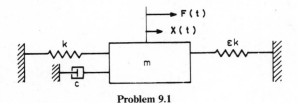

Problem 9.1

9.2. A cantilever with a concentrated mass attached to its tip is subjected to random loading with autocorrelation function in the form of ideal white noise with intensity S_0. The cantilever itself is massless. Geometric

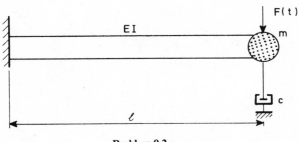

Problem 9.2

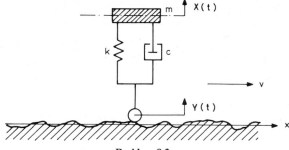

Problem 9.3

dimensions are indicated in the accompanying figure. Find the mean-square displacement and the mean-square velocity.

9.3. A road vehicle travels with uniform velocity v on a rough surface (see figure) and in the process is subjected to a time-varying displacement excitation. The roughness of the profile is a random function of the coordinate x, hence both the excitation and response of the vehicle are random. Derive an expression for the response mean-square value if the autocorrelation function of the profile is $R_Y(x_1, x_2) = d^2 \exp\{-\alpha|x_2 - x_1|\}$.

9.4. Consider a nonsymmetric two-degrees-of-freedom system under ideal white-noise excitation $F_1(t)$.

(a) Verify that the error incurred by disregarding the cross-correlations in evaluating $E(|X_1|^2)$ reaches 40% when ε tends to zero.

(b) Verify that at $\varepsilon = 1$ the cross-correlation terms can still be omitted but $E(|X_1|^2)$ differs from $E(|X_2|^2)$, unlike the example of a symmetric system considered in Sec. 9.4.

9.5. The system shown in the figure is subjected to random loading with the autocorrelation function

$$R_F(\tau) = d^2 e^{-\alpha|\tau|}\left(\cos \beta\tau + \frac{\alpha}{\beta} \sin \beta|\tau|\right)$$

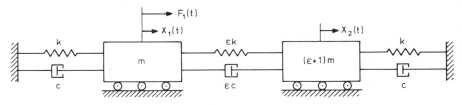

Problem 9.4

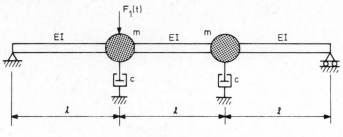

Problem 9.5

The masses are attached to a massless beam simply supported at its ends, with stiffness modulus EI. Find $E(|X_1|^2)$ and $E(|X_2|^2)$.

9.6. A beam is clamped at its ends, but free sliding is permitted in the axial direction. The beam itself is massless. The stiffness modulus of the beam between the masses is εEI, where ε is a nonnegative parameter. $F_1(t)$ represents an ideal white-noise excitation with intensity S_0.

(a) Examine the dependence of $E(|X_1|^2)$ versus the parameter ε.

(b) Verify that for $\varepsilon \to 0$ the result of Prob. 9.2 is obtained.

9.7. Find $E(|X_1|^2)$ and $E(|X_2|^2)$ for the system shown in the figure, where $R_F(\tau) = e^{-\alpha^2 \tau^2}$. Use the approximate method for determining the mean-square values.

9.8. The system shown in the figure is subjected to ideal white-noise excitation.

(a) Verify that the modal matrix is

$$[v] = \frac{1}{2\sqrt{m}} \begin{bmatrix} 1 & \sqrt{2} & 1 \\ \sqrt{2} & 0 & -\sqrt{2} \\ 1 & -\sqrt{2} & 1 \end{bmatrix}$$

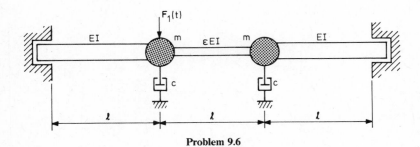

Problem 9.6

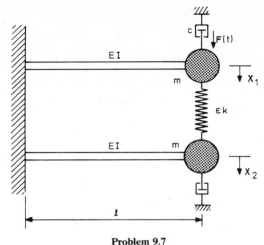

Problem 9.7

where the corresponding natural frequencies are

$$\omega_1^2 = (2 - \sqrt{2})\frac{k}{m}, \quad \omega_2^2 = 2\frac{k}{m}, \quad \omega_3^2 = (2 + \sqrt{2})\frac{k}{m}$$

(b) Verify that $E(|X_1|^2) \simeq E(|X_3|^2)$ and $E(|\dot{X}_1|^2) \simeq E(|\dot{X}_3|^2)$.

9.9. An n-degrees-of-freedom system is subjected to ideal white-noise loading with intensity S_0. Each mass is connected both with the ground and with the other masses, the springs being identical with stiffness k. Damping proportional to the mass is provided by a dashpot attached to each mass, the damping coefficient being c.

(a) Verify that this system possesses $n - 1$ identical natural frequencies.

(b) Estimate $E(|X_1|^2)$ in the particular case $n = 4$ (for $n = 3$ the problem reduces to Example 9.11.)

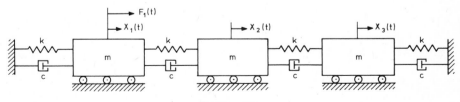

Problem 9.8

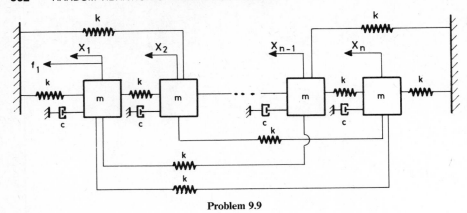

Problem 9.9

CITED REFERENCES

Barnoski, R. L., and Maurer, J. R., "Mean-Square Response of Simple Mechanical Systems to Nonstationary Random Excitation," *ASME J. Appl. Mech.*, **36**, 221–227 (1969).

Bendat, J. S., and Piersol, A. G., *Random Data: Analysis and Measurement Procedures*, Wiley-Interscience, New York, 1971, Chap. 9.

Bolotin, V. V., *Statistical Methods in Structural Mechanics*, State Pub. House for Building, Architecture and Building Materials, Moscow, 1965 (translated into English by S. Aroni), Holden-Day, San Francisco, 1969, p. 122.

Caughey, T. K., "Classical Normal Modes in Damped Linear Dynamic Systems," *ASME J. Appl. Mech.*, **27**, 269–271 (1960).

Caughey, T. K., and Stumpf, H. J., "Transient Response of a Dynamic System under Random Excitation," *ASME J. Appl. Mech.*, **28**, 563–566 (1961).

Crandall, S. H., and Mark, W. D., *Random Vibration in Mechanical Systems*, Academic Press, New York, 1963, p. 79.

Crandall, S. H., and Wittig, L., "Chladni's Patterns for Random Vibration of a Plate," in G. Herrmann and N. Perrone, Eds., *Dynamic Response of Structures*, Pergamon Press, New York, 1972, pp. 55–71.

Holman, R. E., and Hart, G. C., "Structural Response to Segmented Nonstationary Random Excitation," *AIAA J.*, **10**, 1473–1478 (1972).

Hurty, W. C., and Rubinstein, M. F., *Dynamics of Structures*, Prentice-Hall, Englewood Cliffs, NJ, 1964, pp. 313–337.

Meirovich, L., *Elements of Vibration Analysis*, McGraw-Hill, New York, 1975.

Priestley, M. B., "Power Spectral Analysis of Nonstationary Random Processes," *J. Sound Vibration*, **6**, 86–97 (1967).

Warburton, G. B., *Dynamic Behaviour of Structures*, 2nd ed., Pergamon Press, Oxford, 1976, p.33.

RECOMMENDED FURTHER READING

Caughey, T. K., "Nonlinear Theory of Random Vibrations," *Advan. Appl. Mech.*, **11**, 209–253 (1971).

Corotis, R. B., and Vanmarcke, E. H., "Time-Dependent Spectral Content of System Response," *J. Eng. Mech. Div., Proc. ASCE*, **101**, 623–637 (1975).

Eringen, A. C., "Response of Tall Buildings to Random Earthquakes," *Proc. 3rd U.S. Nat. Cong. Appl. Mech.*, ASME, 1958, pp. 141–151.

Hammond, J. K., "On the Response of Single and Multidegree of Freedom Systems to Non-Stationary Random Excitations," *J. Sound Vibration*, **7**, 393–416 (1968).

Iwan, W. D., "Response of Multi-Degree-of-Freedom Yielding Systems", *J. Eng. Mech. Div., Proc. ASCE*, **94**, 421–437 (1968).

Lin, Y. K., *Probabilistic Theory of Structural Dynamics*, McGraw-Hill, New York, 1967. Chap. 5: Linear Structures with Single Degree of Freedom, pp. 110–154; Chap. 6: Linear Structures with Finitely Many Degrees of Freedom, pp. 155–202; Chap. 8: Nonlinear Structures, pp. 253–292.

Madsen, P. H., and Krenk, S., "Stationary and Transient Response Statistics," The Danish Center for Applied Mathematics and Mechanics, Rept. No. 194, Oct. 1980.

Newland, D. E., *An Introduction to Random Vibrations and Spectral Analysis*, Longman, London, 1975. Chap. 6: Excitation–Response Relations for Linear Systems, pp. 53–66; Chap. 7: Transmission of Random Vibration, pp. 67–81.

Robson, J. D., "The Random Vibration Response of a System Having Many Degrees of Freedom," *Aeronaut. Quart.*, **17**, 21–30 (1966).

Sagirow, P., *Stochastic Methods in the Dynamics of Satellites*, Intern. Centre Mech. Sci., Udine, Course No. 57, Springer, Vienna, 1970.

Vanmarcke, E. H., "Some Recent Developments in Random Vibration," *Appl. Mech. Rev.*, **32**(10), 1197–1202 (1979).

chapter 10

Random Vibration of Continuous Structures

Having so far confined ourselves to discrete structures, it is natural now to proceed to continuous structures in which excitation and response are generally random not only in time but also in space. The solution via the normal mode method will be derived, and the effect of cross correlation between different modes will be shown again to be of marked importance.

10.1 RANDOM FIELDS

In Chapter 9 we referred to a family of random variables depending on a single deterministic parameter t as a random function $X(t)$. In applications t is usually time, in which case $X(t)$ is referred to as a random process.

A family of random variables depending on more than one deterministic parameter will be called a *random field*. Examples of such functions of space and time are the ordinates of the sea surface, stresses induced by turbulent boundary-layer pressure, and so on. We will confine ourselves to at most four deterministic parameters: time t and a point in space $\mathbf{r} = x\mathbf{i}$ for a one-dimensional case, $\mathbf{r} = x\mathbf{i} + y\mathbf{j}$ for a two-dimensional case, and $\mathbf{r} = x\mathbf{i} + y\mathbf{j} + z\mathbf{k}$ for the most general, three-dimensional case. In analogy to the definition of a stationary function, in the wide sense, we will consider a random field as *homogeneous* if its mathematical expectation is constant and the autocorrelation function*

$$R_x(\mathbf{r}_1, t_1; \mathbf{r}_2, t_2) = E\left[X(\mathbf{r}_1, t_1) X(\mathbf{r}_2, t_2) \right] \tag{10.1}$$

---

*In this chapter the random functions are denoted by lowercase symbols, whereas capitals are reserved for the corresponding random spectra.

depends only on the differences $\rho = \mathbf{r}_2 - \mathbf{r}_1$ and $\tau = t_2 - t_1$.

$$R_x(\mathbf{r}_1, t_1; \mathbf{r}_2, t_2) = R_x(\mathbf{r}_2 - \mathbf{r}_1, t_2 - t_1) = R_x(\rho, \tau) \qquad (10.2)$$

where

$$\rho = \mathbf{r}_2 - \mathbf{r}_1 = (x_2 - x_1)\mathbf{i} + (y_2 - y_1)\mathbf{j} + (z_2 - z_1)\mathbf{k} = \xi_1\mathbf{i} + \xi_2\mathbf{j} + \xi_3\mathbf{k}$$

$$(10.3)$$

For a random field that is stationary in the wide sense in time, we define the cross-spectral density

$$S_x(\omega, \mathbf{r}_1, \mathbf{r}_2) = \frac{1}{2\pi} \int_{-\infty}^{\infty} R_x(\tau, \mathbf{r}_1, \mathbf{r}_2) e^{-i\omega\tau}\, d\tau \qquad (10.4)$$

where

$$R_x(\tau, \mathbf{r}_1, \mathbf{r}_2) = \int_{-\infty}^{\infty} S_x(\omega, \mathbf{r}_1, \mathbf{r}_2) e^{i\omega\tau}\, d\omega \qquad (10.5)$$

For a random field that is both stationary in the wide sense in time and homogeneous in the wide sense in space,

$$S_x(\omega, \kappa) = \frac{1}{(2\pi)^4} \int_{-\infty}^{\infty}\!\!\!\int\!\!\int\!\!\int R_x(\tau, \rho) \exp(-i\kappa \cdot \rho - i\omega\tau)\, d\rho\, d\tau$$

$$(10.6)$$

$$R_x(\tau, \rho) = \int_{-\infty}^{\infty}\!\!\!\int\!\!\int\!\!\int S_x(\omega, \kappa) \exp(i\kappa \cdot \rho + i\omega\tau)\, d\kappa\, d\omega \qquad (10.7)$$

where

$$\kappa = \kappa_1\mathbf{i} + \kappa_2\mathbf{j} + \kappa_3\mathbf{k}$$

$$\rho = \xi_1\mathbf{i} + \xi_2\mathbf{j} + \xi_3\mathbf{k} \qquad (10.8)$$

and $\kappa \cdot \rho$ is an inner product,

$$\kappa \cdot \rho = \kappa_1\xi_1 + \kappa_2\xi_2 + \kappa_3\xi_3$$

For the one- and two-dimensional cases, the integrals in Eqs. (10.6) and (10.7) are double and triple, respectively.

Example 10.1

Consider a plate infinite in both directions under distributed loading $q(t, x, y)$, where t denotes time and x and y the space coordinates of a point in the middle surface of the plate. Assume that $q(t, x, y)$ is a homogeneous field with zero mathematical expectation. We represent the loading by its expansion

$$q(t, x, y) = \int\int\int_{-\infty}^{\infty} \exp\{i(\omega t + \kappa_1 x + \kappa_2 y)\}Q(\omega, \kappa_1, \kappa_2)\, d\omega\, d\kappa_1\, d\kappa_2$$

(10.9)

The autocorrelation function (coincident with the autocovariance function at zero mathematical expectation) reads

$$R_q(t_1, x_1, y_1; t_2, x_2, y_2) = E[q^*(t_1, x_1, y_1)q(t_2, x_2, y_2)]$$

$$= \int\int\int\int\int\int_{-\infty}^{\infty} E[Q^*(\omega, \kappa_1, \kappa_2)Q(\omega', \kappa_1', \kappa_2')]$$

$$\times \exp[-i(\omega t_1 + \kappa_1 x_1 + \kappa_2 y_1)]$$

$$\times \exp[i(\omega' t_2 + \kappa_1' x_2 + \kappa_2' y_2)]$$

$$\times d\omega\, d\kappa_1\, d\kappa_2\, d\omega'\, d\kappa_1'\, d\kappa_2'$$

(10.10)

We note that this function depends on the differences $\tau = t_2 - t_1, \xi_1 = x_2 - x_1,$ $\xi_2 = y_2 - y_1$, providing the following relation is valid:

$$E[Q^*(\omega, \kappa_1, \kappa_2)Q(\omega', \kappa_1', \kappa_2')]$$

$$= S_q(\omega, \kappa_1, \kappa_2)\delta(\omega' - \omega)\delta(\kappa_1' - \kappa_1)\delta(\kappa_2' - \kappa_2)$$ (10.11)

Then

$$R_q(\tau, \xi_1, \xi_2) = \int\int\int_{-\infty}^{\infty} S_q(\omega, \kappa_1, \kappa_2)\exp[i(\omega\tau + \kappa_1\xi_1 + \kappa_2\xi_2)]\, d\omega\, d\kappa_1\, d\kappa_2$$

(10.12)

which is analogous to Eq. (10.7). The spectral density $S_q(\omega, \kappa_1, \kappa_2)$ becomes

$$S_q(\omega, \kappa_1, \kappa_2) = \frac{1}{(2\pi)^3} \int\int\int_{-\infty}^{\infty} R_q(\tau, \xi_1, \xi_2)$$

$$\times \exp[-i(\omega\tau + \kappa_1\xi_1 + \kappa_2\xi_2)]\, d\tau\, d\xi_1\, d\xi_2$$ (10.13)

The displacement of the plate $w(t, x, y)$ is governed by the differential equation

$$D\left(\frac{\partial^4 w}{\partial x^4} + 2\frac{\partial^4 w}{\partial x^2\, \partial y^2} + \frac{\partial^4 w}{\partial y^4}\right) + \rho h\frac{\partial^2 w}{\partial t^2} = q(t, x, y)$$ (10.14)

where

$$D = \frac{Eh^3}{12(1 - \nu^2)} \tag{10.15}$$

h is the thickness of the plate, ν Poisson's ratio, ρ the mass density, and E Young's modulus, treated as a complex quantity

$$E = E_r(1 + i\mu) \tag{10.16}$$

incorporating the structural damping effect. (Compare the treatment of the stiffness coefficient k in Example 9.6.)

For the displacement $w(t, x, y)$ we have the following expansion

$$w = \int\!\!\!\int\!\!\!\int_{-\infty}^{\infty} \exp\{i(\omega t + \kappa_1 x + \kappa_2 y)\} W(\omega, \kappa_1, \kappa_2) \, d\omega \, d\kappa_1 \, d\kappa_2 \tag{10.17}$$

Substitution of Eqs. (10.9) and (10.17) in (10.14) yields

$$W(\omega, \kappa_1, \kappa_2) = \frac{Q(\omega, \kappa_1, \kappa_2)}{D(\kappa_1^2 + \kappa_2^2)^2 - \rho h \omega^2} \tag{10.18}$$

Thanks to Eqs. (10.11) and (10.18), the autocorrelation function of the displacements can be put in the following form:

$$R_w(\tau, \xi_1, \xi_2) = \int\!\!\!\int\!\!\!\int_{-\infty}^{\infty} \frac{\exp\left[i(\omega\tau + \kappa_1\xi_1 + \kappa_2\xi_2)\right]}{\left|D(\kappa_1^2 + \kappa_2^2)^2 - \rho h \omega^2\right|^2} S_q(\omega, \kappa_1, \kappa_2) \, d\omega \, d\kappa_1 \, d\kappa_2 \tag{10.19}$$

For the autocorrelation function involving the time lag τ only, we have

$$R_w(\tau) = R_w(\tau, 0, 0)$$

$$= \int\!\!\!\int\!\!\!\int_{-\infty}^{\infty} \frac{\exp(i\omega\tau)}{\left|D(\kappa_1^2 + \kappa_2^2)^2 - \rho h \omega^2\right|^2} S_q(\omega, \kappa_1, \kappa_2) \, d\omega \, d\kappa_1 \, d\kappa_2 \tag{10.20}$$

Consider the case where the loading is represented by spacewise white noise, that is,

$$R_q(\tau, \xi_1, \xi_2) = R_q(\tau)\delta(\xi_1)\delta(\xi_2) \tag{10.21}$$

Equation (10.13) then yields

$$S_q(\omega, \kappa_1, \kappa_2) = S_q(\omega) = \frac{1}{2\pi} \int_{-\infty}^{\infty} R_q(\tau) e^{-i\omega\tau} d\tau$$

where $S_q(\omega)$ is a spectral density. Then Eq. (10.20) becomes

$$R_w(\tau) = \int_{-\infty}^{\infty} S_q(\omega) e^{i\omega\tau} A(\omega) \, d\omega$$

$$A(\omega) = \int\int_{-\infty}^{\infty} \frac{d\kappa_1 \, d\kappa_2}{\left| D(\kappa_1^2 + \kappa_2^2)^2 - \rho h \omega^2 \right|^2} \tag{10.22}$$

Introducing the new variables

$$\kappa_1 = r \cos\theta \qquad \kappa_2 = r \sin\theta \qquad z = r^2$$

then

$$A(\omega) = \tfrac{1}{2} \int_0^{2\pi} d\theta \int_0^{\infty} \frac{dz}{D_r^2(1 + \mu^2)z^4 - 2\rho h \omega^2 D_r z^2 + (\rho h \omega^2)^2}$$

For evaluation of the integral we use the formula (Gradshteyn and Ryzhik, p. 293)

$$\int_0^{\infty} \frac{dz}{a + bz^2 + cz^4} = \frac{\pi \cos(\alpha/2)}{2cd^3 \sin\alpha} \qquad d = \left(\frac{a}{c}\right)^{1/4} \qquad \cos\alpha = -\frac{b}{2(ac)^{1/2}}$$

and the final result (due to Pal'mov) is

$$A(\omega) = \frac{\pi^2}{2\sqrt{2}} \frac{\left[1 + (1 + \mu^2)^{1/2}\right]^{1/2}}{\rho h |\omega^3 \mu| (D_r \rho h)^{1/2}} \simeq \frac{\pi^2}{2\rho h |\omega^3 \mu| (D_r \rho h)^{1/2}} \tag{10.23}$$

where $D_r = h^3 E_r/12(1 - \nu^2)$ and $\mu \ll 1$. The autocorrelation function becomes

$$R_w(\tau) = \int_{-\infty}^{\infty} \frac{\pi^2 S_q(\omega) \exp(i\omega\tau) \, d\omega}{2\rho h |\omega^3 \mu| (D_r \rho h)^{1/2}} \tag{10.24}$$

and depends on the particular form of $S_q(\omega)$. As is seen from Eq. (10.24), the vibration intensity is a function of damping, mass density, and stiffness and moreover decreases as these parameters increase.

10.2 NORMAL MODE METHOD

We illustrate this method on an example of a uniform beam under distributed random loading $q(x, t)$, stationary in the wide sense in time, so that its mathematical expectation is independent of time and the autocorrelation function depends only on the time lag τ. The relevant differential equation reads

$$EI\frac{\partial^4 w}{\partial x^4} + c\frac{\partial w}{\partial t} + \rho A\frac{\partial^2 w}{\partial t^2} = q(x, t) \tag{10.25}$$

where E is Young's modulus, I the moment of inertia, c the viscous damping coefficient, ρ mass density per unit area, A the cross-sectional area, and $w(x, t)$ the displacement.

This equation is supplemented by the boundary conditions at the ends of the beam. For a free end, both the bending moment M_z and the shear force V_y vanish:

$$M_z = EI\frac{\partial^2 w}{\partial x^2} = 0 \qquad V_y = EI\frac{\partial^3 w}{\partial x^3} = 0 \tag{10.26}$$

For a simply supported end, the displacement w and the bending moment M_z vanish:

$$w = 0 \qquad M_z = EI\frac{\partial^2 w}{\partial x^2} = 0 \tag{10.27}$$

Finally, for a clamped end, the displacement and the slope of the displacement curve vanish:

$$w = 0 \qquad \frac{\partial w}{\partial x} = 0 \tag{10.28}$$

The boundary conditions can be put as

$$P_i(w) = 0 \quad \begin{cases} i = 1,2 & \text{for } x = 0 \\ i = 3,4 & \text{for } x = l \end{cases} \tag{10.29}$$

where l is the span of the beam and $P_i(\cdots)$ are some operators. For example, for a beam simply supported at its ends, we have

$$P_1(w) = P_3(w) = 1 \cdot w = w$$

$$P_2(w) = P_4(w) = EI\frac{\partial^2 w}{\partial x^2} \tag{10.30}$$

Additional conditions are the initial values at $t = 0$, which for convenience we take as zero:

$$w(x, t)\Big|_{t=0} = 0 \qquad \frac{\partial w(x, t)}{\partial t}\Big|_{t=0} = 0 \qquad (10.31)$$

10.2.1 Free Undamped Vibration. We first consider the free undamped vibration problem. Accordingly, we put $c \equiv 0$ and $q(x, t) \equiv 0$ in Eq. (10.25), to obtain

$$EI\frac{\partial^4 w}{\partial x^4} + \rho A\frac{\partial^2 w}{\partial t^2} = 0$$

We put also $w(x, t) = W(x, \omega)e^{i\omega t}$, which results in

$$EI\frac{d^4 W}{dx^4} - \rho A\omega^2 W = 0 \qquad (10.32)$$

This is an ordinary differential equation, with a solution of the form

$$W(x, \omega) = C_1\sin rx + C_2\cos rx + C_3\sinh rx + C_4\cosh rx \qquad (10.33)$$

where

$$r = \left(\frac{\rho A\omega^2}{EI}\right)^{1/4}$$

This solution must satisfy the boundary conditions (10.26), and since C_j are functions only of ω we have the set

$$C_1 P_j(\sin rx) + C_2 P_j(\cos rx) + C_3 P_j(\sinh rx) + C_4 P_j(\cosh rx) = 0,$$

$$j = 1, 2, 3, 4$$

The above are the homogeneous equations of the system in terms of C_1, C_2, C_3, C_4, and a nontrivial solution is conditional on the following determinant vanishing:

$$\begin{vmatrix} P_1(\sin rx) & P_1(\cos rx) & P_1(\sinh rx) & P_1(\cosh rx) \\ P_2(\sin rx) & P_2(\cos rx) & P_2(\sinh rx) & P_2(\cosh rx) \\ P_3(\sin rx) & P_3(\cos rx) & P_3(\sinh rx) & P_3(\cosh rx) \\ P_4(\sin rx) & P_4(\cos rx) & P_4(\sinh rx) & P_4(\cosh rx) \end{vmatrix} = 0 \quad (10.34)$$

For example, for the beam simply supported at its ends, we have in accordance with Eq. (10.27),

$$\begin{vmatrix} 0 & 1 & 0 & 1 \\ 0 & -r^2 & 0 & r^2 \\ s & c & S & C \\ -r^2 s & -r^2 c & r^2 S & r^2 C \end{vmatrix} = 0 \qquad (10.35)$$

where $s = \sin rl$, $c = \cos rl$, $S = \sinh rl$, and $C = \cosh rl$.

Evaluation of the determinant yields $s = 0$, which means $rl = j\pi$, $j = 1, 2, \dots$, with

$$\omega = \left(\frac{EI}{\rho A} \right)^{1/2} \frac{j^2 \pi^2}{l^2} \equiv \omega_j \qquad (10.36)$$

where ω_j is the jth natural frequency. The corresponding mode shape $W(x, \omega_j)$ turns out to be

$$W(x, \omega_j) = \sin \frac{j\pi x}{l} \qquad (10.37)$$

so that j is the number of half-waves over the span of the beam. The first three mode shapes for this case are shown in Fig. 10.1.

As another example, consider the beam clamped at both its ends. Then, instead of Eq. (10.35) we have

$$\begin{vmatrix} 0 & 1 & 0 & 1 \\ r & 0 & r & 0 \\ s & c & S & C \\ rc & -rs & rC & rS \end{vmatrix} = 0$$

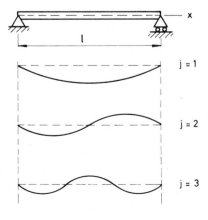

Fig. 10.1. Mode shapes of beam simply supported at its ends.

which is equivalent to

$$1 - cC = 0 \tag{10.38}$$

The solutions of this equation are

$$r_1l = 4.7300 \qquad r_2l = 7.8532 \qquad r_3l = 10.9956 \qquad r_4l = 14.1371$$
$$r_5l = 17.2787 \qquad r_6l = 20.4203 \qquad r_jl \simeq \left(j + \tfrac{1}{2} \right)\pi$$

with the natural frequencies

$$\omega = r_j^2 \left(\frac{EI}{\rho A} \right)^{1/2} \equiv \omega_j \tag{10.39}$$

The appropriate mode shapes are

$$W(x, \omega_j) = \sin r_j x - \sinh r_j x + A_j \left(\cos r_j x - \cosh r_j x \right)$$

$$A_j = \frac{\cos r_j l - \cosh r_j l}{\sin r_j l + \sinh r_j l} \tag{10.40}$$

The first three mode shapes for this case are shown in Fig. 10.2.

Hereinafter we will denote the mode shapes by $\psi_j(x)$. Note that in accordance with Eq. (10.32) we have

$$EI \frac{d^4 \psi_j(x)}{dx^4} - \rho A \omega_j^2 \psi_j(x) = 0 \tag{10.41}$$

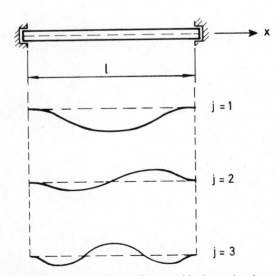

Fig. 10.2. Mode shapes of beam with clamped ends.

We now seek the orthogonality condition for the different mode shapes

$$\int_0^l \rho A \psi_j(x) \psi_k(x) \, dx = 0, \quad j \neq k \tag{10.42}$$

To do this, we multiply Eq. (10.41) by $\psi_k(x)$ and integrate with respect to x over the span:

$$\omega_j^2 \int_0^l \rho A \psi_j(x) \psi_k(x) \, dx = \int_0^l \psi_k(x) EI \frac{d^4 \psi_j(x)}{dx^4} \, dx \tag{10.43}$$

Double integration of the right-hand side of this equation by parts yields

$$\omega_j^2 \int_0^l \rho A \psi_j(x) \psi_k(x) \, dx = \psi_k EI \frac{d^3 \psi_j(x)}{dx^3} \bigg|_0^l - \int_0^l \frac{d\psi_k(x)}{dx} EI \frac{d^3 \psi_j(x)}{dx^3} \, dx$$

$$= \left\{ \psi_k(x) EI \frac{d^3 \psi_j(x)}{dx^3} - \frac{d\psi_k(x)}{dx} EI \frac{d^2 \psi_j(x)}{dx^2} \right\} \bigg|_0^l$$

$$+ \int_0^l EI \frac{d^2 \psi_k(x)}{dx^2} \frac{d^2 \psi_j(x)}{dx^2} \, dx$$

$$= \left\{ \psi_k(x) V_y(x) - \frac{d\psi_k(x)}{dx} M_z(x) \right\} \bigg|_0^l$$

$$+ \int_0^l EI \frac{d^2 \psi_k(x)}{dx^2} \frac{d^2 \psi_j(x)}{dx^2} \, dx \tag{10.44}$$

Since the nonintegral terms for all combinations of free, simply supported and clamped ends at $x = 0$ and $x = l$ vanish, the latter equation reduces to

$$\omega_j^2 \int_0^l \rho A \psi_j(x) \psi_k(x) \, dx = \int_0^l EI \frac{d^2 \psi_k(x)}{dx^2} \frac{d^2 \psi_j(x)}{dx^2} \, dx \tag{10.45}$$

Analogously, we have the kth natural frequency

$$\omega_k^2 \int_0^l \rho A \psi_j(x) \psi_k(x) \, dx = \int_0^l EI \frac{d^2 \psi_j(x)}{dx^2} \frac{d^2 \psi_k(x)}{dx^2} \, dx \tag{10.46}$$

Comparison of the last two equations indicates that, because of the nonequality of ω_j and ω_k Eq. (10.42), the orthogonality condition, is valid.

10.2.2 Deterministic Response via the Normal Mode Method. Let us now consider the deterministic version of Eq. (10.25). To do this we expand the given distributed loading $q(x, t)$ in series in terms of the mode shapes of undamped free vibration:

$$q(x, t) = \sum_{j=1}^{\infty} q_j(t)\psi_j(x) \tag{10.47}$$

Multiplying the above by $\psi_k(x)$ and integrating over the span of the beam, we arrive at an expression which, with k replaced by j, becomes

$$q_j(t) = v_j^{-2} \int_0^l q(x, t)\psi_j(x)\, dx, \quad v_j^2 = \int_0^l \psi_j^2(x)\, dx \tag{10.48}$$

We also expand the displacement $w(x, t)$ in a familiar series:

$$w(x, t) = \sum_{j=1}^{\infty} w_j(t)\psi_j(x) \tag{10.49}$$

Substitution of Eqs. (10.47) and (10.49) in (10.25) yields

$$\sum_{j=1}^{\infty} \left[EIw_j(t)\frac{d^4\psi_j(x)}{dx^4} + c\frac{dw_j(t)}{dt}\psi_j(x) \right.$$

$$\left. + \rho A \frac{d^2 w_j(t)}{dt^2}\psi_j(x) - q_j(t)\psi_j(x) \right] = 0$$

With Eq. (10.41) in mind, we obtain

$$\sum_{j=1}^{\infty} \left[\rho A \omega_j^2 w_j(t) + c\frac{dw_j(t)}{dt} + \rho A \frac{d^2 w_j(t)}{dt^2} - q_j(t) \right] \psi_j(x) = 0$$

and since this equation is valid for any x, the expression in square brackets vanishes for every j and

$$\frac{d^2 w_j(t)}{dt^2} + 2\zeta_j \omega_j \frac{dw_j(t)}{dt} + \omega_j^2 w_j(t) = \frac{1}{\rho A} q_j(t), \quad j = 1, 2, \ldots$$

$$\tag{10.50}$$

where

$$\zeta_j = \frac{c}{2\rho A \omega_j} \tag{10.51}$$

$q_j(t)$ being generalized deterministic forces. Equation (10.50) for a beam has precisely the same structure as Eqs. (9.125) for a discrete system. Accordingly, the solution to Eqs. (10.50) may be put in the form

$$w_j(t) = \frac{1}{\rho A} \int_{-\infty}^t q_j(\tau) h_j(t - \tau) \, d\tau \qquad (10.52)$$

where $h_j(t)$ is the impulse response function associated with the jth mode as per Eq. (9.127), and Eq. (10.49) becomes

$$w(x, t) = \frac{1}{\rho A} \sum_{j=1}^{\infty} \psi_j(x) \int_{-\infty}^t q_j(\tau) h_j(t - \tau) \, d\tau$$

or, with Eq. (10.48) in mind,

$$w(x, t) = \frac{1}{\rho A} \sum_{j=1}^{\infty} v_j^{-2} \psi_j(x) \int_{-\infty}^t h_j(t - \tau) \, d\tau \int_0^l q(\xi, \tau) \psi_j(\xi) \, d\xi$$

or, denoting $t - \tau = \lambda$,

$$w(x, t) = \frac{1}{\rho A} \sum_{j=1}^{\infty} v_j^{-2} \psi_j(x) \int_0^\infty h_j(\lambda) \, d\lambda \int_0^l q(\xi, t - \lambda) \psi_j(\xi) \, d\xi$$

Since shifting of the lower limit for 0 to $-\infty$ does not affect the integral over λ, we have

$$w(x, t) = \frac{1}{\rho A} \sum_{j=1}^{\infty} v_j^{-2} \psi_j(x) \int_{-\infty}^\infty h_j(\tau) \, d\tau \int_0^l q(\xi, t - \tau) \psi_j(\xi) \, d\xi \qquad (10.53)$$

For the generalized forces, representable in the form

$$q_j(t) = \int_{-\infty}^\infty Q_j(\omega) e^{i\omega t} \, d\omega$$

$$Q_j(\omega) = \frac{1}{2\pi} \int_{-\infty}^\infty q_j(t) e^{-i\omega t} \, dt \qquad (10.54)$$

we seek the displacement in an analogous manner

$$w_j(t) = \int_{-\infty}^\infty W_j(\omega) e^{i\omega t} \, d\omega$$

$$W_j(\omega) = \frac{1}{2\pi} \int_{-\infty}^\infty w_j(t) e^{-i\omega t} \, dt \qquad (10.55)$$

Substituting Eqs. (10.54) and (10.55) in Eq. (10.50), we find

$$W_j(\omega) = H_j(\omega)Q_j(\omega) \tag{10.56}$$

$$H_j(\omega) = \frac{1}{L_j(\omega)}, \quad L_j(\omega) = \rho A \big(\omega_j^2 - \omega^2 + 2\zeta_j \omega_j i\omega\big) \tag{10.57}$$

where $H_j(\omega)$ is the complex frequency response associated with the jth mode, and $L_j(\omega)$ the mechanical impedance of this mode.

10.2.3 Random Response via the Normal Mode Method. The formal solutions derived in the preceding section represent the response space-time function $w(x, t)$ due to a particular excitation function $q(x, t)$. In the case of random vibration, we regard the excitation as an ensemble of space-time functions, while the ensemble of response functions constitutes a random field depending on the excitation and the system. The mathematical expectation

$$E[q(x, t)] = m_q(x, t)$$

and the autocorrelation function

$$E[q(x_1, t_1)q(x_2, t_2)] = R_q(x_1, x_2; t_1, t_2)$$

of the force are supposed to be given. The problem consists in determining the mathematical expection and the autocorrelation function of the displacements.
  The mathematical expectation of the displacements is

$$E[w(x, t)] = \frac{1}{\rho A} \sum_{j=1}^{\infty} v_j^{-2} \psi_j(x) \int_{-\infty}^{\infty} h_j(\tau)\,d\tau \int_0^l m_q(\xi, t - \tau)\psi_j(\xi)\,d\xi$$

If the excitation is stationary, then

$$m_q(\xi, t - \tau) = m_q(\xi)$$

that is, m_q is independent of the time coordinate, whence

$$E[w(x, t)] = \frac{1}{\rho A} \sum_{j=1}^{\infty} v_j^{-2} \psi_j(x) \int_{-\infty}^{\infty} h_j(\tau)\,d\tau \int_0^l m_q(\xi)\psi_j(\xi)\,d\xi$$

For random excitation, $q_j(t), j = 1, 2, \ldots$, is replaced by an infinite-dimensional vector, so that

$$E[q_j(t)] = \int_0^l m_q(x, t)\psi_j(x)\,dx \tag{10.58}$$

so that if $m_q(x, t)$ is zero, so is also $E[q_j(t)]$. For simplicity, we assume the latter to be the case. The cross correlation between $q_j(t_1)$ and $q_k(t_2)$ is, for $j, k = 1, 2, \ldots$,

$$R_{q_j q_k}(t_1, t_2) = E[q_j(t_1) q_k(t_2)]$$

$$= v_j^{-2} v_k^{-2} \int_0^l \int_0^l E[q(x_1, t_1) q(x_2, t_2)] \psi_j(x_1) \psi_k(x_2) \, dx_1 \, dx_2$$

$$= v_j^{-2} v_k^{-2} \int_0^l \int_0^l R_q(x_1, x_2; t_1, t_2) \psi_j(x_1) \psi_k(x_2) \, dx_1 \, dx_2$$

where x_1 and x_2 are dummy variables representing two distinct points of the beam $0 \leqslant x_1, x_2 \leqslant l$. If the force is stationary in the wide sense in time, $R_q(x_1, x_2; t_1, t_2)$ depends on the time lag $\tau = t_2 - t_1$ rather than on t_1 and t_2 themselves, and so do the quantities $R_{q_j q_k}(t_1, t_2)$. Thus

$$R_{q_j q_k}(\tau) = v_j^{-2} v_k^{-2} \int_0^l \int_0^l R_q(x_1, x_2, \tau) \psi_j(x_1) \psi_k(x_2) \, dx_1 \, dx_2$$

$$(10.59)$$

The autocorrelation function of the displacement is

$$R_w(x_1, x_2; t_1, t_2) = E[w(x_1, t_1) w(x_2, t_2)]$$

$$= E\left[\sum_{j=1}^{\infty} w_j(t_1) \psi_j(x_1) \sum_{k=1}^{\infty} w_k(t_2) \psi_k(x_2) \right]$$

$$= \sum_{j=1}^{\infty} \sum_{k=1}^{\infty} E[w_j(t_1) w_k(t_2)] \psi_j(x_1) \psi_k(x_2)$$

$$= \sum_{j=1}^{\infty} \sum_{k=1}^{\infty} R_{w_j w_k}(t_1, t_2) \psi_j(x_1) \psi_k(x_2) \qquad (10.60)$$

where

$$R_{w_j w_k}(t_1, t_2) = E\left[\int_{-\infty}^{\infty} h_j(\tau) q_j(t_1 - \tau) \, d\tau \int_{-\infty}^{\infty} h_k(\kappa) q_k(t_2 - \kappa) \, d\kappa \right]$$

$$= \int_{-\infty}^{\infty} \int_{-\infty}^{\infty} h_j(\tau) h_k(\kappa) E[q_j(t_1 - \tau) q_k(t_2 - \kappa)] \, d\tau \, d\kappa$$

$$= \int_{-\infty}^{\infty} \int_{-\infty}^{\infty} h_j(\tau) h_k(\kappa) R_{q_j q_k}(t_1 - \tau, t_2 - \kappa) \, d\tau \, d\kappa$$

Now, if $q(x, t)$ is stationary in the wide sense in time, $R_{q_j q_k}(t_1 - \tau, t_2 - \kappa)$

depends only on the time lag $t_2 - \kappa - t_1 + \tau$, so that

$$R_{w_j w_k}(t_1, t_2) = \int_{-\infty}^{\infty} \int_{-\infty}^{\infty} h_j(\tau) h_k(\kappa) R_{q_j q_k}(t_2 - t_1 + \tau - \kappa) \, d\tau \, d\kappa$$

Denoting $t_2 - t_1 = \theta$, we have

$$R_{w_j w_k}(\theta) = \int_{-\infty}^{\infty} \int_{-\infty}^{\infty} h_j(\tau) h_k(\kappa) R_{q_j q_k}(\theta + \tau - \kappa) \, d\tau \, d\kappa \qquad (10.61)$$

Consequently, $R_{w_j w_k}(\theta)$ is also stationary in the wide sense in time, and the autocorrelation function of the displacements becomes

$$R_w(x_1, x_2; t_1, t_2) = \sum_{j=1}^{\infty} \sum_{k=1}^{\infty} R_{w_j w_k}(t_1, t_2) \psi_j(x_1) \psi_k(x_2)$$

$$= \sum_{j=1}^{\infty} \sum_{k=1}^{\infty} R_{w_j w_k}(t_2 - t_1) \psi_j(x_1) \psi_k(x_2)$$

$$= \sum_{j=1}^{\infty} \sum_{k=1}^{\infty} R_{w_j w_k}(\theta) \psi_j(x_1) \psi_k(x_2) \qquad (10.62)$$

where $R_{w_j w_k}(\theta)$ is as per Eq. (10.61).

An alternative treatment of the problem is possible through Eqs. (10.54)–(10.57). Indeed, combining Eqs. (10.47) and (10.54) we have

$$q(x, t) = \sum_{j=1}^{\infty} \psi_j(x) \int_{-\infty}^{\infty} Q_j(\omega) e^{i\omega t} \, d\omega \qquad (10.63)$$

The autocorrelation function of $q(x, t)$ is

$$R_q(x_1, x_2; t_1, t_2) = E\left[\sum_{j=1}^{\infty} \psi_j(x_1) \int_{-\infty}^{\infty} Q_j^*(\omega_1) e^{-\omega_1 t_1} \, d\omega_1 \right.$$

$$\left. \times \sum_{k=1}^{\infty} \psi_k(x_2) \int_{-\infty}^{\infty} Q_k(\omega_2) e^{i\omega_2 t_2} \, d\omega_2 \right]$$

$$= \sum_{j=1}^{\infty} \sum_{k=1}^{\infty} \psi_j(x_1) \psi_k(x_2) \int_{-\infty}^{\infty} \int_{-\infty}^{\infty} E\left[Q_j^*(\omega_1) Q_k(\omega_2) \right]$$

$$\times \exp\{ i(-\omega_1 t_1 + \omega_2 t_2) \} \, d\omega_1 \, d\omega_2$$

For $q(x, t)$ stationary in the wide sense, $R_q(x_1, x_2; t_1, t_2)$ depends only on

$\tau = t_2 - t_1$. This is the case if

$$E\left[Q_j^*(\omega_1)Q_k(\omega_2)\right] = S_{Q_jQ_k}(\omega_1)\delta(\omega_2 - \omega_1) \tag{10.64}$$

which then yields

$$R_q(x_1, x_2, \tau) = \sum_{j=1}^{\infty}\sum_{k=1}^{\infty} \psi_j(x_1)\psi_k(x_2)\int_{-\infty}^{\infty} S_{Q_jQ_k}(\omega)e^{i\omega\tau}\,d\omega$$

$$\tag{10.65}$$

For the cross-spectral density $S_q(x_1, x_2, \omega)$ of the loading, we have (see Eq. 10.5)

$$R_q(x_1, x_2, \tau) = \int_{-\infty}^{\infty} S_q(x_1, x_2, \omega)e^{i\omega\tau}\,d\omega$$

$$S_q(x_1, x_2, \omega) = \frac{1}{2\pi}\int_{-\infty}^{\infty} R_q(x_1, x_2, \tau)e^{-i\omega\tau}\,d\tau \tag{10.66}$$

so that

$$S_q(x_1, x_2, \omega) = \sum_{j=1}^{\infty}\sum_{k=1}^{\infty} S_{Q_jQ_k}(\omega)\psi_j(x_1)\psi_k(x_2) \tag{10.67}$$

which is a double Fourier series in terms of the mode shapes. To obtain $S_{Q_jQ_k}(\omega)$, we multiply Eq. (10.67) by $\psi_\alpha(x_1)\psi_\beta(x_2)$ and integrate twice over the span of the beam. With Eq. (10.42) the result is (with the formal replacement $\alpha \to j$, $\beta \to k$)

$$S_{Q_jQ_k}(\omega) = \nu_j^{-2}\nu_k^{-2}\int_0^l\int_0^l S_q(x_1, x_2, \omega)\psi_j(x_1)\psi_k(x_2)\,dx_1\,dx_2$$

$$\tag{10.68}$$

An equation for the displacement analogous to (10.63) follows from Eqs. (10.49) and (10.55):

$$w(x, t) = \sum_{j=1}^{\infty} \psi_j(x)\int_{-\infty}^{\infty} W_j(\omega)e^{i\omega t}\,d\omega$$

The corresponding autocorrelation function reads

$$R_w(x_1, x_2; t_1, t_2) = E\left[\sum_{j=1}^{\infty} \psi_j(x_1)\int_{-\infty}^{\infty} W_j^*(\omega_1)e^{-i\omega_1 t_1}\,d\omega_1\right.$$

$$\left.\times \sum_{k=1}^{\infty} \psi_k(x_2)\int_{-\infty}^{\infty} W_k(\omega_2)e^{i\omega_2 t_2}\,d\omega_2\right]$$

With Eq. (10.56), we have

$$R_w(x_1, x_2; t_1, t_2)$$

$$= E\left[\sum_{j=1}^{\infty} \psi_j(x_1) \int_{-\infty}^{\infty} H_j^*(\omega_1) Q_j^*(\omega_1) e^{-i\omega_1 t_1} d\omega_1 \right.$$

$$\left. \times \sum_{k=1}^{\infty} \psi_k(x_2) \int_{-\infty}^{\infty} H_k(\omega_2) Q_k(\omega_2) e^{i\omega_2 t_2} d\omega_2 \right]$$

$$= \sum_{j=1}^{\infty} \sum_{k=1}^{\infty} \psi_j(x_1) \psi_k(x_2) \int\int_{-\infty}^{\infty} H_j^*(\omega_1) H_k(\omega_2) E\left[Q_j^*(\omega_1) Q_k(\omega_2) \right]$$

$$\times \exp\{ -i\omega_1 t_1 + i\omega_2 t_2 \} \, d\omega_1 \, d\omega_2$$

Taking into account Eq. (10.64), we have

$$R_w(x_1, x_2; t_1, t_2)$$

$$= \sum_{j=1}^{\infty} \sum_{k=1}^{\infty} \psi_j(x_1) \psi_k(x_2) \int_{-\infty}^{\infty} S_{Q_j Q_k}(\omega) H_j^*(\omega) H_k(\omega) e^{i\omega\tau} d\omega$$

$$\equiv R_w(x_1, x_2, \tau) \tag{10.69}$$

so that the response also turns out to be stationary in the wide sense in time. For the cross-spectral density of the displacements, we have, in perfect analogy with Eq. (10.66),

$$R_w(x_1, x_2, \tau) = \int_{-\infty}^{\infty} S_w(x_1, x_2, \omega) e^{i\omega\tau} d\omega \tag{10.70a}$$

$$S_w(x_1, x_2, \omega) = \frac{1}{2\pi} \int_{-\infty}^{\infty} R_w(x_1, x_2, \tau) e^{-i\omega\tau} d\tau \tag{10.70b}$$

Therefore,

$$S_w(x_1, x_2, \omega) = \sum_{j=1}^{\infty} \sum_{k=1}^{\infty} S_{Q_j Q_k}(\omega) H_j^*(\omega) H_k(\omega) \psi_j(x_1) \psi_k(x_2)$$

$$\tag{10.71}$$

The cross-spectral density can be put in the form

$$S_w(x_1, x_2, \omega) = \sum_{j=1}^{\infty} \sum_{k=1}^{\infty} \left(\frac{\psi_j(x_1) \psi_k(x_2)}{L_j^*(\omega) L_k(\omega)} \right) \left(\frac{S_q(\omega) l^2}{v_j^2 v_k^2} \right) A_{Q_j Q_k}(\omega)$$

$$S_q(\omega) = S_q(0, 0, \omega) \tag{10.72}$$

where $A_{Q_jQ_j}(\omega)$ is called the *joint acceptance* and is defined as

$$A_{Q_jQ_j}(\omega) = \frac{S_{Q_jQ_j}(\omega)v_j^2 v_j^2}{S_q(\omega)l^2} = \int_0^l \int_0^l \overline{S}_q(x_1, x_2, \omega)\psi_j(x_1)\psi_j(x_2)\, dx_1\, dx_2$$

$$\overline{S}_q(x_1, x_2, \omega) = \frac{S_q(x_1, x_2, \omega)}{S_q(\omega)l^2} \tag{10.73}$$

while $A_{Q_jQ_k}(\omega), j \neq k$ is called the *cross acceptance* and is defined as

$$A_{Q_jQ_k}(\omega) = \frac{S_{Q_jQ_k}(\omega)v_j^2 v_k^2}{S_q(\omega)l^2} = \int_0^l \int_0^l \overline{S}_q(x_1, x_2, \omega)\psi_j(x_1)\psi_k(x_2)\, dx_1\, dx_2$$

$$\tag{10.74}$$

The cross-spectral density at point x is given by

$$S_w(x, x, \omega) = \sum_{j=1}^{\infty} \frac{\psi_j^2(x)}{|L_j(\omega)|^2} \frac{S_q(\omega)l^2}{v_j^4} A_{Q_jQ_j}(\omega)$$

$$+ \sum_{j=1}^{\infty} \sum_{\substack{k=1 \\ j \neq k}}^{\infty} \frac{\psi_j(x)\psi_k(x)}{L_j^*(\omega)L_k(\omega)} \frac{S_q(\omega)l^2}{v_j^2 v_k^2} A_{Q_jQ_k}(\omega) \tag{10.75}$$

The first sum is associated with the modal autocorrelations, identical modes being involved, and the second, with the modal cross correlations, since nonidentical modes are involved.

The mean-square value of the displacements is

$$d_w^2(x) = R_w(x, x, 0) = \int_{-\infty}^{\infty} S_w(x, x, \omega)\, d\omega$$

$$= \sum_{j=1}^{\infty} \psi_j^2(x)d_{jj} + \sum_{j=1}^{\infty} \sum_{\substack{k=1 \\ j \neq k}}^{\infty} \psi_j(x)\psi_k(x)d_{jk} \tag{10.76}$$

where

$$d_{jj} = \int_{-\infty}^{\infty} \frac{S_q(\omega)l^2 A_{Q_jQ_j}(\omega)}{|L_j(\omega)|^2 v_j^4}\, d\omega$$

$$d_{jk} = \int_{-\infty}^{\infty} \frac{S_q(\omega)l^2 A_{Q_jQ_k}(\omega)}{L_j^*(\omega)L_k(\omega)v_j^2 v_k^2}\, d\omega \tag{10.77}$$

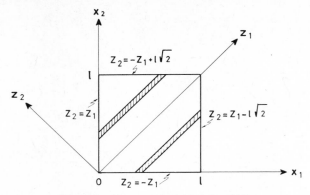

Fig. 10.3. Change of variables (Sec. 10.3).

Of interest is also the *span average* s^2 of the mean-square values $d_w^2(x)$:

$$s^2 = \frac{1}{l} \int_0^l d_w^2(x)\, dx$$

$$= \sum_{j=1}^{\infty} d_{jj} \frac{1}{l} \int_0^l \psi_j^2(x)\, dx + \sum_{j=1}^{\infty} \sum_{\substack{k=1 \\ j \neq k}}^{\infty} d_{jk} \frac{1}{l} \int_0^l \psi_j(x)\psi_k(x)\, dx$$

$$(10.78)$$

The second sum vanishes identically, by orthogonality of the mode shapes (10.42), and we have

$$s^2 = \frac{1}{l} \sum_{j=1}^{\infty} d_{jj} \nu_j^2 \qquad (10.79)$$

that is, s^2 depends on the joint acceptances only.

A case where the *exact* mean-square value of the displacement is also independent of the cross acceptances is described in Sec. 10.4.

10.3 DETERMINATION OF JOINT AND CROSS ACCEPTANCES

While the double integrations in Eqs. (10.73) and (10.74) for the joint and cross acceptances mostly call for numerical procedures, in certain specific cases they can be realized in a closed form, depending on the cross-correlation function $S_q(x_1, x_2, \omega)$ and the mode shapes $\psi_j(x)$. We consider here one such common case, namely, where the external loading $q(x, t)$ is a homogeneous random field, that is, the cross-correlation function is a function of the separation $x_2 - x_1$. Here the double integration can be reduced to a single one, irrespective of the mode shapes $\psi_j(x)$. First of all we rewrite Eq. (10.74) as follows:

$$A_{Q_j Q_k}(\omega) = \int_0^l \int_0^l \bar{S}_q(x_2 - x_1, \omega) \psi_j(x_1) \psi_k(x_2) \, dx_1 \, dx_2$$

and introduce new variables

$$z_1 = \frac{x_1 + x_2}{\sqrt{2}} \qquad z_2 = \frac{x_2 - x_1}{\sqrt{2}}$$

(see Fig. 10.3). Then

$$dx_1 \, dx_2 = \mathrm{abs} \left| \frac{\partial(x_1, x_2)}{\partial(z_1, z_2)} \right| dz_1 \, dz_2 = \mathrm{abs} \begin{vmatrix} \dfrac{\partial x_1}{\partial z_1} & \dfrac{\partial x_1}{\partial z_2} \\ \dfrac{\partial x_2}{\partial z_1} & \dfrac{\partial x_2}{\partial z_2} \end{vmatrix} dz_1 \, dz_2$$

$$= \mathrm{abs} \begin{vmatrix} \dfrac{1}{\sqrt{2}} & -\dfrac{1}{\sqrt{2}} \\ \dfrac{1}{\sqrt{2}} & \dfrac{1}{\sqrt{2}} \end{vmatrix} dz_1 \, dz_2 = dz_1 \, dz_2$$

and

$$A_{Q_j Q_k}(\omega) = K_1(\omega) + K_2(\omega)$$

where

$$K_1(\omega) = \int_0^{l/\sqrt{2}} \bar{S}_q(z_2\sqrt{2}, \omega) \, dz_2 \int_{z_2}^{l/\sqrt{2} - z_2} \psi_j \left(\frac{z_1 - z_2}{\sqrt{2}} \right) \psi_k \left(\frac{z_1 + z_2}{\sqrt{2}} \right) dz_1$$

$$K_2(\omega) = \int_{-l/\sqrt{2}}^0 \bar{S}_q(z_2\sqrt{2}, \omega) \, dz_2 \int_{-z_2}^{l/\sqrt{2} + z_2} \psi_j \left(\frac{z_1 - z_2}{\sqrt{2}} \right) \psi_k \left(\frac{z_1 + z_2}{\sqrt{2}} \right) dz_1$$

Introducing further new variables,

$$z_1 = \eta\sqrt{2} \qquad z_2 = \frac{\xi}{\sqrt{2}}$$

we obtain

$$K_1(\omega) = \int_0^l \overline{S}_q(\xi, \omega) \, d\xi \int_{\xi/2}^{l-\xi/2} \psi_j\left(\eta - \frac{\xi}{2}\right) \psi_k\left(\eta + \frac{\xi}{2}\right) d\eta$$

$$K_2(\omega) = \int_0^l \overline{S}_q(-\xi, \omega) \, d\xi \int_{\xi/2}^{l-\xi/2} \psi_j\left(\eta + \frac{\xi}{2}\right) \psi_k\left(\eta - \frac{\xi}{2}\right) d\eta$$

and finally

$$A_{Q_jQ_k}(\omega) = \int_0^l \overline{S}_q(\xi, \omega) M_{jk}(\xi) \, d\xi$$

$$M_{jk}(\xi) = \int_{\xi/2}^{l-\xi/2}\left[\psi_j\left(\eta - \frac{\xi}{2}\right)\psi_k\left(\eta + \frac{\xi}{2}\right) + \psi_j\left(\eta + \frac{\xi}{2}\right)\psi_k\left(\eta - \frac{\xi}{2}\right)\right] d\eta$$

$$(10.80)$$

Example 10.2

Consider a beam simply supported at both its ends, so that

$$\psi_j(x) = \sin\frac{j\pi x}{l}$$

Equation (10.80) yields ($\zeta = x/l$)

$$A_{Q_jQ_j}(\omega) = 4\int_0^1 \frac{S_q(\zeta l, \omega)}{S_q(\omega)}\left[(1 - \zeta)\cos j\pi\zeta + \frac{1}{j\pi}\sin j\pi\zeta\right] d\zeta \qquad (10.81)$$

$$A_{Q_jQ_k}(\omega) = \frac{4\left[1 - (-1)^{j+k+1}\right]}{\pi(k^2 - j^2)}\int_0^1 \frac{S_q(\zeta l, \omega)}{S_q(\omega)}(k\sin j\pi\zeta - j\sin k\pi\zeta) \, d\zeta,$$

$$j \neq k \quad (10.82)$$

Equation (10.81) is analogous to Powell's formula, and Eq. (10.82) to Lin's. From Equation (10.82), it follows that the modes symmetric about the half-span cross section $x = l/2$ of the beam (those with an even number of half-waves) do not correlate with the antisymmetric ones (those with an odd number of half-waves):

$$A_{Q_jQ_k}(\omega) = 0 \quad j + k \text{ odd}$$

Very often, in applications, the cross-spectral density has the form

$$S_q(\zeta, \omega) = S_q(\omega)e^{-A|\zeta|}\cos B\zeta \qquad (10.83)$$

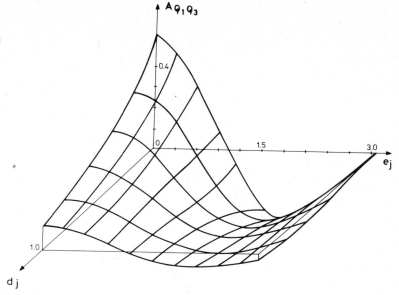

Fig. 10.4. Cross acceptance $A_{Q_1Q_3}$ as a function of nondimensional parameters e_j and d_j.

where A and B are positive nondimensional quantities. Then

$$A_{Q_jQ_j} = \frac{4d_j}{\pi j}\frac{E_j}{R_j} + \frac{32}{\pi^2}\frac{d_je_j}{j^2R_j^2}\left(E_j^2 - 2e_j^2\right)(-1)^j e^{-A}\sin B$$

$$+ \frac{8}{\pi^2 j^2 R_j^2}\left[E_j^2 - 4e_j^2\left(1 + 2d_j^2\right)\right]\left[1 - (-1)^j e^{-A}\cos B\right] \qquad (10.84)$$

$$A_{Q_jQ_k} = \frac{4\left[1 - (-1)^{j+k+1}\right]}{\pi\left(k^2 - j^2\right)}\left\{\frac{k}{j\pi}\frac{E_j - 2e_j^2}{R_j} - (-1)^j e^{-A}\frac{k}{j\pi R_j}\right.$$

$$\times\left[-2d_je_j\sin B + \left(E_j - 2e_j^2\right)\cos B\right]$$

$$- \frac{j}{k\pi}\frac{E_k - 2e_k^2}{R_k} + (-1)^k e^{-A}\frac{j}{k\pi R_k}$$

$$\left.\times\left[-2d_ke_k\sin B + \left(E_k - 2e_k^2\right)\cos B\right]\right\}, \quad j \neq k,$$

$$(10.85)$$

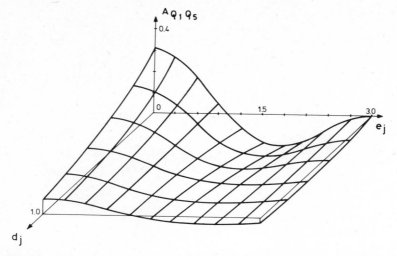

Fig. 10.5.   Cross acceptance $A_{Q_1Q_5}$ as a function of nondimensional parameters e_j and d_j.

where

$$d_j = \frac{A}{j\pi} \qquad e_j = \frac{B}{j\pi}$$

$$E_j = 1 + d_j^2 + e_j^2 \qquad R_j = E_j^2 - 4e_j^2$$

As an example, Figs. 10.4 and 10.5 show $A_{Q_1Q_3}$ and $A_{Q_1Q_5}$ as functions of the nondimensional parameters e_j and d_j.

10.4   CASE CAPABLE OF CLOSED-FORM SOLUTION

Let us consider the case where the autocorrelation function of the loading is given as space-time white noise:

$$R_q(x_1, x_2; t_1, t_2) = \frac{R}{l}\delta(x_2 - x_1)\delta(t_2 - t_1) = \frac{R}{l}\delta(\zeta)\delta(\tau)$$

$$(10.86)$$

where R is some positive constant, ζ the separation of the observation cross sections of $q(x, t)$, and τ the time lag. The beam is simply supported at its ends, so that the mode shapes are as per Eq. (10.37).

The cross-spectral density $S_q(x_1, x_2, \omega)$ is, in accordance with Eq. (10.66.2),

$$S_q(x_1, x_2, \omega) = \frac{1}{2\pi} \int_{-\infty}^{\infty} R_q(x_1, x_2, \tau) e^{-i\omega\tau} \, d\tau$$

$$= \frac{R}{2\pi l} \int_{-\infty}^{\infty} \delta(x_2 - x_1)\delta(\tau) e^{-i\omega\tau} \, d\tau = \frac{R}{2\pi l}\delta(x_2 - x_1)$$

and, by Eq. (10.68),

$$S_{Q_j Q_k} = \frac{4}{l^2} \int_0^l \int_0^l \frac{R}{2\pi l}\delta(x_2 - x_1)\sin\frac{j\pi x_1}{l} \sin\frac{k\pi x_2}{l} dx_1 \, dx_2$$

$$= \frac{R}{\pi l^2}\delta_{jk} \qquad (10.87)$$

where δ_{jk} is Kronecker's delta, so that the cross acceptances vanish identically and all joint acceptances equal $1/2l$. Consequently the second sum in Eq. (10.75) vanishes, and we obtain for zero time lag by Eqs. (10.75) and (10.70a),

$$R_w(x_1, x_2, 0) = \frac{R}{\pi(\rho A l)^2} \sum_{j=1}^{\infty} \sin\frac{j\pi x_1}{l} \sin\frac{j\pi x_2}{l} \int_{-\infty}^{\infty} \frac{d\omega}{\left(\omega_j^2 - \omega^2\right)^2 + 4\zeta_j^2\omega_j^2\omega^2}$$

$$= \frac{R}{\rho A l^2 c} \sum_{j=1}^{\infty} \frac{1}{\omega_j^2} \sin\frac{j\pi x_1}{l} \sin\frac{j\pi x_2}{l}$$

$$= \frac{Rl^2}{\pi^4 EIc} \sum_{j=1}^{\infty} \frac{1}{j^4} \sin\frac{j\pi x_1}{l} \sin\frac{j\pi x_2}{l}$$

$$= \frac{Rl^2}{2\pi^4 EIc} \sum_{j=1}^{\infty} \frac{1}{j^4}\left[\cos\frac{j\pi(x_1 - x_2)}{l} - \cos\frac{j\pi(x_1 + x_2)}{l}\right]$$

$$(10.88)$$

bearing in mind Eq. (9.65.3) for the integral.
Using the following summation formula (Gradshteyn and Ryzhik, p. 39)

$$\sum_{j=1}^{\infty} \frac{\cos jx}{j^4} = \frac{\pi^4}{90} - \frac{\pi^2 x^2}{12} + \frac{\pi x^3}{12} - \frac{x^4}{48}, \quad 0 \leqslant x \leqslant 2\pi$$

we obtain after some algebra

$$R_w(x_1, x_2, 0) = \frac{Rl^2}{24EIc}\left[-(\xi_1 - \xi_2)^2 + (\xi_1 - \xi_2)^3 - \tfrac{1}{4}(\xi_1 - \xi_2)^4\right.$$

$$\left. + (\xi_1 + \xi_2)^2 - (\xi_1 + \xi_2)^3 + \tfrac{1}{4}(\xi_1 + \xi_2)^4\right] \qquad (10.89)$$

where $\xi_1 = x_1/l$, and $\xi_2 = x_2/l$. This formula is due to Eringen. Note that analogous results have been obtained by van Lear and Uhlenbeck for the string without a foundation, whereas the string on foundation was treated by Wedig via his covariance analysis method.

The mean-square value of the displacement is ($\xi_1 = \xi_2 = \xi$)

$$d^2(\xi) = \frac{Rl^2}{6EIc}\xi^2(1 - \xi)^2, \quad \xi = \frac{x}{l} \tag{10.90}$$

with its maximum at the midspan section of the beam,

$$d^2(\tfrac{1}{2}) = \frac{Rl^2}{96EIc} \tag{10.91}$$

whereas the span average is

$$s^2 = \int_0^1 d^2(\xi)\, d\xi = \frac{Rl^2}{180EIc} \tag{10.92}$$

so that the maximum of $d^2(\xi)$ exceeds the span average by factor of $\frac{15}{8}$. The conclusion is that average quantities may not be used as design data for structures subjected to random loading.

10.5 CRANDALL'S PROBLEM*

We now consider a problem which generalizes one of those included in Crandall's recent study of systems driven by a point force, random in time: A beam on an elastic foundation with the force in section $x = a$ (Fig. 10.6). The force can be represented as distributed loading as follows:

$$q(x, t) = F(t)\delta(x - a) \tag{10.93}$$

where $F(t)$ is a random function of time, representing band-limited white noise with cutoff frequency ω_c:

$$E[F(t)] = 0$$

$$S_F(\omega) = \begin{cases} S_0, & |\omega| \leqslant \omega_c \\ 0, & \text{otherwise} \end{cases}$$

---

*This section follows closely the paper by Elishakoff, van Zanten and Crandall (1979), where axisymmetric random vibration of a cylindrical shell under ring loading is considered. Its results are obtainable from the latter by formal substitution $\nu = 0, (L/R)^2 = K$.

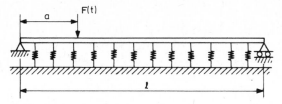

Fig. 10.6. Beam simply supported at both ends, on elastic foundation, under concentrated random force $F(t)$.

The differential equation (10.25) in this case reads

$$EI\frac{\partial^4 w}{\partial x^4} + kw + c\frac{\partial w}{\partial t} + \rho A\frac{\partial^2 w}{\partial t^2} = q(x, t)$$

where k is the stiffness of the foundation. The rest of the notation as in the original equation (10.25). The formalism of normal-mode method described in Sec. 10.2 is fully applicable here as well, with Eq. (10.69) as the final result. We have to find the expression for $S_{Q_j Q_k}(\omega)$ and for the natural frequencies. The mode shape is given as per Eq. (10.37), and

$$\omega_j^2 = \frac{EI}{\rho A}\frac{j^4 \pi^4}{l^4} + \frac{k}{\rho A} \tag{10.94}$$

The autocorrelation function of the loading is

$$R_q(x_1, x_2; t_1, t_2) = E\big[F(t_1)\delta(x_1 - a)F(t_2)\delta(x_2 - a)\big]$$

$$= \delta(x_1 - a)\delta(x_2 - a)E\big[F(t_1)F(t_2)\big]$$

$$= \delta(x_1 - a)\delta(x_2 - a)R_F(t_1, t_2)$$

$$= \delta(x_1 - a)\delta(x_2 - a)R_F(\tau)$$

and

$$S_q(x_1, x_2, \omega) = \delta(x_1 - a)\delta(x_2 - a)S_F(\omega) \tag{10.95}$$

Therefore, in view of Eq. (10.68),

$$S_{Q_j Q_k}(\omega) = \nu_j^{-2}\nu_k^{-2}\int_0^l\int_0^l \delta(x_1 - a)\delta(x_2 - a)S_F(\omega)\psi_j(x_1)\psi_k(x_2)\,dx_1\,dx_2$$

$$= \nu_j^{-2}\nu_k^{-2}\psi_j(a)\psi_k(a)S_F(\omega) \tag{10.96}$$

and Eq. (10.69) yields

$$R_w(x_1, x_2; t_1, t_2) = \sum_{j=1}^{\infty} \sum_{k=1}^{\infty} \psi_j(x_1)\psi_k(x_2)\psi_j(a)\psi_k(a)v_j^{-2}v_k^{-2}$$

$$\times \int_{-\infty}^{\infty} S_F(\omega)H_j^*(\omega)H_k(\omega)e^{i\omega\tau}\,d\omega$$

The mean-square value of the displacements is

$$E[w^2(x, t)] = \sum_{j=1}^{\infty} \sum_{k=1}^{\infty} \psi_j(x)\psi_k(x)\psi_j(a)\psi_k(a)v_j^{-2}v_k^{-2}I_{jk} \quad (10.97)$$

where

$$I_{jk} = \int_{-\infty}^{\infty} S_F(\omega)H_j^*(\omega)H_k(\omega)\,d\omega \quad (10.98)$$

The mean-square value of the velocity $v(x, t) = \partial w(x, t)/\partial t$ is (see Prob. 10.6)

$$E[v^2(x, t)] = \sum_{j=1}^{\infty} \sum_{k=1}^{\infty} \psi_j(x)\psi_k(x)\psi_j(a)\psi_k(a)v_j^{-2}v_k^{-2}I'_{jk} \quad (10.99)$$

where

$$I'_{jk} = \int_{-\infty}^{\infty} \omega^2 S_F(\omega)H_j^*(\omega)H_k(\omega)\,d\omega \quad (10.100)$$

For the displacement integral,

$$I_{jk} = \frac{S_0}{(\rho A)^2} \int_{-\omega_c}^{\omega_c} \frac{d\omega}{(\omega_j^2 - \omega^2 - 2i\beta\omega)(\omega_k^2 - \omega^2 + 2i\beta\omega)}$$

$(\beta = 2\zeta_j\omega_j = c/\rho A)$ we obtain

$$I_{jk} = \frac{S_0}{(\rho A)^2} \frac{\Phi(\omega_j, \omega_k; \omega_c) + \Phi(\omega_k, \omega_j; \omega_c)}{(\omega_j^2 - \omega_k^2)^2 + 2\beta^2(\omega_j^2 + \omega_k^2)} \quad (10.101)$$

where

$$\Phi(\omega_j, \omega_k; \omega_c) = \frac{\omega_j^2 - \omega_k^2 - \beta^2}{2(\omega_j^2 - \beta^2/4)^{1/2}} \ln \frac{\omega_c^2 + \omega_j^2 - 2\omega_c(\omega_j^2 - \beta^2/4)^{1/2}}{\omega_c^2 + \omega_j^2 + 2\omega_c(\omega_j^2 - \beta^2/4)^{1/2}}$$

$$+ 2\beta \left\{ \tan^{-1}\left[\frac{\omega_c - (\omega_j^2 - \beta^2/4)^{1/2}}{\beta/2} \right] \right.$$

$$+ \tan^{-1}\left[\frac{\omega_c + (\omega_j^2 - \beta^2/4)^{1/2}}{\beta/2} \right] \right\}$$

For the velocity integral we obtain

$$I'_{jk} = \frac{S_0}{(\rho A)^2} \frac{\Phi'(\omega_j, \omega_k; \omega_c) + \Phi'(\omega_k, \omega_j; \omega_c)}{\left(\omega_j^2 - \omega_k^2\right)^2 + 2\beta^2\left(\omega_j^2 + \omega_k^2\right)} \qquad (10.102)$$

where

$$\Phi'(\omega_j, \omega_k; \omega_c) = \frac{\omega_j^2\left(\omega_j^2 - \omega_k^2\right) + \beta^2\left(\omega_j^2 + \omega_k^2\right)/2}{2\left(\omega_j^2 - \beta^2/4\right)^{1/2}}$$

$$\times \ln \frac{\omega_c^2 + \omega_j^2 - 2\omega_c\left(\omega_j^2 - \beta^2/4\right)^{1/2}}{\omega_c^2 + \omega_j^2 + 2\omega_c\left(\omega_j^2 - \beta^2/4\right)^{1/2}}$$

$$+ \beta\left(\omega_j^2 + \omega_k^2\right)\left\{\tan^{-1}\left[\frac{\omega_c - \left(\omega_j^2 - \beta^2/4\right)^{1/2}}{\beta/2}\right]\right.$$

$$\left. + \tan^{-1}\left[\frac{\omega_c + \left(\omega_j^2 - \beta^2/4\right)^{1/2}}{\beta/2}\right]\right\}$$

For $j = k$, Eq. (10.101) is analogous to (9.56) for a single-degree-of-freedom system under band-limited white noise, whereas Eq. (10.102) is analogous to Eq. (9.62). As follows from Figs. 9.3 and 9.4, I_{jj} and I'_{jj} are very small at $\omega_j > \omega_c$. The conclusion is that only terms with $\omega_j \leqslant \omega_c$ and $\omega_k \leqslant \omega_c$ have to be taken into account in summations (10.97) and (10.99). The number N_c of modes in the excitation band is defined as the largest value of j such that $\omega_j \leqslant \omega_c$. As a result, expression (10.97) and (10.99) can be approximated as

$$E\left[w^2(x, t)\right] \simeq \sum_{j=1}^{N_c} \sum_{k=1}^{N_c} \psi_j(x)\psi_k(x)\psi_j(a)\psi_k(a)\nu_j^{-2}\nu_k^{-2}I_{jk}$$

$$E\left[v^2(x, t)\right] \simeq \sum_{j=1}^{N_c} \sum_{k=1}^{N_c} \psi_j(x)\psi_k(x)\psi_k(a)\psi_k(a)\nu_j^{-2}\nu_k^{-2}I'_{jk}$$

$$(10.103)$$

Let us consider the expression for $E[v^2(x, t)]$ in more detail. It resolves into two sums,

$$E\left[v^2(x, t)\right] \simeq s_1(x, t) + s_2(x, t)$$

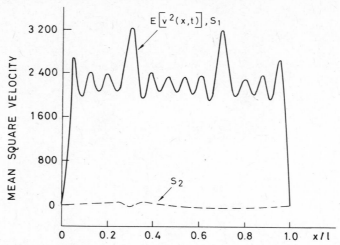

Fig. 10.7. Mean-square velocity $E[v^2(x, t)]$ and component sums s_1 and s_2 for beam without elastic foundation, $H = 0.01$, $K = 0$, $\Omega_c = 2\pi$, $B = 0.02$.

where

$$s_1(x, t) = 4 \sum_{j=1}^{N_c} \sin^2 \frac{j\pi x}{l} \sin^2 \frac{j\pi a}{l} \frac{I'_{jj}}{l^2}$$

$$s_2(x, t) = 4 \sum_{\substack{j=1 \\ j \neq k}}^{N_c} \sum_{k=1}^{N_c} \sin \frac{j\pi x}{l} \sin \frac{k\pi x}{l} \sin \frac{j\pi a}{l} \sin \frac{k\pi a}{l} \frac{I'_{jk}}{l^2}$$

$$= 8 \sum_{\substack{j=1 \\ j < k}}^{N_c} \sum_{k=1}^{N_c} \sin \frac{j\pi x}{l} \sin \frac{k\pi x}{l} \sin \frac{j\pi a}{l} \sin \frac{k\pi a}{l} \frac{I'_{jk}}{l^2}$$

$$(10.104)$$

$s_1(x, t)$ is associated with the modal autocorrelations, and $s_2(x, t)$ with the modal cross correlations. Direct calculation shows that

$$s_1(a, t) = s_1(l - a, t) \qquad (10.105)$$

meaning that the part of the response associated with the modal autocorrelations is symmetric about the half-span $x = l/2$ of the beam. This result is in perfect analogy with the two-degree-of-freedom system considered in Sec. 9.4.

In cases where the natural frequencies are far apart, the contribution of modal cross correlations is negligible and $E[v^2(x, t)] \simeq s_1(x, t)$. This implies that in these circumstances

$$E[v^2(a, t)] \simeq E[v^2(l - a, t)] \qquad (10.106)$$

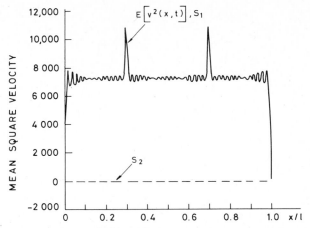

Fig. 10.8. Mean-square velocity $E[v^2(x, t)]$ and component sums s_1 and s_2, for beam without elastic foundation, $H = 0.01$, $K = 0$, $\Omega_c = 20\pi$, $B = 0.02$.

that is, that this symmetry is also a feature of the mean-square velocity. Such a case, involving a beam without an elastic foundation, is shown in Fig. 10.7. The beam cross section is chosen to be rectangular, with depth h and width b. The nondimensional parameters $H = h/l$, $K = kl^2/EA$, $\Omega_c = \omega_c l/(E/\rho)^{1/2}$, $B = \beta l/(E/\rho)^{1/2}$, $\eta = a/l$ are fixed at $H = 0.01$, $K = 0$, $\Omega_c = 2\pi$, $B = 0.02$, $\eta = 0.3$. The number of modes in the band with $\Omega_c = 2\pi$ is $N_c = 14$. Figure 10.8 shows $E[v^2(x, t)]$ for $\Omega_c = 20\pi$, $\xi = x/l$ ($N_c = 46$). As is seen, the response is largely uniform over the beam except for zones of enhanced response* in the vicinity of cross sections $x = a$ and $x = l - a$.

Note that the expression I'_{jj} in Eq. (10.104) varies slowly with $j \leqslant N_c$ and may very well be approximated as $I'_{jj} = \pi S_0/\rho Ac$; then

$$s_1(x, t) = V_1^2 G\left(\frac{x}{l}, \frac{a}{l}, N_c\right) \tag{10.107}$$

where

$$V_1^2 = \frac{\pi S_0}{\rho Acl^2}, \quad G\left(\frac{x}{l}, \frac{a}{l}, N_c\right) = \sum_{j=1}^{N_c} 4\sin^2\frac{j\pi x}{l} \sin^2\frac{j\pi a}{l} \tag{10.108}$$

This sum can be evaluated in closed form (Crandall and Wittig):

$$G\left(\frac{x}{l}, \frac{a}{l}, N_c\right) = g\left(\frac{x}{l}, N_c\right) + g\left(\frac{a}{l}, N_c\right) - \frac{1}{2}g\left(\frac{x - a}{l}, N_c\right)$$

$$- \frac{1}{2}g\left(\frac{x - l + a}{l}, N_c\right) \tag{10.109}$$

*This interesting phenomenon was first observed by Crandall and Wittig.

where

$$g(\xi, N_c) = N_c + \frac{1}{2} - \frac{\sin(2N_c + 1)\pi\xi}{2\sin \pi\xi} \qquad (10.110)$$

If N_c is large and a not too close to zero or to l, the sum oscillates in the neighborhood of N_c for most values of x in the interval $0 \leqslant x \leqslant l$, except for the pairs of points $x = 0$ and $x = l$, where it vanishes, and $x = a$ and $x = l - a$, where it approximates to $3N_c/2$, that is, the intensification factor is $3N_c/2 : N_c = \frac{3}{2}$, and there are zones of 50% enhancement of the response.

For a beam on an elastic foundation, the picture is rather different. Figure 10.9 shows the mean-square velocity response for $H = 0.01$, $K = 4$, $B = 0.02$, and $\Omega_c = 2\pi$, with $N_c = 14$. In this case the cross correlation sum s_2 introduces a substantial degree of asymmetry into the distribution of $E[v^2(x, t)]$, although the zones of enhanced response at $x = a$ and $x = l - a$ are still recognizable. Figure 10.10 shows the result for a beam with $H = 0.001$, $K = 4$, $\Omega_c = 2\pi$, $B = 0.02$; the excitation cutoff frequency is the same as in Fig. 10.9, but N_c is increased due to clustering of the frequencies at the low end of the spectrum. This is also seen from Table 10.1, where the natural frequencies are listed for different H values. Similar asymmetry due to clustering was demonstrated in Sec. 9.4 for the two-degree-of-freedom system; here $N_c = 45$ for $\Omega_c = 2\pi$. The cross correlation sum s_2 now dominates the response at the driving cross section $x = a$; a trace of enhanced response is still noticeable at $x = l - a$. Finally, the beam in Fig. 10.11 ($K = 4$, $H = 0.0003$, $\Omega_c = 2\pi$, $B = 0.02$) has a number of natural frequencies within the excitation band, $N_c = 83$. The contribution of s_2 is now more than three times that of s_1 at the driving section, and the enhanced response at $x = l - a$ is all but erased.

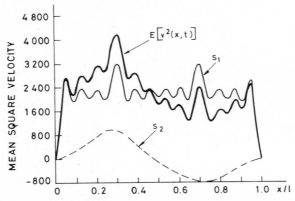

Fig. 10.9. Mean-square velocity $E[v^2(x, t)]$ and component sums s_1 and s_2 for beam on elastic foundation, $H = 0.01$, $K = 4$, $\Omega_c = 2\pi$, $B = 0.02$.

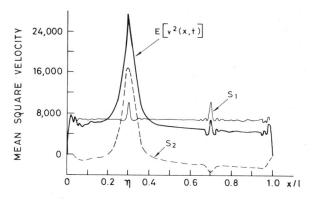

Fig. 10.10. Mean-square velocity $E[v^2(x, t)]$ and component sums s_1 and s_2 for beam on elastic foundation $H = 0.001$, $K = 4$, $\Omega_c = 2\pi$, $B = 0.02$.

TABLE 10.1 Values of Natural Frequencies for Beam on an Elastic Foundation with $K = 4$ and Different H Ratios

	Ω_j	
j	$H = 0.01$	$H = 0.001$
1	2.000203	2.000002
2	2.003244	2.000032
3	2.016371	2.000164
4	2.051294	2.000520
5	2.123049	2.001268
6	2.247669	2.002628
7	2.439056	2.004867
8	2.706455	2.008295
9	3.053824	2.013271
10	3.481009	2.020192
11	3.985564	2.029494
12	4.564240	2.041647
13	5.213845	2.057144
14	5.931602	2.076497
15	6.715240	2.100225

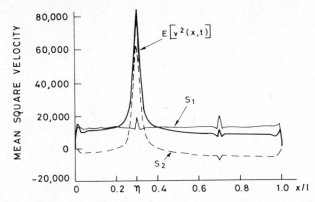

Fig. 10.11.   Mean-square velocity $E[v^2(x, t)]$ and component sums for beam on elastic foundation with $H = 0.0003$, $K = 4$, $\Omega_c = 2\pi$, $B = 0.02$.

TABLE 10.2   Components of Driving-Section Response for Beams with $K = 4$ and $\Omega_c = 2\pi$, and Percentage Error Incurred by Disregarding $s_2(a, t)$

H	N_c	$s_1(a, t)$	$s_2(a, t)$	$\eta, \%$
0.01	14	3,211.4	1,000.0	23.7%
0.005	20	4,699.4	2,482.4	34.6%
0.001	45	10,876.4	16,688.2	60.5%
0.0005	64	14,955.9	35,854.9	70.6%
0.0003	83	19,574.4	62,086.0	76.0%

The contributions to the driving-section mean-square response in the five beams with $K = 4$ are summarized in Table 10.2. The values of $s_1(a, t)$ and $s_2(a, t)$ at $x = a$ are tabulated. As the parameter H decreases from 0.01 to 0.0003, the percentage error incurred by disregarding the contribution of the cross correlation sum increases from 23.7% to 76.0%.

10.6   RANDOM VIBRATION DUE TO BOUNDARY-LAYER TURBULENCE

We now consider a case incapable of exact solution, where the normal-mode approach has to be combined with approximate techniques—namely, a system excited by pressure fluctuations in turbulent flow.

Boundary-layer turbulence sets in when a vehicle travels at high speed (see Fig. 10.12). In the process, the excited system transmits the noise (random sound pressure) by radiation to the surrounding medium, including the cabin,

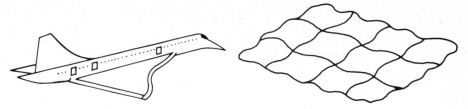

Fig. 10.12. Supersonic transport in turbulent flow.

thereby causing discomfort to the passengers. The acoustic-elastic problem as such is beyond the scope of this book*; what concerns us here is the random vibration due to the external load consisting of pressure fluctuations in the turbulent boundary layer and pressure perturbation depending on the deformation of the structure.

Ordinarily, random vibration of elastic bodies in a turbulent boundary layer is analyzed as though it were of the forced type, but the contribution of the surrounding medium is not taken into account. Such a contribution does exist, however. For example, part of the vibration energy may be dissipated in the medium in the form of radiation. Incidentally, it should be noted that as a rule the damping characteristics of structural systems are determined in air rather than in a vacuum, so that the energy radiated into medium may be assumed to be automatically accounted for. This radiative dissipation (per cycle) in aircraft structures is relatively small compared with that lost in the structure itself—normally of the order of 10^{-3} or less, sometimes about 10^{-2}—out of a total of 0.05–0.10. At supersonic flow velocities, and with a flexible skin, however, the picture may change and aeroelastic effects may be observed. The frequency characteristics of the structure are changed considerably, and at sufficiently high velocities, *flutter* (i.e., dynamic instability due to interaction with the flow) of the skin may set in. In this context mention should be made of the *piston theory* (Ashley and Zartarian; Ilyushin), extensively used in aeroelastic design in the supersonic range. Here the pressure perturbation on a surface moving in the direction of the normal, with the supersonic flow oriented along the x axis, is approximated by

$$p \simeq \frac{\kappa p_\infty}{c_\infty} \left(\frac{\partial w}{\partial t} + U \frac{\partial w}{\partial x} \right) \qquad (10.111)$$

where w is the displacement of the surface, p_∞ the unperturbed pressure in the medium, c_∞ the unperturbed velocity in it, U the flow velocity, and κ the

---

*For the determination of noise levels inside soundproofed shells enclosing an acoustic medium, see paper by Bolotin and Elishakoff.

polytropic exponent. Equation (10.111) (which is equivalent to the assumption that each small element of the vibratory surface may be treated as a piston applying a one-directional perturbation) is analogous to the piston approximation in acoustics. In fact, recalling that

$$c_\infty = \left(\frac{\kappa p_\infty}{\rho_\infty} \right)^{1/2}$$

where ρ_∞ is the unperturbed density of the medium, we rewrite equation (10.111) in the form

$$p \simeq \rho_\infty c_\infty \left(\frac{\partial w}{\partial t} + U \frac{\partial w}{\partial x} \right) \tag{10.112}$$

The product $\rho_\infty c_\infty$ represents the active component of the acoustic impedance of the piston at wavelength sufficiently large compared with its characteristic dimension (Morse). Accordingly, Eqs. (10.111) and (10.112) allow for the dissipation in the medium; at sufficiently high velocities, Eq. (10.111) yields a correct estimate of the critical flutter value (Dowell).

In beams, the correlation scale of the external load is much larger, and the standard deviation of the displacements much smaller, than the radius of inertia of the cross section, so that the problem may be considered in a probabilistically linear setting. In these circumstances, the equation of forced vibration of the system is given by Eq. (10.25), where x is flow direction. For the next step, the pressure perturbation (10.111) and the fluctuating components of the load, q are summed, whereby equation (10.25) becomes

$$EI \frac{\partial^4 w}{\partial x^4} + c \frac{\partial w}{\partial t} + \rho A \frac{\partial^2 w}{\partial t^2} = q(x, t) - \frac{\kappa p_\infty}{c_\infty} \left(\frac{\partial w}{\partial t} + U \frac{\partial w}{\partial x} \right)$$

$$\tag{10.113}$$

Strictly speaking, this summation is unjustified, as the "piston theory" represents an approximation to an asymptotically limiting case in supersonic aerodynamics, with the interaction between the boundary layer and the surface neglected. As a rule, the fluctuations component $q(x, t)$ is determined from measurements on rigid walls. Skin vibrations have some influence on fluctuations on the boundary layer; conversely, the presence of the latter must, in principle, be taken into account in determining the influence of the unperturbed supersonic flow on the vibrations. Equation (10.113) should be regarded as a first engineering approximation in studying the interaction of the vibrating panel and the external flow at supersonic velocities. It is indirectly justified by the fact that the critical panel flutter velocities determined from Eq. (10.111) are consistent with the experimental data, despite the presence of a turbulent boundary layer.

10.7 ANALYTIC APPROXIMATIONS FOR PRESSURE FLUCTUATIONS IN A TURBULENT BOUNDARY LAYER

As established experimentally, pressure fluctuations in a turbulent boundary layer may be approximated by a stationary function, provided the characteristic thickness of the layer is approximately constant and the gradient of the mean-square pressure sufficiently small. Almost no published data are available for experiments on natural objects, although there are a considerable number of works dealing with model tests in a wind tunnel, with analytic approximations of the results. We will list some of the results applicable for a two-dimensional system.

Dyer used the following simple expression for the autocorrelation function:

$$E\left[q(x_1, x_2, t)q(x_1', x_2', t + \tau)\right] = ad_q^2\delta(\xi - U_{\text{conv}}\tau)\delta(\eta)\exp\left(-\frac{|\tau|}{\theta}\right)$$

where a is the correlation area; d_q^2 the mean-square pressure, U_{conv} the mean convection velocity in the flow direction; $\xi = x_1' - x_1$, $\eta = x_2' - x_2$ the separations along the x_1 and x_2 axes, respectively; τ the time delay; and θ the life expectancy of the turbulent eddies.

Most popular is Corcos' concept of cross-spectral density of turbulent fluctuations:

$$S_q(\xi, \eta, \omega) = S_q(\omega)A(S_\xi)B(S_\eta)\exp(-iS_\xi) \qquad (10.114)$$

where $S_\xi = \omega\xi/U_{\text{conv}}$ and $S_\eta = \omega\eta/U_{\text{conv}}$ are the longitudinal and lateral Strouhal numbers, respectively, and $A(S_\xi)$ and $B(S_\eta)$ the longitudinal and lateral correlations presented graphically and approximated by exponential functions.

Wilby proposed the following approximation for $S_q(\xi, \eta, \omega)$:

$$S_q(\xi, \eta, \omega) = d_q^2 S_q(\omega)f_q(\xi, \eta, \omega)\cos\left(\frac{\omega\xi}{U_{\text{conv}}}\right)$$

where

$$f_q(\xi, 0, \omega) = \exp(-a_\xi|\xi|)$$

$$f_q(0, \eta, \omega) = \exp(-b_\eta|\eta|) \tag{10.115}$$

and

$$a_\xi = 0.1 \frac{\omega}{U_{\text{conv}}} \quad \text{and} \quad b_\eta = 0.715 \frac{\omega}{U_{\text{conv}}} \tag{10.116}$$

Measurements of $f_q(\xi, \eta, \omega)$ at $30°$ and $60°$ to the direction of flow showed that in the high-frequency range the following product-type expression does not conflict with experimental data:

$$f_q(\xi, \eta, \omega) = f_q(\xi, 0, \omega) f_q(0, \eta, \omega)$$

All these approximations have a common drawback in that they represent nondifferentiable random fields. This is the consequence of the "coarse" approach to the eddies, and also of the desire to simplify the approximating functions as far as possible.

Willmarth and Roos proposed, for the longitudinal and lateral correlation, an analytic approximation that corresponds to a differentiable random process and for our purpose here differs immaterially from Corcos' approximation:

$$A(S_\xi) = \exp(-0.1445|S_\xi|) + 0.1445|S_\xi|\exp(-2.5|S_\xi|)$$

$$B(S_\eta) = \exp(-0.092|S_\eta|) + 0.70\exp(-0.789|S_\eta|)$$

$$+ 0.145\exp(-2.916|S_\eta|) + 0.99|S_\eta|\exp(-4.0|S_\eta|)$$

Corcos' formulation was criticized in certain works. It was noted that in the low-frequency range, namely,

$$\frac{\omega\delta_*}{U_{\text{conv}}} < 0.2$$

where δ_* is the displacement thickness of the boundary layer, Eqs. (10.116) are inadequate for a qualitative picture of turbulent fluctuations; for $\omega \to 0$, they correspond to loads with infinitely increasing scales of correlation. Accordingly, the following approximations may be used, in the low-frequency range, for the empirical functions a_ξ and b_η (Crocker):

$$a_\xi = \frac{0.1\omega}{U_{\text{conv}}} + \frac{0.265}{\delta_0} \qquad b_\eta = \frac{0.715\omega}{U_{\text{conv}}} + \frac{2.0}{\delta_0} \tag{10.117}$$

where δ_0 is the thickness of the boundary layer.

The convective velocity U_{conv} is a function of the frequency and of the parameters of the boundary layer. As a first approximation, at supersonic velocities $U_{\text{conv}} = 0.95U$, U being the free stream velocity.

In conclusion, it should be recalled that these experimental data refer to a rigid wall. In the case of an elastic body, strong vibration may change the vortex pattern in the boundary layer. Still, approximations of the type discussed may be valid at not too large displacements, namely, at mean-square deviations not exceeding the thickness of the boundary layer.

10.8 FLUTTER AND RANDOM VIBRATION OF BEAMS—APPROXIMATE SOLUTION*

The differential equation (10.113) of motion of the beam is rewritten as

$$EI\frac{\partial^4 w}{\partial x^4} + 2\varepsilon_0 m\frac{\partial w}{\partial t} + gmU\frac{\partial w}{\partial x} + m\frac{\partial^2 w}{\partial t^2} = q(x,t) \qquad (10.118)$$

where

$$m = \rho A \qquad \varepsilon_0 = \varepsilon + \frac{g}{2}, \qquad g = \frac{\kappa p_\infty}{mc_\infty}$$

We confine ourselves to a beam clamped at both ends, so that the boundary conditions are as per (10.28). The mode shapes are given by the functions (10.40); the natural frequencies are the roots of Eq. (10.38) and have approximated values as per (10.39). The fluctuating components of the external force $q(x, t)$ and the beam displacement $w(x, t)$ are expressed as

$$q(x,t) = \sum_{j=1}^{N} \int_{-\infty}^{\infty} Q_j(\omega)\psi_j(x)e^{i\omega t}\,d\omega$$

$$w(x,t) = \sum_{j=1}^{N} \int_{-\infty}^{\infty} W_j(\omega)\psi_j(x)e^{i\omega t}\,d\omega \qquad (10.119)$$

where $Q_j(\omega)$ and $W_j(\omega)$ are random complex spectra and N is the number of modes taken into account. Substituting Eqs. (10.119) in Eq. (10.118), we apply the approximative technique of Galerkin (see Appendix E, for a short review of this method). Namely, we multiply both sides of the resulting equation by $\psi_k(x)$, $k = 1, 2, \ldots, N$ (N being number of terms taken into account) and

*Based on Elishakoff (1977), from which Figs. 10.13–10.19 are also reproduced, with the permission of the publisher.

integrate over the span, to obtain

$$
\begin{bmatrix}
a_{11} & a_{12} & \cdots & a_{1N} \\
a_{21} & a_{22} & \cdots & a_{2N} \\
\vdots & & & \\
a_{N1} & a_{N2} & \cdots & a_{NN}
\end{bmatrix}
\begin{Bmatrix}
W_1 \\
W_2 \\
\vdots \\
W_N
\end{Bmatrix}
= \frac{1}{m}
\begin{Bmatrix}
Q_1 \\
Q_2 \\
\vdots \\
Q_N
\end{Bmatrix}
\qquad (10.120a)
$$

$$
a_{jk} = \left(\omega_j^2 - \omega^2 + 2\varepsilon_0 i\omega \right)\delta_{jk} + gUb_{jk}
\qquad (10.120b)
$$

$$
b_{jk} = v_j^{-2} \int_0^l \frac{d\psi_k(x)}{dx}\psi_j(x)\,dx, \quad v_j^2 = \int_0^l \psi_j^2(x)\,dx, \quad j,k = 1,2,\ldots,N
$$

$$(10.120c)$$

For system parameters such that the matrix $[A] = [a_{jk}]$ is nonsingular, the solution of set (10.120) is

$$
W_k = \frac{1}{m}\sum_{\delta=1}^{N}(A^{-1})_{k\delta}Q_\delta(\omega)
$$

$(A^{-1})_{k\delta}$ being elements of the inverse matrix $[A^{-1}]$. The cross-spectral densities of the beam displacements are given by

$$
S_w(x_1, x_1', \omega) = \frac{1}{m^2}\sum_{j=1}^{N}\sum_{k=1}^{N}\sum_{\gamma=1}^{N}\sum_{\delta=1}^{N}(A^{-1})_{j\gamma}^*(A^{-1})_{k\delta}S_{Q_\gamma Q_\delta}\psi_j(x_1)\psi_k(x_1')
$$

The cross-spectral density of the displacement at point x is given by

$$
S_w(x, x, \omega) = S_{w,\mathrm{I}} + S_{w,\mathrm{II}}
$$

$$
S_{w,\mathrm{I}} = \frac{1}{m^2}\sum_{\gamma=1}^{N}S_{Q_\gamma Q_\gamma}\left|\sum_{j=1}^{N}(A^{-1})_{j\gamma}\psi_j(x)\right|^2
$$

$$
S_{w,\mathrm{II}} = 2\,\mathrm{Re}\,\frac{1}{m^2}\sum_{\substack{\gamma=1 \\ \gamma<\delta}}^{N}\sum_{\delta=1}^{N}S_{Q_\gamma Q_\delta}\sum_{j=1}^{N}(A^{-1})_{j\gamma}^*\psi_j(x)\sum_{k=1}^{N}(A^{-1})_{k\delta}\psi_\delta(x)
$$

$$(10.121)$$

where $\mathrm{Re}(\cdots)$ denotes the real part, and $S_{Q_\gamma Q_\delta}$ is

$$
S_{Q_\gamma Q_\delta} = v_\gamma^{-2}v_\delta^{-2}\int_0^l\int_0^l S_q(x_1, x_1', \omega)\psi_\gamma(x_1)\psi_\delta(x_1')\,dx_1\,dx_1'
$$

When the flow velocity is such that the matrix $[A]$ is singular (in which case, by definition, $U = U_{cr}$), we have instability of the beam. Consequently, the equations (10.121) are valid only in the range $U < U_{cr}$.

Consider the variation interval of U, in which the set (10.120) is uniquely compatible. For it to be incompatible, the condition is $\det[A(\omega)] = 0$. Since $[A]$ has complex coefficients, necessary and sufficient conditions for incompatibility are that both the real and the imaginary parts of the determinant $\det[A]$ vanish:

$$\text{Re}\{\det[A(\omega)]\} = 0, \quad \text{Im}\{\det[A(\omega)]\} = 0 \qquad (10.122)$$

A simple check shows that at $U = 0$ the above conditions are uniquely compatible, and so is the system in the half-interval $U \in [0, U_{cr})$.

We consider now how the probabilistic response characteristics are influenced by the number of modes used.

10.8.1   Single-Term Approximation.   For this case, the determinant of coefficients of the spectral equation is

$$\det[A(\omega)] = \omega_1^2 - \omega^2 + 2\varepsilon_0 i\omega + gUb_{11}$$

The critical velocity is given by

$$\text{Re}\{\det[A(\omega)]\} = \omega_1^2 - \omega^2 + gUb_{11} = 0$$

$$\text{Im}\{\det[A(\omega)]\} = 2\varepsilon_0 i\omega = 0$$

The total damping coefficient ε_0 differs from zero, and the latter equation holds only for $\omega = 0$; that is, flutter-type instability is apparently impossible. At the same time, $\text{Re}\{\det[A(\omega)]\} = 0$, because $b_{11} = 0$ is also impossible. It is thus seen that the single-term approximation, frequently used in random vibration analysis, does not cover the complete physical picture and actually leads to the incorrect conclusion of nonexistence of flutter-type instability.

10.8.2   Two-Term Approximation.   The spectral equations (10.120) take the form

$$a_{11}W_1 + a_{12}W_2 = \frac{1}{m}Q_1 \qquad a_{21}W_1 + a_{22}W_2 = \frac{1}{m}Q_2 \qquad (10.123)$$

The determinant of the system becomes

$$\det[A(\omega)] = \left(\omega_1^2 - \omega^2 + gUb_{11} + 2\varepsilon_0 i\omega\right)$$

$$\times \left(\omega_2^2 - \omega^2 + gUb_{22} + 2\varepsilon_0 i\omega\right) - g^2U^2b_{12}b_{21} \qquad (10.124)$$

Its real and imaginary parts are

$$\text{Re}\{\det[A(\omega)]\} = c_0\omega^4 - c_2\omega^2 + c_4$$

$$\text{Im}\{\det[A(\omega)]\} = -c_1\omega^3 + c_3\omega \tag{10.125}$$

where

$$c_0 = 1, \qquad c_1 = 4\varepsilon_0 \qquad c_2 = \omega_1^2 + \omega_2^2 + gU(b_{11} + b_{22}) + 4\varepsilon_0^2$$

$$c_3 = 2\varepsilon_0(c_2 - 4\varepsilon_0^2), \qquad c_4 = (\omega_1^2 + gUb_{11})(\omega_2^2 + gUb_{22}) - g^2U^2b_{12}b_{21}$$

To clarify the incompatibility condition for Eqs. (10.123), we seek the common root of $\text{Re}\{\det[A(\omega)]\} = 0$, $\text{Im}\{\det[A(\omega)]\} = 0$. From the latter we have

$$\omega_{1,\,\text{cr}} = 0, \quad \omega_{2,\,\text{cr}}^2 = \frac{c_3}{c_1} = \frac{1}{2}\left[\omega_1^2 + \omega_2^2 + gU(b_{11} + b_{22})\right]$$

for which

$$\text{Re}\{\det[A(\omega_{1,\,\text{cr}})]\} = \omega_1^2\omega_2^2 - g^2U^2b_{12}b_{21}$$

$$\text{Re}\{\det[A(\omega_{2,\,\text{cr}})]\} = \left[-c_3(c_1c_2 - c_0c_3) + c_4c_1^2\right]c_1^{-2}$$

The critical flow velocity associated with $\omega_{1,\,\text{cr}}$, which is identically zero, indicates static instability, or *divergence*, while that associated with nonzero critical frequency indicates dynamic instability, or *flutter*. $U_{\text{divergence}}$ and U_{flutter} are given by the following conditions, respectively:

$$\omega_1^2\omega_2^2 - g^2U^2b_{12}b_{21} = 0 \tag{10.126}$$

$$- c_3(c_1c_2 - c_0c_3) + c_4c_1^2 = 0 \tag{10.127}$$

It may be shown that $b_{12}b_{21} < 0$. Consequently, Eq. (10.126) has no real roots and static instability is also ruled out under this approximation.

The critical flutter velocity is obtained from Eq. (10.127)

$$U_{\text{flutter}} = \frac{1}{2g}\left\{-\frac{1}{b_{12}b_{21}}\left[(\omega_2^2 - \omega_1^2)^2 + 8\varepsilon_0(\omega_1^2 + \omega_2^2)\right]\right\}$$

with b_{jk} as per (10.120).

For $U < U_{\text{flutter}}$, the matrix $[A]$ is nonsingular, and the spectral equations have the following solution:

$$W_1(\omega) = \frac{a_{22}}{\Delta}Q_1 - \frac{a_{12}}{\Delta}Q_2 \qquad W_2(\omega) = \frac{a_{11}}{\Delta}Q_2 - \frac{a_{21}}{\Delta}Q_2 \tag{10.128}$$

The mean-square value of the displacements may be formulated as

$$E\left[w^2(x, t)\right] = \int_{-\infty}^{\infty}\left[S_{Q_1Q_1}\varphi_{11} + S_{Q_2Q_2}\varphi_{22} - 2\,\mathrm{Re}\!\left(S_{Q_1Q_2}\varphi_{12}\right)\right]m^{-2}\Delta^{-2}\,d\omega$$

$$(10.129)$$

where

$$\varphi_{11} = |a_{22}|^2\psi_1^2(x) + |a_{21}|^2\psi_2^2(x) + 2\,\mathrm{Re}(a_{21}^*a_{22})\psi_1(x)\psi_2(x)$$

$$\varphi_{22} = |a_{12}|^2\psi_1^2(x) + |a_{11}|^2\psi_2^2(x) + 2\,\mathrm{Re}(a_{11}^*a_{12})\psi_1(x)\psi_2(x)$$

$$\varphi_{12} = a_{22}^*a_{12}\psi_1^2(x) + a_{21}^*a_{11}\psi_2^2(x) + (a_{21}^*a_{12} + a_{22}^*a_{11})\psi_1(x)\psi_2(x)$$

Higher approximations are realized in a similar manner. Numerical results are shown in Figs. 10.13–10.19, given the following data and flow parameters: $l = 0.55$ m, $\rho = 2700$ kg/m^3, $E = 7.05 \times 10^{10}$ N/m^2, $\nu = 0.3$, $\varepsilon = 0.1$, $c_\infty = 303.3$ m/s, $\rho_\infty = 0.459$ kg/m^3, altitude $= 10{,}000$ m. The cross-spectral density of the external forces is given as

$$S_q(x_1, x_1', \omega) = d_q^2 S_q(\omega)\exp\left[\left(-\frac{0.1\omega l}{U_{\mathrm{conv}}} + 0.265\frac{l}{\delta_0}\alpha\right)|\xi_1 - \xi_1'| \right.$$

$$\left. - \frac{i\omega l}{U_{\mathrm{conv}}}(\xi_1 - \xi_1')\right] \qquad (10.130)$$

where d_q^2 is the mean-square fluctuation in the turbulent boundary layer, $S_q(\omega)$ the frequency spectral density, ω the frequency, U_{conv} the convection velocity, $\xi_1 = x_1/l$ and $\xi_1' = x_1'/l$ nondimensional coordinates, and δ_0 the thickness of the boundary layer, its nondimensional value being $\delta_0/l = 0.32$. For $\alpha = 0$ and $\alpha = 1$, Eq. (10.130) reduces, respectively, to Wilby's and Crocker's approximations. Note that the nondimensional cross-spectral density $S_q(x_1, x_1', \omega)/d_q^2 S_q(\omega)$ obtainable from Eq. (10.130) is probably a good empirical expression for the normalized cross-spectral density at supersonic as well as subsonic speeds. d_q^2 is about $0.003q_0$ for Mach numbers $2 \leqslant M \leqslant 5$, where q_0 is the dynamic pressure of free flow and M is the ratio of the free stream velocity to the speed of the sound. M_{flutter} equals, respectively, 1.545, 1.976, and 2.08 by the two-, three-, and four-term approximations. The last value may be regarded in practice as an exact solution.

The nondimensional cross-spectral density of the maximal normal stresses

$$\bar{S}_\sigma = \frac{m^2h^4}{36(EI)^2}\frac{S_\sigma(\omega)}{d_q^2 S_q(\omega)}$$

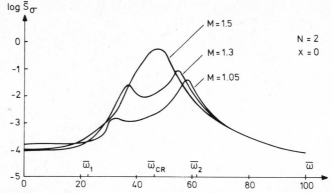

Fig. 10.13. Nondimensional cross-spectral density $\log \bar{S}_\sigma$ versus nondimensional frequence $\bar{\omega}$, $x = 0$, $N = 2$.

is calculated as a function of the nondimensional frequency

$$\bar{\omega} = \omega l^2 \left(\frac{m}{EI} \right)^{1/2}$$

The following nondimensional quantities are also used:

$$\bar{\omega}_\alpha = \omega_\alpha f \qquad \bar{\varepsilon} = \varepsilon f \qquad \bar{\varepsilon}_0 = \varepsilon_0 f$$

$$\bar{g} = gf \qquad \bar{b}_{jk} = lb_{jk} \qquad \bar{U} = \frac{Uf}{l} \qquad \bar{\omega}_j = \left(r_j l \right)^2 \qquad \bar{x} = \frac{x}{l}$$

with $f = l^2 (m/EI)^{1/2}$ and r_j as per Eq. (10.39).

In Figs. 10.13 and 10.14, $\bar{S}_\sigma$ is plotted against $\bar{\omega}$ at $\bar{x} = 0$ in the two-term approximation. At $M = 1.05$ and $M = 1.3$, the plots are seen to contain two

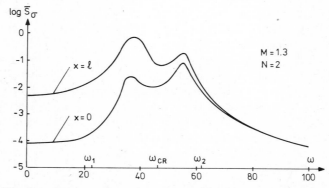

Fig. 10.14. Nondimensional cross-spectral density $\log \bar{S}_\sigma$ versus nondimensional frequency, for different cross sections, $N = 2$.

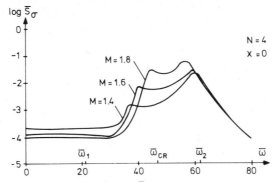

Fig. 10.15. Variation of $\log \bar{S}_\sigma$ for different Mach numbers.

distinct maxima: For $M = 1.05$, the first maximum appears at $\bar{\omega} = 32$, the first nondimensional natural frequency being $\bar{\omega}_1 = 22.37$; the second maximum is at $\bar{\omega} = 56$, the second nondimensional natural frequency being $\bar{\omega}_2 = 61.67$. As the Mach number increases and approaches the flutter level, they merge into a single maximum close to the critical flutter frequency $\bar{\omega}_{cr} = 46.4$. At $M = M_{flutter}$, the cross-spectral density of the maximal normal stresses (valid only in the preflutter range) has a pole at $\bar{\omega}_{cr}$.

Figures 10.15 and 10.16 present the nondimensional cross-spectral densities of the maximal normal stresses in the four-term approximation for $M = 1.4$, 1.6, and 1.8. In the two-term approximation, the two last Mach numbers fall within the post-flutter range, again illustrating how approximations below the minimum safe number of terms N^* (in the present case, four) are of purely qualitative significance. Figure 10.16 shows four maxima, the first of them occurring considerably above the first natural frequency and the remaining

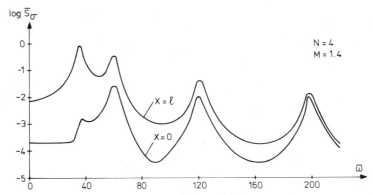

Fig. 10.16. Nondimensional cross-spectral density $\log \bar{S}_\sigma$ versus nondimensional frequency, for different cross sections, $N = 4$.

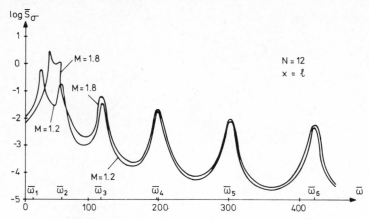

Fig. 10.17. Results of 12-mode approximation for log $\bar{S}_\sigma$.

three close to the corresponding natural frequencies. As the Mach number increases, the first two maxima draw together and eventually merge, leaving a total of three maxima (as though the system had only three degrees of freedom!). This degenerative tendency may serve as an advance indication of possible flutter-type instability.

Figure 10.14 and 10.16 show that for a given flow velocity, the mean-square stress at the cross section $x = l$,

$$E\left[\sigma^2(l, t)\right] = \int_{-\infty}^{\infty} S_\sigma(l, l, \omega)\, d\omega \qquad (10.131)$$

exceeds its counterpart $E[\sigma^2(0, t)]$ at $x = 0$, provided that the function $S_q(\omega)$ is sufficiently smooth. Figures 10.13 and 10.15, moreover, show that in the four-term approximation the cross-spectral density is insensitive to change of the free flow velocity in the interval outside of the first two natural frequencies.

On comparing the true maximal stress spectral density with its partial counterpart based on terms corresponding to like modes of vibration alone (joint acceptances), it turns out that recourse to the partial counterpart results in 40–45% underestimation of some frequencies. Consequently, when deformational pressure perturbations are considered, the terms containing cross acceptances may not be neglected in calculating the probabilistic response.

The convergence analysis, undertaken to determine the minimum number of terms to be taken into account, showed that $N = N^*$ is still too small: At least 12 terms are necessary for satisfactory estimation of log $\bar{S}_\sigma$ in the frequency interval $\bar{\omega} \leqslant 500$. The logarithm is plotted in Figs. 10.17 and 10.18 against $\bar{\omega}$ at 12-term approximation. For $M = 1.2$, the first maximum appears at $\bar{\omega} = 32$ and the second at $\bar{\omega} = 60$, while for $M = 1.8$ the first maximum shifts to $\bar{\omega} = 44$, and the second to $\bar{\omega} = 59$. The mean square of the maximal stress at $x = l$ again exceeds that at $x = 0$ (Fig. 10.18). In Fig. 10.19, log $\bar{S}_\sigma$ is plotted

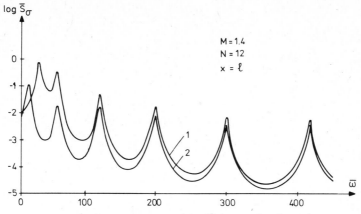

Fig. 10.18.   Comparison of log $\bar{S}_\sigma$ for $x = 0$ and $x = l$.

against $\bar{\omega}$, with the last term in Eq. (10.120b) formally taken as zero, that is, with only part of the deformational pressure perturbations considered. As may be seen from Fig. 10.19, such an approximation leads to an underestimate of the true maximal stress spectral density, especially at the low frequencies.

PROBLEMS

10.1.   An infinite plate is subjected to random loading represented by a spacewise fully correlated random function, $R_q(\tau, \xi_1, \xi_2) = R_q(\tau)$. Find the mean-square displacement and acceleration.

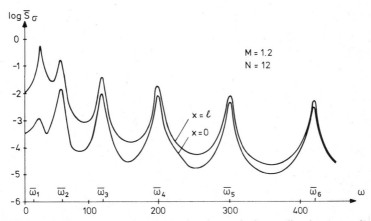

Fig. 10.19.   Disregard of the pressure perturbation due to the beam vibration (curve *2*) may lead to an underestimate of the true maximal stress spectral density (curve *1*).

10.2. An infinite plate is subjected to random loading, represented by spacewise white noise and time-cut white noise. Find the mean-square acceleration.

10.3. Apply the normal mode method for a beam under a distributed random load, simply supported at one end and clamped at the other.

10.4. Put Eq. (10.53) in the form

$$w(x, t) = \int_{-\infty}^{\infty} d\tau \int_{0}^{l} q(x, t - \tau) h(x, \tau, \xi) \, d\xi$$

where

$$h(x, \tau, \xi) = \frac{1}{\rho A} \sum_{j=1}^{\infty} \psi_j(x) \psi_j(\xi) h_j(\tau) v_j^{-2}$$

and show that $h(x, \tau, \xi)$ is the response to excitation in the form $\delta(x - \xi)\delta(t - \tau)$. [$h(x, \tau, \xi)$ is called the *unit impulse response function* or *Green function* of the system.]

10.5. A cantilever beam is subjected to random loading with autocorrelation function as in Eq. (10.80). Find a closed expression for the mean-square value of the displacement at the free end.

10.6. Derive expression (10.93).

10.7. Show that the span average of $G(x/l, a/l, N_c)$ in Sec. 10.4 is $g(a/l, N_c)$, and the average with respect to the loading positions of the span average is N_c. That is,

$$\frac{1}{l^2} \int_0^l dx \int_0^l G\left(\frac{x}{l}, \frac{a}{l}, N_c\right) da = N_c$$

10.8. Use the normal mode approach to derive the mean-square values of the displacement, velocity, and acceleration of a beam of span l, clamped at both ends and subjected to a concentrated load $F(t)$ at the cross section $x = a$. $F(t)$ is band-limited white noise with ω_c as the cutoff frequency. Examine the behavior of the responses as ω_c increases and tends to infinity.

10.9. A nonuniform beam is simply supported at both ends and subjected to a concentrated force at the cross section $x = a$, $a < l$. $F(t)$ is a

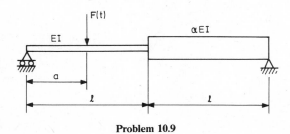

Problem 10.9

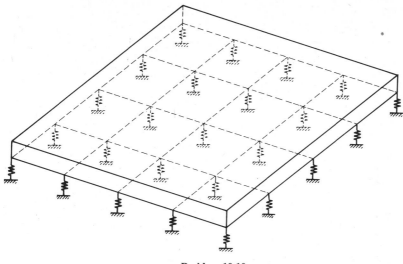

Problem 10.10

band-limited white noise with ω_c as the cutoff frequency. Find the mean-square velocity as a function of the parameter α. Evaluate the span average mean-square velocity as a function of α (see figure).

10.10. Generalize the results of Sec. 10.4 to a plate on an elastic foundation (see figure).

CITED REFERENCES

Ashley, H., and Zartarian, G., "Piston Theory—A New Aerodynamic Tool for Aeroelastician," *J. Aeronaut. Sci.*, **23**, 1109–1118 (1956).

Bolotin, V. V., and Elishakoff, I., "Random Vibrations of Elastic Shells Containing an Acoustic Medium," *Mech. Solids*, **6**, 99–107 (1971), Allerton Press, Inc.

Corcos, E. M., "Resolution of Pressure in Turbulence," *J. Acoust. Soc. Am.*, **35**, 192–199 (1963).

Crandall, S. H., and Wittig, L. E., "Chladni's Patterns for Random Vibration of a Plate," in G. Herrmann and N. Perrone, Eds., *Dynamic Response of Structures*, Pergamon Press, New York, 1972, pp. 55–71.

Crocker, M. J., "The Response of a Supersonic Transport Fuselage to Boundary Layer and to Reverberant Noise," *J. Sound Vibration*, **9**, 6–20 (1969).

Dowell, E. H., "Panel Flutter: a review of Aeroelastic Stability of Plates and Shells," *AIAA J.*, **8**, 385–399 (1970).

Dyer, I., "Response of Plates to a Decaying and Convecting Random Pressure Field," *J. Acoust. Soc. Am.*, **31**, 922–928 (1959).

Elishakoff, I., "Flutter and Random Vibrations in Plates," in B. L. Clarkson, Ed., *Stochastic Problems in Dynamics*, Pitman, London, 1977, pp. 390–410.

Elishakoff, I., "Simulation of Space-Random Fields for Solution of Stochastic Boundary-Value Problems," *J. Acoust. Soc. Am.*, **65** (2), 399–403 (1979).

Elishakoff, I., van Zanten, A. Th., and Crandall, S. H., "Wide-Band Random Axisymmetric Vibration of Cylindrical Shells," *ASME J. Appl. Mech.*, **46**, 417–422 (1979).

Eringen, A. C., "Response of Beams and Plates to Random Loads," *ASME. J. Appl. Mech.*, **24**, 46–52 (1957).

Gradshteyn, I. S., and Ryzhik, I. M., *Tables of Integrals, Series and Products*, Academic Press, New York, 1965.

Ilyushin, A. A., "Law of Plane Sections at High Supersonic Velocities," *PMM Appl. Math. Mech.*, **20**(6), 1956.

Lin, Y. K., *Probabilistic Theory of Structural Dynamics*, McGraw-Hill, New York, 1967, p. 221.

Morse, P. M., *Vibration and Sound*, 2nd ed., McGraw-Hill, New York, 1948.

Pal'mov, V. A., "Thin Plates Under the Wide-Band Random Loading," *Proc. Leningrad Polytech. Inst.*, **252**, 97–106 (1965).

Powell, A., "An Introduction to Acoustic Fatigue," in W. J. Trapp and D. M. Forney, Jr., Eds., *Acoustical Fatigue in Aerospace Structures*, Syracuse Univ. Press, 1965, pp. 1–15.

Van Lear, G. A., Jr., and Uhlenbeck, G. E., "Brownian Motion of Strings and Elastic Rods," *Phys. Rev.*, **38**, 1583–1598 (1931).

Wedig, W., "Zufallschwingungen von querangestroemten Saiten", *Ingenieur-Archiv*, **48**, 325–335 (1979).

Wilby, J. F., "Turbulent Boundary Layer Pressure Fluctuations and Their Effect on Adjacent Structures," *Jahrb. 1964 Wiss. Ges. Luft Raumfahrt*.

Willmarth, W. W., and Roos, F. W., "Resolution and Structure of the Wall Pressure Field Beneath a Turbulent Boundary Layer," *J. Fluid Mech.*, **22**, 81–94 (1965).

RECOMMENDED FURTHER READING

Bolotin, V. V., *Random Vibrations of Elastic Bodies*, "Nauka" Publishing House, Moscow, 1979.

Crandall, S. H., "Random Vibration of One- and Two-Dimensional Structures," R. P. Krishnaiah, Ed., *Developments in Statistics*, Vol. 2, Academic Press, New York, 1979, pp. 1–82.

Dimentberg, M. F., *Nonlinear Stochastic Problems of Mechanical Vibrations*, "Nauka" Publishing House, Moscow, in press.

Heinrich, W. and Hennig, K., *Random Vibrations of Mechanical Systems*, Vieweg, Braunschweig, 1978.

Lin, Y. K., *Probabilistic Theory of Structural Dynamics*, McGraw-Hill, New York, 1976; Chap. 7: Linear Continuous Structures, pp. 203–252.

Sobczyk, K., *Methods in Statistical Dynamics*, Polish Scientific Publ., Warsaw, 1973.

Svetlitskii, V. A., *Random Vibrations of Mechanical Systems*, "Mashinostroenie" Publishing House, Moscow, 1976.

chapter **11**

Monte Carlo Method

Up to now we have been concerned with problems capable of exact or effective approximate solution. These, however, constitute only a small fraction of the probabilistic problems encountered in the different branches of engineering. In cases where exact solution is cumbersome or impossible, the Monte Carlo method is a logical remedy, especially with the advent of high-speed digital computers. Even if it may seem a "last resort" at present, it is bound to become more and more attractive in the future.

11.1 DESCRIPTION OF THE METHOD

The name "Monte Carlo method" dates from about 1944 and is due to von Neumann and Ulam, who introduced it as a code name for their secret work on neutron diffusion problems (encountered in work on the atomic bomb) at the Los Alamos Scientific Laboratory. The name itself was chosen because roulette (with which the casino town Monte Carlo is traditionally associated) is one of the simplest tools for generating random numbers. Systematic development dates from 1949, with publication of the paper by Metropolis and Ulam.

The method may be described as a means of solving problems numerically in mathematics, physics, and other sciences through sampling experiments. The problem may be posed in either probabilistic or deterministic form. In the probabilistic case the actual random variable or function appearing in the problem is simulated, whereas in the deterministic case an artificial random variable or function is first constructed and then simulated. The simulation process is usually computerized, although a roulette wheel and pencil and paper may sometimes suffice. The method normally consists of three steps: (*1*) simulation of the random variable function, (*2*) solution of the deterministic problem for a large number of realizations of the latter, and (*3*) statistical analysis of the results.

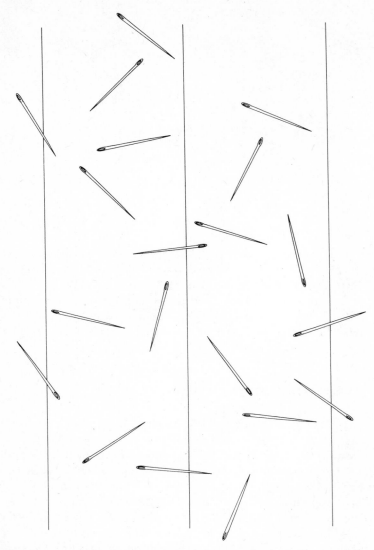

Fig. 11.1. Illustration of needle experiment (From D. D. McCracken, "The Monte Carlo Method." Copyright © 1955 by *Scientific American*, Inc. All rights reserved.)

The earliest known application of the Monte Carlo method is apparently Buffon's classic "needle problem," which we reproduce here as an illustrative example (Fig. 11.1): *Given a board ruled in parallel lines with uniform spacing a, on which a needle of length l is dropped at random. What is the probability of the needle falling across one of the lines?*

Let X be the random variable representing the distance between the midpoint M of the needle and the nearest line, and Θ the random variable representing the acute angle between the needle (or its extension) and the line. x and θ denote any particular values of X and Θ, it being apparent that the former can take on any value between 0 and $a/2$ and the latter, any value between 0 and $\pi/2$.

Since X and Θ are independent, the joint probability density equals

$$f_{X\Theta}(x, \theta) = f_X(x)f_\Theta(\theta)$$

where

$$f_X(x) = \frac{2}{a}, \quad f_\Theta(\theta) = \frac{2}{\pi}$$

and therefore

$$f_{X\Theta}(x, \theta) = \frac{4}{a\pi} \qquad (11.1)$$

Figure 11.2 shows that the condition for the needle to hit a line is $AM \leqslant CM \sin \theta$, or $X \leqslant (l/2)\sin \theta$. The probability we seek is accordingly given by

$$P(A) = \frac{4}{a\pi} \int_0^{\pi/2} \int_0^{(l/2)\sin\theta} d\theta \, dx = \frac{2l}{a\pi} \qquad (11.2)$$

A remarkable feature of this result is that it contains the number π, thereby providing an experimental means for its estimation. For this purpose, we drop the needle a sufficiently large number of times and record the hits. The relative

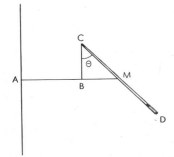

Fig. 11.2. Needle misses all lines if $AM > CM \sin \theta$, $\theta = MCB$. (From D. D. McCracken, "The Monte Carlo Method." Copyright © 1955, by *Scientific American*, Inc. All rights reserved.)

frequency of these hits, $\hat{P}(A)$, stabilizes near the theoretical probability, so that equating them we obtain an approximation for π:

$$\pi \simeq \frac{2l}{\hat{P}(A)a} \qquad (11.3)$$

Such a series of experiments was carried out between 1849 and 1853 by the Swiss astronomer Wolf, using a ruled board with $a = 45$ mm and a needle 36 mm long. The needle was thrown 5000 times, and 2532 hits were scored, yielding a relative frequency of 0.5064. The corresponding approximation for π equals

$$\pi \simeq \frac{2 \cdot 36}{0.5064 \cdot 45} = 3.1596$$

deviating from the real value $(3.1416\ldots)$ by less than 0.02.

11.2   GENERATION OF RANDOM NUMBERS

Consider the random variable E, which can take on 10 values $0, 1, 2, \ldots, 9$ with uniform probability .1. Numbers generated by repeated independent selection from these values are referred to as *random digits*. The generation of random digits is equivalent to placing 10 balls numbered 0 to 9 in an urn and drawing one of them at a time and noting its number, replacing it and shaking the urn thoroughly before drawing the next one.

Any group of consecutive random digits is known as a *random number*. We will show, following Sobol', that the uniformly distributed random variable X in the interval $(0, 1)$ may be put in the form

$$X = 0.E_1 E_2 \ldots E_k \ldots \qquad (11.4)$$

where $E_1, E_2, \ldots, E_k, \ldots$ are independent random digits. Expression (11.4) is equivalent to

$$X = \sum_{k=1}^{\infty} E_k \times 10^{-k}$$

An arbitrary number from the interval $(0, 1)$ may be put in the form

$$x = 0.a_1 a_2 \ldots a_k \ldots$$

If $X < x$, then in the expansion (11.4) either $E_1 < a_1$ or $E_1 = a_1$ and $E_2 < a_2$; either $E_1 = a_1$, $E_2 = a_2$ and $E_3 < a_3$; etc. Therefore,

$$P(X < x) = \sum_{k=1}^{\infty} P\{E_1 = a_1, \ldots, E_{k-1} = a_{k-1}, E_k < a_k\}$$

As $E_1, E_2, \ldots, E_k, \ldots$ are independent, we find

$$P(X < x) = \sum_{k=1}^{\infty} P(E_1 = a_1) \cdots P(E_{k-1} = a_{k-1}) P(E_k < a_k)$$

Noting that $P(E_k < a_k) = a_k \times 10^{-1}$, since in this case the values taken on by E_k are $0, 1, \ldots, a_{k-1}$, we have

$$P(X < x) = \sum_{k=1}^{\infty} 10^{-(k-1)} a_k \times 10^{-1} = \sum_{k=1}^{\infty} a_k \times 10^{-k} = x \qquad (11.5)$$

which is the result we sought. Hereafter a uniform random number will be called simply a *random number*. After N independent experiments, we are in possession of N random digits. These we can "deposit" in a table and use the latter for "withdrawals" as necessity may arise, either singly or in a group as a random number, as desired, $X = 0. E_{\alpha} E_{\alpha+1} \ldots E_{\alpha+N-1}$. Since randomness is already inherent in the table, there is no need to observe a random selection procedure, and we are completely free to start anywhere and draw digits either in running order (in any direction) or sporadically, with recourse to any predetermined algorithm independent of the actual values of the digits.

Several such comprehensive tables are available in the literature for pencil-and-paper Monte Carlo processes. By contrast, computerized processes call for perfectly deterministic generation of numbers and the product, not being truly random, is referred to accordingly as *pseudo-random* or *quasi-random*.

The first presentation of pseudo-random numbers (due to Metropolis and von Neumann) involved the *midsquare method*: An arbitrary number α_0, comprising an even number $2k$ of binary digits and serving as starting point of a recurrence process, was squared to yield a $4k$-digit number; then the central group of $2k$ binary digits of the latter was taken as the next number α_1 and the process repeated, and so on. The attempt proved unsatisfactory, owing to the cyclic character of the sequences so constructed.

Eventually, the *congruential* method proved the best for the purpose in question, for which Lehmer proposed the relation

$$x_i = ax_{i-1} (\text{mod } m) \qquad (11.6)$$

where x_i is the remainder when ax_{i-1} is divided by m; $a = 23$, $m = 10^8 + 1$. Once the initial value x_0 is specified, the result is a sequence of eight-digit decimal numbers with period 5,882,352. Since such a sequence repeats itself after at most m steps, the period must be required to exceed the number of random numbers involved in the experiment under consideration; this is why m was set at a large power of 10.

The above relation was later replaced by the more general one

$$x_i = ax_{i-1} + b \ (\text{mod } m)$$

with the quotients x_i/m themselves constituting the sequence of pseudo-random numbers, m again being a large integer determined by the design of the computer.

11.3   SIMULATION OF CONTINUOUS RANDOM VARIABLES

A random variable with continuous distribution may be simulated with the aid of the *inverse-function method* described in Example 4.7. Let a random variable Y be determined in the interval (a, b) and possess the probability density function $f_Y(y)$. If it is to satisfy the equation

$$F_Y(Y) = X \tag{11.7}$$

where X is a uniformly distributed random variable, it must have the distribution $F_Y(y)$. Indeed, since the latter is strictly increasing in the interval (a, b) [i.e., between $F_Y(a) = 0$ and $F_Y(b) = 1$], Eq. (11.7) has a single root for each realization of X, and

$$P\{y \leqslant Y \leqslant y + dy\} = P\{F_Y(y) < X \leqslant F_Y(y + dy)\} \tag{11.8}$$

but as X has a uniform distribution in the interval $(0, 1)$, then

$$P\{y \leqslant Y \leqslant y + dy\} = F_Y(y + dy) - F_Y(y) = f_Y(y)\, dy \tag{11.9}$$

which is the desired result.

For example, for a random variable with Rayleigh distribution, we have (see Sec. 4.5)

$$f_Y(y) = \frac{y}{a^2} \exp\left(-\frac{y^2}{2a^2}\right), \qquad y \geqslant 0$$

$$F_Y(y) = 1 - \exp\left(-\frac{y^2}{2a^2}\right), \qquad y \geqslant 0$$

where a is the distribution parameter. This random variable can be obtained from the uniformly distributed random variable X as

$$Y = a\left[-2\ln(1 - X)\right]^{1/2} = a(-2\ln X)^{1/2}$$

since X and $1 - X$ have identical distributions.

Analogously, for an exponentially distributed variable, we use Eq. (4.71):

$$Y = \frac{1}{a}\ln(1 - X) = -\frac{1}{a}\ln X$$

For a normally distributed variable $N(0, 1)$, we have to solve

$$\frac{1}{\sqrt{2\pi}} \int_{-\infty}^{Y} e^{-t^2/2} \, dt = X \tag{11.10}$$

which involves determination of the inverse of the error function. This is unfeasible in closed form and calls for a numerical procedure, subroutines for which (based, for example, on piecewise approximation through Chebyshev polynomials) are readily available.

For Y with a finite interval (a, b), von Neumann's *acceptance-rejection method* may be used. This consists in choosing, from the generator of random numbers in the interval $(0, 1)$, a pair of independent uniformly distributed random numbers X_1, X_2 and forming the pair

$$Y_1 = a + (b - a) X_1, \quad Y_2 = f X_2$$

where f is a maximum of $f_Y(y)$. If

$$Y_2 \leqslant f_Y(Y_1) \tag{11.11}$$

we accept $y_1^* \equiv y_1$ as the realization of the random variable Y; otherwise we choose a new pair x_1 and x_2 and proceed as before.

To justify this procedure, we note that the pair X_1, X_2 may be considered as coordinates of random points in a plane, uniformly distributed along the axes y and $f_Y(y)$ within the rectangle $aa'b'b$, where a' and b' have the coordinates (a, f) and (b, f), respectively. Similarly the pair Y_1, Y_2, satisfying condition (11.11) are the coordinates of random points of the plane, uniformly distributed along the axes within the lower part of the rectangle bounded by the $f_Y(y)$. The probability of the event of a random point below $f_Y(y)$ falling within the elementary strip $(y, y + \Delta y)$ is proportional to $f_Y(y)$. The overall probability of it lying anywhere in the lower part of the rectangle is, by (11.11), unity.

In addition to the above, specific methods are available for simulating some random variables. In particular, for simulating a normally distributed variable, approximative use may be made of the central limit theorem (Sec. 7.4), by virtue of which the sum of a large number of random variables tends asymptotically to a normal distribution, subject to certain general conditions. Accordingly, we take

$$Y = X_1 + X_2 + \cdots + X_n - \tfrac{1}{2}n$$

where

$$E(Y) = 0 \qquad \sigma_Y = \tfrac{1}{2}\left(\frac{n}{3}\right)^{1/2}$$

Setting $n = 12$, we find $\sigma_Y = 1$, hence $N(0, 1)$.

For other methods of simulation, the reader is referred to the books of Hammersley and Handscomb, and of Rubinstein.

Example 11.1

Consider a simply supported I-beam with span $l = 1000$ mm, cross section depth $h = 32$ mm (moment of inertia 28,800 mm^4), and allowable stress $\sigma_{allow} = 250$ N/mm^2, under a random concentrated load G at midspan. The load is normally distributed with mean $a_G = 1500$ N and standard deviation $\sigma_G = 300$ N. We shall determine the reliability of the beam in closed form and compare the result with that of the Monte Carlo method.

The exact solution is obtained as follows. The maximum bending moment appears in the midspan cross section and equals

$$M_{max} = \tfrac{1}{4}Gl$$

so that the nonfailure condition reads

$$-\sigma_{allow} \leqslant \frac{M_{max}}{S} \leqslant \sigma_{allow} \tag{11.12}$$

where S is the section modulus of the beam

$$S = \frac{I}{\tfrac{1}{2}h} = \frac{28,800}{16} = 1800 \text{ mm}^3$$

The reliability is then given by the probability of the random event (11.12),

$$R = P\left(-\sigma_{allow} \leqslant \frac{Gl}{4S} \leqslant \sigma_{allow}\right)$$

$$= F_G\left(\frac{4S\sigma_{allow}}{l}\right) - F_G\left(-\frac{4S\sigma_{allow}}{l}\right) \tag{11.13}$$

or, because of the normality of G,

$$R = \text{erf}\left(\frac{4S\sigma_{allow}/l - a_G}{\sigma_G}\right) - \text{erf}\left(-\frac{4S\sigma_{allow}/l + a_G}{\sigma_G}\right)$$

$$= \text{erf}(1) + \text{erf}(11) = 0.34134 + 0.5 = 0.84134 \tag{11.14}$$

Hence, for an ensemble of 1000 beams, the number of nonfailures has to stabilize, in accordance with the statistical interpretation of probability, in the vicinity of 841.

The Monte Carlo method proceeds in this case as follows: The moment $M_{max} = \tfrac{1}{4}Gl$ is calculated, for each realization of the simulated load the nonfailure condition (11.12) is checked, and the beams that satisfy it are counted. The results for ten series of beams as above are 832, 821, 847, 872, 840, 874, 845, 858, 851, so that the maximal percentage difference between the exact reliability R and its simulated counterpart R^* is 0.03266. For an ensemble of 10,000 beams, we have $R^* = 0.8496$, which differs from the exact value by 0.00826.

Unlike the foregoing relatively simple case, for which the exact solution is readily obtainable, that of a set of random concentrated forces is almost intractable analytically. By contrast, it lends itself to straightforward solution by the Monte Carlo method, provided we know how to simulate the vector of random loads. This is done in the next section.

11.4 SIMULATION OF RANDOM VECTORS

Consider a random vector with joint probability density $f(y_1, y_2, \ldots, y_n)$. In the general case where the random variables $y_1, y_2, \ldots, y_n$ are dependent, the (unconditional) joint probability density can be expressed as the product of the (conditional) component densities, using Eq. (6.35):

$$f(y_1, y_2, \ldots, y_n) = f(y_1)f(y_2|y_1)f(y_3|y_1, y_2) \cdots f(y_n|y_1, y_2, \ldots, y_{n-1})$$

$$(11.15)$$

Conversely, all conditional densities can be derived from the joint density, as shown in Sec. 6.4.

We will now show that if $X_1, X_2, \ldots, X_n$ represent independent random numbers, the random variables $Y_1, Y_2, \ldots, Y_n$ obtained from the set of equations

$$
\begin{aligned}
F(Y_1) &= X_1 \\
F(Y_2|Y_1) &= X_2 \\
&\vdots
\end{aligned}
$$

$$(11.16)$$

$$F(Y_n|Y_1, Y_2, \ldots, Y_{n-1}) = X_n$$

have the joint probability density $f(y_1, y_2, \ldots, y_n)$.

Indeed, if the values $Y_1 = y_1, \ldots, Y_{k-1} = y_{k-1}$ are given, then the random variable $\{Y_k|Y_1, Y_2, \ldots, Y_{k-1}\}$ can be obtained by Eq. (11.7):

$$F(Y_k|Y_1, Y_2, \ldots, Y_{k-1}) = X_k$$

and the corresponding probability of the random event $y_k \leqslant Y_k \leqslant y_k + dy_k$ is

$$P\{y_k \leqslant Y_k \leqslant y_k + dy_k|y_1, y_2, \ldots, y_{k-1}\} = f(y_k|y_1, y_2, \ldots, y_{k-1})\, dy_k$$

Therefore,

$$P\{y_1 \leqslant Y_1 \leqslant y_1 + dy_1, \ldots, y_n \leqslant Y_n \leqslant y_n + dy\}$$

$$= P\{y_1 \leqslant Y_1 \leqslant y_1 + dy_1\}P\{y_2 \leqslant Y_2 \leqslant y_2 + dy_2|Y_1 = y_1\} \times \cdots$$

$$\times P\{y_n \leqslant Y_n \leqslant y_n + dy_n|Y_1 = y_1, Y_2 = y_2, \ldots, Y_{n-1} = y_{n-1}\}$$

$$= f(y_1)f(y_2|y_1) \cdots f(y_n|y_1, y_2, \ldots, y_{n-1}) \, dy_1 \, dy_2 \ldots dy_n$$

$$= f(y_1, y_2, \ldots, y_n) \, dy_1 \, dy_2 \cdots dy_n$$

Example 11.2

Consider a random vector $\{Y\} = (Y_1, Y_2)$ with joint probability density

$$f(y_1, y_2) = \tfrac{2}{3}(y_1 + 2y_2), \quad 0 < y_1 < 1, \quad 0 < y_2 < 1$$

We have

$$f(y_1) = \int_0^1 \tfrac{2}{3}(y_1 + 2y_2) \, dy_2 = \tfrac{2}{3}(1 + y_1)$$

$$f(y_2|y_1) = \frac{f(y_1, y_2)}{f(y_1)} = \frac{y_1 + 2y_2}{1 + y_1}$$

Corresponding probability distributions are

$$F(y_1) = \int_0^{y_1} f(y_1) \, dy_1 = \tfrac{1}{3}y_1(2 + y_1)$$

$$F(y_2|y_1) = \int_0^{y_2} f(y_2|y_1) \, dy_2 = \frac{y_2(y_1 + y_2)}{1 + y_1}$$

The set (11.6) yields equations for the simulation of Y_1 and Y_2:

$$\tfrac{1}{3}Y_1(2 + Y_1) = X_1$$

$$\frac{Y_2(Y_1 + Y_2)}{1 + Y_1} = X_2$$

where X_1 and X_2 are independent random numbers.

11.5 METHOD OF LINEAR TRANSFORMATION

The preceding section presented a general tool for simulation of a random vector possessing a given joint probability density function. Let us now attempt a more limited objective, a random vector $\{Y\}$ with given mean vector

$\{m\}$ and variance-covariance matrix $[V]$, which, in the case of a normal distribution, are the sole characteristics of the random vector. Furthermore, $\{E(Y)\}$ is assumed to coincide with the zero vector; this entails no loss of generality, because when $\{Y\}$ is simulated as below, the vector $\{Y + m\}$ has the mean $\{m\}$ and the variance-covariance matrix $[V] = [v_{jk}]_{N \times N}$.

As is known, any linear transformation of the n-dimensional vector is the equivalent of its premultiplication by some matrix $[C]$ (see Sec. 6.11):

$$\{Y\} = [C]\{Z\} \tag{11.17}$$

where

$$\{Y\}^T = [Y_1 \quad Y_2 \quad \cdots \quad Y_N]$$

$$\{Z\}^T = [Z_1 \quad Z_2 \quad \cdots \quad Z_N]$$

$$[C] = \begin{bmatrix} c_{11} & c_{12} & \cdots & c_{1N} \\ c_{21} & c_{22} & \cdots & c_{2N} \\ \vdots & & & \\ c_{N1} & c_{N2} & \cdots & c_{NN} \end{bmatrix}$$

$[C]$ is chosen to be lower triangular, so that

$$Y_1 = c_{11}Z_1$$
$$Y_2 = c_{21}Z_1 + c_{22}Z_2$$
$$\vdots$$
$$Y_N = c_{N1}Z_1 + c_{N2}Z_2 + \cdots + c_{NN}Z_N$$

and its elements are

$$E(Y_jY_k) = v_{jk}, \quad E(Z_jZ_k) = \delta_{jk} \tag{11.18}$$

where δ_{jk} is the Kronecker delta and $Z_1, Z_2, \ldots, Z_n$ are independent standard normal variables (i.e., with zero mean and unity variance). Then, from the equalities

$$E(Y_1^2) = c_{11}^2 = v_{11}$$

$$E(Y_1Y_2) = c_{11}c_{21} = v_{12}$$

$$E(Y_2^2) = c_{21}^2 + c_{22}^2 = v_{22}$$

we obtain

$$
c_{11} = v_{11}^{1/2} \qquad c_{21} = \frac{v_{12}}{v_{11}^{1/2}}, \qquad c_{22} = \left(v_{22} - \frac{v_{12}^2}{v_{11}} \right)^{1/2}
$$

Analogously we can find

$$
c_{31} = \frac{v_{13}}{v_{11}^{1/2}} \qquad c_{32} = \frac{v_{23} - v_{12}v_{13}/v_{11}}{\left(v_{22} - v_{12}^2/v_{11} \right)^{1/2}}
$$

$$
c_{33} = \left(v_{33} - \frac{v_{13}^2}{v_{11}} - \frac{(v_{23} - v_{12}v_{13}/v_{11})^2}{v_{22} - v_{12}^2/v_{11}} \right)^{1/2}
$$

the general expression being

$$
c_{j1} = \frac{v_{j1}}{\sqrt{v_{11}}}, \quad 1 \leqslant j \leqslant N
$$

$$
c_{jj} = \left(v_{jj} - \sum_{i=1}^{j-1} c_{ji}^2 \right)^{1/2}, \quad 1 \leqslant j \leqslant N
$$

$$
c_{jk} = \frac{v_{jk} - \sum_{i=1}^{k-1} c_{ji}c_{ki}}{c_{kk}}, \quad 1 < k < j \leqslant N
$$

$$
c_{jk} = 0, \quad j < k \leqslant N \tag{11.19}
$$

Since, in accordance with Eq. (11.18),

$$
E\left(\{Y\}\{Y\}^T \right) = [V], \quad E\left(\{Z\}\{Z\}^T \right) = [I]
$$

where $[I]$ is an identity matrix, we find from Eq. (11.17) that

$$
[C][C]^T = [V] \tag{11.20}
$$

This expression represents decomposition of the positive-definite symmetric matrix $[V]$ into a product of the lower triangular matrix $[C]$ and the upper triangular matrix $[C]^T$ and is known as the *Cholesky decomposition*. Vector $\{Z\}$, as is seen from Eq. (11.18), has unity variances. Consequently, to simulate a random vector $\{Y\}$ with mean $\{m\}$, we use the expression

$$
\{Y\} = [C]\{Z\} + \{m\} \tag{11.21}
$$

where the components of $\{Z\}$ are the independent standard normal variables.

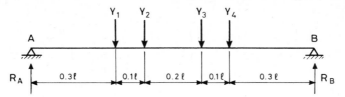

Fig. 11.3.   Beam under four concentrated loads.

Example 11.3

We seek the reliability of the beam shown in Fig. 11.3, with span $l = 1000$ mm,
section modulus $S = 1800$ mm^3, and allowable stress $\sigma_{\text{allow}} = 250$ N/mm^2,
subjected to four concentrated loads. The mean vector of the forces and the
variance-covariance matrix are, respectively,

$$\{m\}^T = [800 \quad 400 \quad 400 \quad 800] \text{ N}$$

$$[V] = 100^2 \begin{bmatrix} 1 & \frac{1}{2} & \frac{1}{3} & \frac{1}{4} \\ \frac{1}{2} & \frac{1}{3} & \frac{1}{4} & \frac{1}{5} \\ \frac{1}{3} & \frac{1}{4} & \frac{1}{5} & \frac{1}{6} \\ \frac{1}{4} & \frac{1}{5} & \frac{1}{6} & \frac{1}{7} \end{bmatrix} \text{N}^2 \tag{11.22}$$

The reaction at the left-hand support is

$$R_A = 0.1(7Y_1 + 6Y_2 + 4Y_3 + 3Y_4)$$

The expression for the bending moment is

$$M(x) = -0.1(7Y_1 + 6Y_2 + 4Y_3 + 3Y_4)x + Y_1\langle x - 0.3l\rangle$$

$$+ Y_2\langle x - 0.4l\rangle + Y_3\langle x - 0.6l\rangle + Y_4\langle x - 0.7l\rangle$$

The bending moments in the cross sections with the concentrated forces are

$$M_1 = M(0.3l) = -0.01l(21Y_1 + 18Y_2 + 12Y_3 + 9Y_4)$$

$$M_2 = M(0.4l) = -0.01l(18Y_1 + 24Y_2 + 16Y_3 + 12Y_4)$$

$$M_3 = M(0.6l) = -0.01l(12Y_1 + 16Y_2 + 24Y_3 + 18Y_4)$$

$$M_4 = M(0.7l) = -0.01l(9Y_1 + 12Y_2 + 18Y_3 + 12Y_4) \tag{11.23}$$

The reliability is, then, the probability of the normal stresses $\sigma_j = M_j/S$ in
these cross sections not exceeding in absolute value the allowable stress σ_{allow},

that is,

$$R = P\left\{ \bigcap_{j=1}^{4} \left(|\sigma_j| \leq \sigma_{\text{allow}} \right) \right\}$$
(11.24)

The bending moments in Eq. (11.23) have to be simulated through the set Y_1, Y_2, Y_3, Y_4.

The Cholesky decomposition, in accordance with Eqs. (11.19), yields the lower triangular matrix $[C]$:

$$[C] = 100 \begin{bmatrix} 1 & 0 & 0 & 0 \\ \frac{1}{2} & \frac{\sqrt{3}}{6} & 0 & 0 \\ \frac{1}{3} & \frac{\sqrt{3}}{6} & \frac{\sqrt{5}}{30} & 0 \\ \frac{1}{4} & \frac{3\sqrt{3}}{20} & \frac{\sqrt{5}}{20} & \frac{\sqrt{7}}{140} \end{bmatrix}$$

and the vector $\{Y\}$ is accordingly simulated as follows:

$$\begin{Bmatrix} Y_1 \\ Y_2 \\ Y_3 \\ Y_4 \end{Bmatrix} = 100 \begin{bmatrix} 1 & 0 & 0 & 0 \\ \frac{1}{2} & \frac{\sqrt{3}}{6} & 0 & 0 \\ \frac{1}{3} & \frac{\sqrt{3}}{6} & \frac{\sqrt{5}}{30} & 0 \\ \frac{1}{4} & \frac{3\sqrt{3}}{20} & \frac{\sqrt{5}}{20} & \frac{\sqrt{7}}{140} \end{bmatrix} \begin{Bmatrix} Z_1 \\ Z_2 \\ Z_3 \\ Z_4 \end{Bmatrix} + \begin{Bmatrix} 800 \\ 400 \\ 400 \\ 800 \end{Bmatrix}$$
(11.25)

The procedure is as follows. First, we create a large ensemble of realizations of the independent random variables Z_1, Z_2, Z_3, and Z_4, and Eqs. (11.25) yield those of the random forces. For each of the latter we determine the bending moments M_1, M_2, M_3, and M_4 and check whether the normal stresses do not exceed in absolute value the allowable stress. The fraction of beams for which Eq. (11.24) is satisfied provides the estimate of the reliability.

For this example, an ensemble of 100,000 beams was "created." In 10 series of 10,000, nonfailures totaled 8881, 8888, 8880, 8897, 8868, 8816, 8823, 8858, 8848, 8876, so that the reliability estimate in series of 100,000 beams is $R^* = 0.88635$.

11.6 SIMULATION OF RANDOM FUNCTIONS

Consider a general nonlinear two-point boundary-value problem as follows. Let $F(x)$ be the function (with given mean and autocovariance functions) representing randomness of a structure, and let the corresponding deterministic linear boundary-value problem have eigenfunction $\psi_j(x)$, $j = 1, 2, \ldots$; $x \in [0, l]$; or, in less obligatory terms, let the eigenfunctions of a familiar deterministic problem be known and likewise denoted $\psi_j(x)$. The latter represent a complete set of orthogonal functions on $[0, l]$.

The problem consists in simulating the random function $F(x)$, which can be expanded in a series in terms of the eigenfunctions $\psi_j(x)$:

$$F(x) = \sum_{j=1}^{\infty} A_j \psi_j(x) \tag{11.26}$$

where A_j is a random variable for every fixed j. In the sequel, the series (11.26) is truncated to some N (number of eigenfunctions taken into consideration) in accordance with the degree of accuracy required for the probabilistic characteristics of the sought output random function. Equation (11.26) is replaced by

$$F(x) = \sum_{j=1}^{N} A_j \psi_j(x) \tag{11.27}$$

In order to find the expression for A_j we multiply expression (11.27) by $\psi_k(x)$ and integrate over the span l, to yield

$$\int_0^l F(x) \psi_k(x) \, dx = \sum_{j=1}^{N} A_j \int_0^l \psi_j(x) \psi_k(x) \, dx$$

Due to the orthogonality property,

$$\int_0^l \psi_j(x) \psi_k(x) \, dx = \delta_{jk} \nu_k^2, \quad \nu_k^2 = \int_0^l \psi_k^2(x) \, dx$$

where δ_{jk} is the Kronecker delta, we find

$$A_k = \nu_k^{-2} \int_0^l F(x) \psi_k(x) \, dx$$

The mean value of A_j is found by replacing $k \to j$ and taking the mathematical expectation:

$$E(A_j) = \nu_j^{-2} \int_0^l E[F(x)] \psi_j(x) \, dx \tag{11.28}$$

The autocovariance function of $F(x)$,

$$C(x_1, x_2) = E\{[F(x_1) - EF(x_1)][F(x_2) - EF(x_2)]\}$$

is given by

$$C(x_1, x_2) = \sum_{j=1}^{N} \sum_{k=1}^{N} v_{jk} \psi_j(x_1) \psi_k(x_2) \tag{11.29}$$

where

$$v_{jk} = E\{[A_j - E(A_j)][A_k - E(A_k)]\} \tag{11.30}$$

Multiplying both sides of Eq. (11.29) by $\psi_\alpha(x_1)\psi_\beta(x_2)$, $\alpha, \beta = 1, 2, \ldots, N$, integrating, and replacing $\alpha \to j$, $\beta \to k$ in the final result, we obtain

$$v_{jk} = v_j^{-2} v_k^{-2} \int_0^l \int_0^l C(x_1, x_2)\psi_j(x_1)\psi_k(x_2) \, dx_1 \, dx_2 \tag{11.31}$$

The problem now reduces to simulation of the random vector $\{A\}^T = [a_1 \quad a_2 \quad \cdots \quad a_N]$ with mean as per (11.28) and with the variance-covariance matrix $[V] = [v_{jk}]_{N \times N}$ defined by Eq. (11.31). This is the problem we considered in the preceding section. We decompose matrix $[V]$ into the product

$$[V] = [C][C]^T \tag{11.32}$$

as in Eq. (11.20), and vector $\{A\}$ is simulated as

$$\{A\} = [C]\{Z\} + \{E(A)\} \tag{11.33}$$

just as in Eq. (11.21).

Note that expression (11.31) for the elements v_{jk} of the variance-covariance matrix closely resembles formula (10.74) for the cross acceptances in the random vibration problem. Thus, if $F(x)$ is a segment of homogeneous random function with autocovariance function $C(x_1, x_2) = C(\xi)$, $\xi = x_2 - x_1$, we can use a simple representation analogous to (10.80):

$$v_{jk} = v_j^{-2} v_k^{-2} \int_0^l C(\xi) M_{jk}(\xi) \, d\xi \tag{11.34}$$

where

$$M_{jk} = \int_{\xi/2}^{l-\xi/2} \left[\psi_j\left(\eta - \frac{\xi}{2}\right)\psi_k\left(\eta + \frac{\xi}{2}\right) + \psi_j\left(\eta + \frac{\xi}{2}\right)\psi_k\left(\eta - \frac{\xi}{2}\right) \right] d\eta \tag{11.35}$$

Example 11.4

Let the boundary conditions associated with a two-point boundary value problem be

$$w = 0 \qquad \frac{d^2 w}{dx^2} = 0, \quad x = 0, \quad x = l$$

Then

$$\psi_j(x) = \sin \frac{j\pi x}{l}$$

For this case, in analogy with Eqs. (10.81) and (10.82), we have

$$v_{jj} = 4\int_0^1 \left[(1 - \zeta)\cos j\pi\zeta + \frac{1}{j\pi} \sin j\pi\zeta\right] C(\zeta l)\, d\zeta, \qquad\qquad j = k$$

$$v_{jk} = \frac{4\left[1 - (-1)^{j+k+1}\right]}{\pi(k^2 - j^2)} \int_0^1 (k\sin j\pi\zeta - j\sin k\pi\zeta)C(\zeta l)\, d\zeta, \quad j \neq k$$

$$(11.36)$$

Let the autocovariance function be

$$C(\zeta l) = \frac{1}{2\alpha} \exp\left(-\frac{|\zeta l|}{\alpha}\right)$$

where α is some positive constant. The elements of the variance-covariance matrix are

$$v_{jk} = \frac{2\delta_{jk}}{1 + J^2} + \frac{2JK\alpha}{(1 + J^2)(1 + K^2)}\left\{1 + (-1)^{j+k} - \left[(-1)^j + (-1)^k\right]e^{-1/\alpha}\right\}$$

where $J = j\pi\alpha$, and $K = k\pi\alpha$. The elements of the lower triangular matrix c_{jk} are determined as above, by triangular decomposition of $[V]$. It is worth noting that for $\alpha \to 0$, $C(\zeta l)$ tends to Dirac's delta function $\frac{1}{2}\delta(\zeta)$. The variance-covariance matrix becomes diagonal $[V] = [\ 2\ ]$, and the triangular decomposition is readily effected ($c_{jk} = \sqrt{2}\,\delta_{jk}$).

Example 11.5

Let the boundary conditions associated with some two-point boundary value problem be

$$w = 0 \qquad \frac{dw}{dx} = 0, \quad \text{for } x = 0, \quad x = l$$

For the sake of determinacy, consider a clamped-clamped beam with some random property or random external force to be simulated. The vibrational mode shapes of a homogeneous beam given in Eq. (10.40) may be taken as the eigenfunctions $\psi_j(x)$ in expansion (11.27).

If $F(x)$ is a segment of a homogeneous random function with the autocovariance function $C(\zeta l)$, it can be shown that under the boundary conditions in question the symmetric and antisymmetric modes do not correlate.

We next consider the numerical solution for a space-random function with zero mean and autocovariance function:

$$C(\zeta l) = e^{-A|\zeta|}\cos B\zeta$$

with $A = 4$ and $B = 8$. The elements of the variance-covariance matrix, obtained through approximate evaluation of Eq. (11.31) by numerical quadrature, are

$v_{11} =$ 0.130571	$v_{22} =$ 0.198181	$v_{31} = -0.029745$
$v_{33} =$ 0.185467	$v_{42} =$ 0.011332	$v_{44} =$ 0.125444
$v_{51} = -0.014961$	$v_{53} =$ 0.046897	$v_{55} =$ 0.043972
$v_{62} =$ 0.026416	$v_{64} = -0.027749$	$v_{66} =$ 0.052857
$v_{71} =$ 0.028091	$v_{73} = -0.113921$	$v_{75} =$ 0.061534
$v_{77} =$ 0.068084	$v_{82} = -0.075171$	$v_{84} =$ 0.053030
$v_{86} =$ 0.021419	$v_{88} =$ 0.016944	

while the nonzero elements of $[C]$ are obtained from Eq. (11.19):

$c_{11} =$ 0.361346	$c_{22} =$ 0.445175	$c_{31} = -0.082318$
$c_{33} =$ 0.422719	$c_{42} = -0.025456$	$c_{44} =$ 0.353265
$c_{51} = -0.041404$	$c_{53} =$ 0.102876	$c_{55} =$ 0.177971
$c_{62} =$ 0.059338	$c_{64} = -0.074274$	$c_{66} =$ 0.209332
$c_{71} =$ 0.077740	$c_{73} = -0.254357$	$c_{75} =$ 0.510870
$c_{77} =$ 0.068084	$c_{82} = -0.168858$	$c_{84} =$ 0.137945
$c_{86} =$ 0.199134	$c_{88} =$ 0.016944	

Application of the simulation procedure for random functions is illustrated in next section. Exact solution of the problem is unfeasible and therefore the Monte Carlo method is called for.

11.7 BUCKLING OF A BAR ON A NONLINEAR FOUNDATION*

11.71 Introductory Remarks. It is now generally recognized that initial imperfections play a dominant role in reducing the buckling load of certain structures. As is well known, a thin shell is often highly imperfection-sensitive

*This section is an extension of Elishakoff (1979b), from which Figs. 11.5–11.9 are also reproduced.

in this context (see, for example, the survey by Hutchinson and Koiter and the article by Budiansky). However, despite the accepted theoretical explanation of the buckling behavior of different structures, use of the concept of imperfection sensitivity in engineering practice is still in the ad hoc stage and engineers prefer to rely on the *knockdown factor* (see NASA SP-8007), chosen so that its product by the classical buckling load yields a lower bound for all available experimental data for the configuration in question. This apparent reluctance to take advantage of theoretical findings stems from the fact that most imperfection studies are conditional on detailed advance knowledge of the geometric imperfections of the particular structure, which is rarely available. At best, the imperfections can be measured experimentally and incorporated in the theoretical analysis to predict the buckling loads. This approach, however, while justified for single prototypal structures, is impracticable as a general means of behavior prediction: Information on the type and magnitude of imperfections of a particular structure would be too specific and thus invalid for other realizations of the same structure, even those obtained by the same manufacturing process.

In the light of these considerations, and bearing in mind the scatter of the experimental results, it is clear that practicability of the imperfection-sensitivity theories is conditional on their being combined with probabilistic analysis of the imperfections and buckling loads. The notion of randomness of initial imperfections was given considerable attention in the literature, and a bibliography can be found in Amazigo's paper. A probabilistic approach was first suggested in a study of imperfection-sensitive structures by Bolotin, who postulated, in brief, that the buckling load α^* of a structure can be expressed as a function of a finite number of random variables $\overline{\xi}_i$ representing the initial imperfections.

$$\alpha^* = \varphi\left(\overline{\xi}_1, \overline{\xi}_2, \ldots, \overline{\xi}_N\right) \qquad (11.37)$$

Given a particular function φ, the probability density of α^* can be derived from the joint probability density of $\overline{\xi}_i$.

Bolotin applied this method to a cylindrical panel under a uniform compressive load along its curved edges, with the initial imperfections represented by a single normally distributed amplitude parameter. A single-term Galerkin approximation yielded an equation of the type (11.37).

A different approach was used by Boyce in a study of buckling of a finite column with an initial imperfection. He assumed that the initial imperfection can be described by a section of a homogeneous random function with a known autocovariance function

$$C_{\overline{u}}(\eta_1, \eta_2) = E\{[\overline{u}(\eta_1) - E\overline{u}(\eta_1)][\overline{u}(\eta_2) - E\overline{u}(\eta_2)]\} \qquad (11.38)$$

where η_1 and η_2 are the nondimensional coordinates of the column centerline and $\bar{u}(\eta)$ is the initial imperfection function. Two representations were taken for the autocovariance function, namely,

$$C_{\bar{u}}^{(1)}(\eta_1, \eta_2) = A(\eta_1 - \eta_2)^{-1}\sin B(\eta_1 - \eta_2) \tag{11.39}$$

and

$$C_{\bar{u}}^{(2)}(\eta_1, \eta_2) = A \exp\{-B|\eta_1 - \eta_2|\}\cos C(\eta_1 - \eta_2) \tag{11.40}$$

where A, B, and C are some positive constants. Jacquot also used this approach for the same problem, with the following approximation of the autocovariance function of the initial imperfections:

$$C_{\bar{u}}^{(3)}(\eta_1, \eta_2) = A \sin \pi\eta_1 \sin \pi\eta_2 \tag{11.41}$$

We will now show how the Monte Carlo method can be utilized to bridge Bolotin's and Boyce's approaches, using the case of a bar on a nonlinear elastic foundation (which resembles a cylindrical shell in its imperfection-sensitive behavior) as illustration, and with the initial imperfections represented multi-parametrically. A large ensemble of imperfection vectors $(\bar{\xi}_1, \bar{\xi}_2, \ldots, \bar{\xi}_N)$ is created as per Sec. 11.6 from the set of random numbers, and for each "trial" (realization) of these vectors the corresponding buckling load α^* is calculated by a special numerical procedure. We will study the reliability of the structure, the reliability function being, in fact, the ultimate objective in the probabilistic approach to imperfect structures.

11.7.2 Differential Equation of a Finite Bar.

The differential equation for the deflection of an imperfect bar on a nonlinear "softening" foundation (Fig. 11.4) reads:

$$EI\frac{d^4w}{dx^4} + P\frac{d^2w}{dx^2} + k_1w - k_3w^3 = -P\frac{d^2\bar{w}}{dx^2} \tag{11.42}$$

where $\bar{w}(x)$ is the initial imperfection function, $w(x)$ the additional deflection

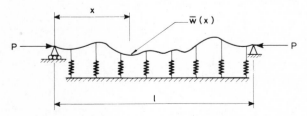

Fig. 11.4.   Column on nonlinear elastic foundation.

due to the axial load P, k_1 and k_3 the linear and nonlinear "spring" coefficients of the foundation, respectively, x the axial coordinate, E Young's modulus, and I the section moment of inertia. The bar is simply supported, so that the boundary conditions are

$$w = \frac{d^2w}{dx^2} = 0, \quad x = 0, \quad x = l \tag{11.43}$$

where l is the length of the bar. If $\bar{w} \equiv 0$ and $k_3 \equiv 0$, we obtain an associated linear equation for a perfect bar on a linear foundation. The following expression satisfies Eqs. (11.42) and (11.43) for this particular case:

$$w = w_0 \sin \frac{n\pi x}{l} \tag{11.44}$$

where n is the number of half-waves. Substituting (11.44) in the equation thus obtained from (11.42), we have

$$\frac{P}{P_E} = n^2 + \frac{k_1 l^4}{EIn^2\pi^4} \equiv \frac{P_n}{P_E}, \quad P_E = \frac{\pi^2 EI}{l^2} \tag{11.45}$$

where P_E is the buckling load of the bar without a foundation. The number $n = n_*$, which minimizes P_n/P_E, determines the number of half-waves during buckling of the linear structure. Consequently,

$$\frac{P_{min}}{P_E} = n_*^2 + \frac{\kappa_1}{n_*^2 \pi^4} \equiv \frac{P_c}{P_E}, \quad \kappa_1 = \frac{k_1 l^4}{EI} \tag{11.46}$$

where P_c is the (classical) buckling load of a bar on a linear elastic foundation. In contrast to the bar without a foundation, n_* does not necessarily equal unity (see also Timoshenko and Gere).

Assume that the nondimensional linear foundation coefficient κ_1 is such that the number of half-waves at buckling equals n_*. Increasing κ_1, we finally arrive at a situation where P_{min}/P_E in Eq. (11.46) is smaller for $n_* + 1$ than for n_*. The limiting value of κ_1, at which transition from n_* to $n_* + 1$ occurs, is found from the condition that the corresponding expression (11.46) should yield the same value for P_{min}/P_E for n_* and for $n_* + 1$, namely,

$$n_*^2 + \frac{\kappa_1}{n_*^2 \pi^4} = (n_* + 1)^2 + \frac{\kappa_1}{(n_* + 1)^2 \pi^4} \tag{11.47}$$

and n_* is determined from

$$n_*^2(n_* + 1)^2 = \frac{\kappa_1}{\pi^4} \tag{11.48}$$

As κ_1 increases, so does the number of half-waves at buckling, and for $n_* \gg 1$, this equation reduces to

$$n_*^4 \simeq \frac{\kappa_1}{\pi^4} \tag{11.49}$$

Substituting the above in Eq. (11.46), we obtain

$$\frac{P_c}{P_E} \simeq \frac{2\kappa_1^{1/2}}{\pi^2} \tag{11.50}$$

The same result is obtained formally by treating n in Eq. (11.46) as a continuous variable (for high values of κ_1), in which case the requirement

$$\frac{d}{dn}\left(\frac{P_n}{P_E}\right) = 0 \tag{11.51}$$

yields the value (11.50), which is in fact a global minimum for P_c/P_E and coincides with the buckling load of an infinite beam; the latter in dimensional terms equals $2(k_1 EI)^{1/2}$.

Introducing, for convenience, the dimensionless quantities

$$u = \frac{w}{\Delta} \qquad \bar{u} = \frac{\bar{w}}{\Delta} \qquad \eta = \frac{x}{l} \qquad \alpha = \frac{P}{P_c}$$

$$\gamma(\kappa_1) = n_*^2 \pi^2 + \frac{\kappa_1}{n_*^2 \pi^2} \qquad \kappa_3 = \frac{k_3 l^4 \Delta^2}{EI} \tag{11.52}$$

in Eqs. (11.42) and (11.43), we obtain

$$\frac{d^4 u}{d\eta^4} + \alpha\gamma(\kappa_1)\frac{d^2 u}{d\eta^2} + \kappa_1 u - \kappa_3 u^3 = -\alpha\gamma(\kappa_1)\frac{d^2\bar{u}}{d\eta^2} \tag{11.53}$$

$$u = \frac{d^2 u}{d\eta^2} = 0, \quad \eta = 0, \quad \eta = 1 \tag{11.54}$$

The problem is formulated as follows: *Given the probabilistic characteristics of $\bar{u}(\eta)$, find those of the buckling load α^* (for specified values of κ_1 and κ_3).*

For a given realization of the imperfection function $\bar{u}(\eta)$, the buckling load α^* is defined from the requirement

$$\frac{d\alpha}{dF} = 0, \quad \alpha = \alpha^* \tag{11.55}$$

where $F = F(u, \bar{u})$ is some functional of the nondimensional displacements increasing with u. The one we use is the end shortening of the bar (the distance of the bar ends move together under load);

$$d = \int_0^l \left[\frac{1}{2}\left(\frac{dw}{dx} \right)^2 + \left(\frac{dw}{dx} \right)\left(\frac{d\bar{w}}{dx} \right) \right] dx \tag{11.56}$$

In terms of the nondimensional quantities, the equation reads

$$F(u, \bar{u}) = \frac{dl}{\Delta^2} = \int_0^1 \left[\frac{1}{2}\left(\frac{du}{d\eta} \right)^2 + \left(\frac{du}{d\eta} \right)\left(\frac{d\bar{u}}{d\eta} \right) \right] d\eta \tag{11.57}$$

11.7.3 Approximate Solution of a Nonlinear Equation. A Galerkin-type solution (for a description of this method see Appendix E) is obtained for Eqs. (11.53)–(11.55) by expanding u and $\bar{u}$ in series in terms of the modes of stability loss of the associated linear structure:

$$\bar{u} = \sum_{k=1}^{\infty} \bar{\xi}_k \sin(k\pi\eta) \tag{11.58}$$

$$u = \sum_{k=1}^{\infty} \xi_k \sin(k\pi\eta) \tag{11.59}$$

Substituting Eqs. (11.58) and (11.59) into Eq. (11.53), we apply the approximate technique by Galerkin. We multiply both sides of the resulting equation by $\sin(m\pi\eta)$, $m = 1, 2, \ldots, N$ (N being the number of retained terms) and integrate over the span, to obtain the following set of coupled nonlinear algebraic equations for ξ_m:

$$\alpha_m\xi_m - \alpha(\xi_m + \bar{\xi}_m) - s\left(\frac{n_*}{m} \right)^2 \sum_{p=1}^{N} \sum_{q=1}^{N} \sum_{r=1}^{N} \xi_p\xi_q\xi_r A_{pqrm} = 0,$$

$$m = 1, 2, \ldots, N \tag{11.60}$$

where

$$\alpha_m = \frac{(m\pi)^2 + \kappa_1(m\pi)^{-2}}{(n_*\pi)^2 + \kappa_1(n_*\pi)^{-2}}, \quad s = \frac{2\kappa_3}{n_*^4\pi^4 + \kappa_1} \tag{11.61}$$

$$A_{pqrm} = \int_0^1 \sin(p\pi\eta)\sin(q\pi\eta)\sin(r\pi\eta)\sin(m\pi\eta)\, d\eta$$

$$= \frac{1}{8}\left[\delta_{p+q, r+m} - \delta_{|p-q|, r+m} - \delta_{p+q, |r-m|} + \delta_{|p-q|, |r-m|} + \delta_{p, q}\delta_{r, m} \right] \tag{11.62}$$

and $\delta_{i,j}$ is the Kronecker delta. Equation (11.60) can be rewritten as

$$\alpha_m \xi_m - \alpha(\xi_m + \bar{\xi}_m) + \frac{s}{8}\left(\frac{n_*}{m}\right)^2 I_m = 0, \quad m = 1, 2, \ldots, N \quad (11.63)$$

with

$$I_m = 8 \sum_{p=1}^{N} \sum_{q=1}^{N} \sum_{r=1}^{N} \xi_p \xi_q \xi_r A_{pqrm} \quad (11.64)$$

If $\kappa_1 \gg 4\pi^4$, then according to Eq. (11.49), $\kappa_1 \simeq n_*^4 \pi^4$, and α_m and s become

$$\alpha_m = \frac{1}{2}\left[\left(\frac{m}{n_*}\right)^2 + \left(\frac{n_*}{m}\right)^2\right] \quad s = \frac{\kappa_3}{\kappa_1} = \frac{k_3 \Delta^2}{k_1} \quad (11.65)$$

$F(u, \bar{u})$ takes the form

$$F(u, \bar{u}) = \frac{1}{4}\sum_{k=1}^{N} \xi_k^2 k^2 \pi^2 + \frac{1}{2}\sum_{k=1}^{N} \xi_k \bar{\xi}_k k^2 \pi^2 \quad (11.66)$$

Since a closed solution to the set of Eqs. (11.63) is unfeasible, they have to be solved numerically. Before doing so, let us consider the analytical solution obtainable in the case of a single equation, such as the one to which the set (11.60)–(11.66) reduces when only the mth term is retained in series (11.58) and (11.59):

$$\alpha = \frac{\xi_m}{\xi_m + \bar{\xi}_m}\left[\xi_m - \frac{3}{8}\left(\frac{n_*}{m}\right)^2 s\xi_m^2\right]$$

$$F = \frac{1}{4}m^2\pi^2\xi_m(\xi_m + 2\bar{\xi}_m) \quad (11.67)$$

with α_m and s as per Eq. (11.61). Differentiating Eq. (11.67) with respect to F and setting

$$\frac{d\alpha}{dF} = 0, \quad \alpha = \alpha^* \quad (11.68)$$

we obtain, after the appropriate algebraic manipulations, the relation between the buckling load α^* and the initial deflection amplitude $\bar{\xi}_m$ (see Eq. 2.18 in Fraser's work):

$$(\alpha_m - \alpha^*)^3 = \frac{81}{32}\left(\frac{n_*}{m}\right)^2 s\bar{\xi}_m^2(\alpha^*)^2 \quad (11.69)$$

From this equation the buckling load α^* is obtainable, given the amplitude $\bar{\xi}_m$

of the initial imperfections. Significantly, Eq. (11.69) is in full agreement with a single-degree-of-freedom model considered in Sec. 5.5, with

$$a = 0 \quad \text{and} \quad b = -\tfrac{3}{8}\left(\frac{n_*}{m}\right)^2 s$$

11.7.4 Multiple-Term Solution.   Here, the method of transformation of an appropriately rewritten set of equations

$$B_m \equiv (\alpha_m - \alpha)\xi_m - \alpha\bar{\xi}_m - \frac{s}{8}\left(\frac{n_*}{m}\right)^2 I_m = 0, \quad m = 1, 2, \ldots, N \qquad (11.70)$$

$$F = \tfrac{1}{4}\sum_{k=1}^{N} k^2\pi^2\xi_k(\xi_k + 2\bar{\xi}_k) \tag{11.71}$$

$$I_m = 8\sum_{p=1}^{N}\sum_{q=1}^{N}\sum_{r=1}^{N} \xi_p\xi_q\xi_r A_{pqrm} \tag{11.72}$$

is applied to the initial-value problem as presented by Qiria and Davidenko. F is treated now as the independent variable, and α and ξ_m as its functions. Differentiating (11.70) with respect to F, we obtain

$$\frac{\partial B_m}{\partial \alpha}\frac{d\alpha}{dF} + \sum_{p=1}^{N}\frac{\partial B_m}{\partial \xi_p}\frac{d\xi_p}{dF} = 0 \tag{11.73}$$

We also refer to the identity $dF/dF - 1 = 0$, in our case in the form

$$\sum_{p=1}^{N}\frac{\partial F}{\partial \xi_p}\frac{d\xi_p}{dF} - 1 = 0 \tag{11.74}$$

Equations (11.73) and (11.74) may be rewritten as

$$-(\xi_m + \bar{\xi}_m)\frac{d\alpha}{dF} + \sum_{p=1}^{N}\left[(\alpha_m - \alpha)\delta_{p,m} - 3\sum_{i=1}^{N}\sum_{j=1}^{N} C_{ijpm}\xi_i\xi_j\right]\frac{d\xi_p}{dF} = 0 \tag{11.75}$$

$$\tfrac{1}{2}\sum_{p=1}^{N}(\xi_p + \bar{\xi}_p)p^2\pi^2\left(\frac{d\xi_p}{dF}\right) - 1 = 0 \tag{11.76}$$

where

$$C_{ijpm} = s\left(\frac{n_*}{m}\right)^2 A_{ijpm} \tag{11.77}$$

The result is a set of $N+1$ differential equations in $N+1$ variables—$\xi_1, \xi_2, \ldots, \xi_N$, and α—subject to initial conditions representing the unstressed state of the bar, namely,

$$\alpha = 0 \qquad \xi_1 = \xi_2 = \cdots = \xi_N = 0, \quad F = 0 \qquad (11.78)$$

The set (11.75), (11.76) in matrix form reads

$$\left[D(\xi, \bar{\xi}, \alpha) \right]\{\xi'\} = \{e\} \qquad [D] = \left[d_{mp} \right]_{(N+1)\times(N+1)} \qquad (11.79)$$

$$\{\xi\}^T = [\xi_i \; \xi_2 \; \cdots \; \xi_N \; \alpha] \qquad \{\xi'\} = \left\{ \frac{d\xi}{dF} \right\} \qquad \{e\}^T = [0 \quad 0 \quad \cdots \quad 0 \quad 1]$$

$$(11.80)$$

$$d_{mp} = (\alpha_m - \alpha)\delta_{m,p} - 3 \sum_{i=1}^{N} \sum_{j=1}^{N} C_{ijpm}\xi_i\xi_j, \quad m, p = 1, 2, \ldots, N \quad (11.81)$$

$$d_{N+1,p} = \tfrac{1}{2}\left(\xi_p + \bar{\xi}_p\right)p^2\pi^2 \qquad d_{m,N+1} = -\left(\xi_m + \bar{\xi}_m\right) \qquad (11.82)$$

Solving (11.79) subject to (11.78), we have the vector $\{\xi\}$ for every α and $\{\bar{\xi}\}$, and hence also the α–F curve. According to Eq. (11.55), the maximum point on the branch that originates at zero nondimensional load represents the nondimensional buckling load α^*.

11.7.5 Generation of Random Initial Imperfection Vectors. In its truncated form, the series in Eq. (11.58) reads

$$\bar{u}(\eta) = \sum_{k=1}^{N} \bar{\xi}_k \sin(k\pi\eta) \qquad (11.83)$$

The initial imperfection mean function $\bar{U}(\eta) = E[\bar{u}(\eta)]$ is supposed to be given. The mean values $m_k = E[\bar{\xi}_k]$ are obtainable as

$$m_k = 2\int_0^1 \bar{U}(\eta)\sin(k\pi\eta)\,d\eta \qquad (11.84)$$

Analogously, the autocovariance function of the imperfections

$$C_{\bar{u}}(\eta_1, \eta_2) = E\{[\bar{u}(\eta_1) - \bar{U}(\eta_1)][\bar{u}(\eta_2) - \bar{U}(\eta_2)]\} \qquad (11.85)$$

$$C_{\bar{u}}(\eta_1, \eta_2) = \sum_{k=1}^{N} \sum_{r=1}^{N} E[(\bar{\xi}_k - m_k)(\bar{\xi}_r - m_r)]\sin(k\pi\eta)\sin(r\pi\eta)$$

$$(11.86)$$

and the elements v_{kr} of the variance-covariance matrix $[V]$ are determined by

$$v_{kr} = E\big[\bar{\xi}_k - m_k\big)(\bar{\xi}_r - m_r)\big]$$

$$= 4 \int_0^1 \int_0^1 C_{\bar{u}}(\eta_1, \eta_2) \sin(k\pi\eta_1)\sin(r\pi\eta_2)\, d\eta_1\, d\eta_2 \qquad (11.87)$$

The problem consists in "creating" a very large ensemble of normal imperfection vectors $\{\bar{\xi}\}$ with given mean vector $\{m\}$ and variance-covariance matrix $[V]$. According to Eq. (11.21),

$$\{\bar{\xi}\} = [C]\{Z\} + \{m\} \qquad (11.88)$$

where $\{Z\}$ is an independent normal vector. We consider the particular case $\{m\} = \{0\}$.

Note that with Eqs. (11.69) and (11.87) available, the single-term approximation ($m = n_*$, $\alpha_{n_*} = 1$) can be brought to conclusion to yield an analytical expression for the reliability function $R(\alpha')$:

$$R(\alpha') = \text{Prob}(\alpha' < \alpha^* \leqslant 1) \qquad (11.89)$$

That is, the probability of the random event of the buckling load falling within the interval $(\alpha', 1]$, in other words, of the structure not buckling prior to the specified nondimensional load α'. In this case $\bar{\xi}_{n_*}$ has a normal distribution $N(0, v_{n_*n_*})$. As can be seen from Eq. (11.69),

$$R(\alpha') = \text{Prob}\big(-\bar{\xi}'_{n_*} < \bar{\xi}_{n_*} < \bar{\xi}'_{n_*}\big) \qquad (11.90)$$

where

$$\bar{\xi}'_{n_*} = \tfrac{4}{9}\left(\frac{2}{s}\right)^{1/2} \frac{(1 - \alpha')^{3/2}}{\alpha'}$$

In conclusion, we have

$$R(\alpha') = 2\text{erf}\left(\frac{\bar{\xi}'_{n_*}}{2v_{n_*n_*}^{1/2}}\right) \qquad (11.91)$$

11.7.6 Numerical Analysis. The numerical examples were worked out by means of the autocovariance function of the initial imperfections as per Eq. (11.40), with the product AB as their variance and with their mean function chosen as zero. The elements v_{kr} of the variance-covariance matrix $[V]$ were approximated from Eq. (11.87) by numerical quadrature, and the elements of matrix $[C]$ obtained by Cholesky's decomposition procedure. Equations (11.79)

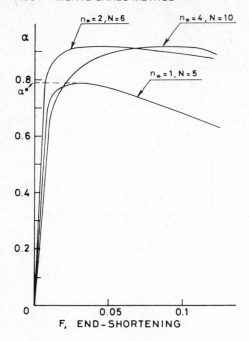

Fig. 11.5. Load versus end-shortening curves.

were integrated for numerous "trial" values of the random Fourier coefficients $\bar{\xi}_1, \bar{\xi}_2, \ldots, \bar{\xi}_N$, using Hamming's predictor-corrector method with varying step size. For this purpose the derivatives vector must be constructed in advance (prior to the integration proper); this was done by solving a set of linear algebraic equations derived from Eq. (11.79). The sign of the quantity $d\alpha/dF$, the $(N + 1)$th component of the unknown vector $\{\bar{\xi}'\}$, was checked throughout the process to yield the interval containing the buckling load, which was in turn identified through the maximal value of α on it. Continuous load/end-shortening curves are shown in Fig. 11.5 for different values of n_*; the "linear" stiffness coefficient was chosen as $\kappa_1 = n_*^4 \pi^4$ with $A = 0.01$ and $B = 1$. The realization of the imperfection vector used in the calculation for Fig. 11.5 was

$$\bar{\xi}_1 = -0.07875 \qquad \bar{\xi}_2 = 0.01304 \qquad \bar{\xi}_3 = -0.025 \qquad \bar{\xi}_4 = 0.006467$$
$$\bar{\xi}_5 = -0.01549 \qquad \bar{\xi}_6 = 0.004305 \qquad \bar{\xi}_7 = 0.01106 \qquad \bar{\xi}_8 = 0.003227$$
$$\bar{\xi}_9 = -0.0086 \qquad \bar{\xi}_{10} = 0.002569$$

$$(11.92)$$

For $n_* = 1$, five modes were retained in determining the load/end-shortening curve, although a single-term approximation would have sufficed for the buckling load. As κ_1 increases, so does the number of terms that must be retained to ensure convergence of the multiple-term solution: Six terms suffice for $n_* = 2$, eight for $n_* = 3$, and 10 for $n_* = 4$.

With the buckling loads known for a sufficient number of realizations, the empirical reliability function

$$R^*(\alpha') = \frac{1}{M}\mu_M(\alpha') \tag{11.93}$$

was constructed, where $\mu_M(\alpha')$ is the number of α^* values exceeding α', and M the ensemble size.

Results as per Eq. (11.91) are presented in Fig. 5.19(a) showing the solution of Eq. (11.69) and 5.19(b) the probability density of the initial imperfection amplitude. The shaded area represents the reliability at nondimensional load $\alpha' = 0.7$ with $v_1 = 0.03192$.

Figure 11.6 shows the influence of the coefficient B on $R^*(\alpha')$, which is seen to decrease as B increases (for constant $A = 0.01$). For example, at load level $\alpha' = 0.7$, the empirical reliability equals 0.69 (i.e., 69% of the bars buckle above that level) for $B = 1$, but only 0.36 for $B = 3$. This is understandable, as a larger B signifies a higher variance of the initial imperfections.

In Fig. 11.7 the reliability function is plotted for different values of A at $n_* = 3$. As expected, the reliability decreases as A increases. For example, at load level $\alpha' = 0.8$, it equals 0.76 for $A = 0.01$ and 0.44 for $A = 0.03$. Figure 11.8 shows the reliability function for $n_* = 4$, $(\kappa_1 = 4\pi^4, B = 1, A = 0.01)$. Curve *1* refers to a three-term, curve *2* to a seven-term approximation. For curve *1*, the third, fourth, and fifth terms were used in Eqs. (11.58) and (11.59):

$$\bar{u}(\eta) = \bar{\xi}_3\sin(3\pi\eta) + \bar{\xi}_4\sin(4\pi\eta) + \bar{\xi}_5\sin(5\pi\eta)$$

$$u(\eta) = \xi_3\sin(3\pi\eta) + \xi_4\sin(4\pi\eta) + \xi_5\sin(5\pi\eta) \tag{11.94}$$

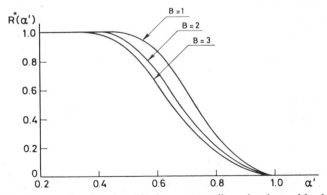

Fig. 11.6. Empirical reliability function $R^*(\alpha')$ versus nondimensional actual load α', $A = 0.01$, $\kappa_1 = \kappa_3 = \pi^4$, $n_* = 1$, $N = 4$.

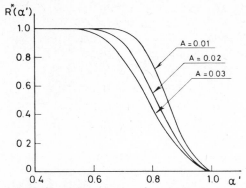

Fig. 11.7. Empirical reliability function $R^*(\alpha')$ versus nondimensional actual load α'; $B = 1$, $\kappa_1 = \kappa_3 = (3\pi)^4$, $n_* = 3$, $N = 4$.

As is seen from the figure, a three-term approximation (let alone a single-term one) would not be at the safe side in determining $R^*(\alpha')$. In this case ($n_* = 4$) the nondimensional buckling loads are plotted as a histogram (Fig. 11.9). Most bars buckle in the load-level intervals (0.80, 0.85] (103 bars), (0.85, 0.90] (145 bars) and (0.90, 0.95] (115 bars). Figures 11.10 and 11.11 are plots of the estimate of the mean buckling loads $E(\alpha^*)$ versus A and the relative nonlinearity ratio κ_3/κ_1, respectively; the latter figure shows that as the relative nonlinearity increases, the mean values of the buckling loads decrease.

11.7.7   Kolmogorov-Smirnov Test for Goodness of Fit.   The reliability functions derived above with the aid of formula (11.93) are empirical. Strictly speaking, $R^*(\alpha')$ represents step functions, which were smoothed out in Figs. 11.6–11.8. An answer to the question of what interrelation exists between

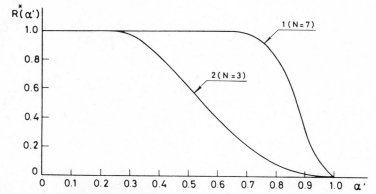

Fig. 11.8. Empirical reliability function $R^*(\alpha')$ versus nondimensional load α'; $B = 1$, $\kappa_1 = \kappa_3 = (4\pi)^4$ (*1*) Seven-term approximation. (*2*) Three-term approximation.

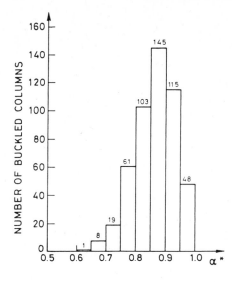

Fig. 11.9. Histogram of nondimensional buck-ling load α^*; $\kappa_1 = \kappa_3 = (4\pi)^4$, $A = 0.01$, $B = 1$, $n_* = 4$, $N = 7$.

$R^*(\alpha')$ and the unknown exact function $R(\alpha')$ is provided by the *Kolmogorov-Smirnov test for goodness of fit*. It states that for a continuous random variable α^*,

$$P\big[\max|F(\alpha') - F^*(\alpha')| < d_\delta(M)\big] = 1 - \delta \qquad (11.95)$$

where $F(\alpha')$ is the exact distribution function of α', $F^*(\alpha')$ is its calculated estimate, and $\max|F(\alpha') - F^*(\alpha')|$ is the maximum absolute difference be-tween them. Since $F(\alpha') = 1 - R(\alpha')$ and $F^*(\alpha') = 1 - R^*(\alpha')$, Eq. (11.95) becomes

$$P\big[\max|R(\alpha') - R^*(\alpha')| < d_\delta(M)\big] = 1 - \delta \qquad (11.96)$$

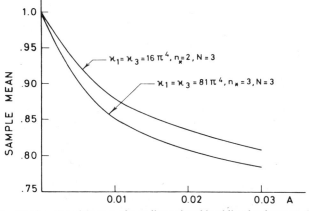

Fig. 11.10. Sample mean of nondimensional buckling load versus A.

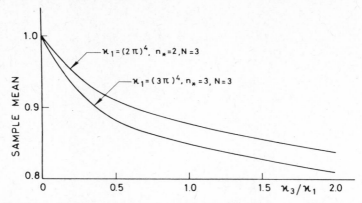

Fig. 11.11. Sample mean of nondimensional buckling load versus ratio κ_3/κ_1.

where M is the ensemble size, and δ a small positive decimal number known as the *level of significance*. For ensemble sizes exceeding $M = 35$, the following formulas are valid for d_δ (Massey):

$$d_{0.2} = \frac{1.07}{\sqrt{M}} \qquad d_{0.15} = \frac{1.14}{\sqrt{M}} \qquad d_{0.1} = \frac{1.22}{\sqrt{M}}$$

$$d_{0.05} = \frac{1.36}{\sqrt{M}} \qquad d_{0.01} = \frac{1.63}{\sqrt{M}} \tag{11.97}$$

For example, for $M = 10{,}000$ at the level of significance $\delta = 0.01$, that is, with the probability $1 - \delta = 0.99$, the maximum absolute difference between the exact and empirical reliability functions is smaller than $d_{0.01}(10{,}000) = 1.63/\sqrt{10{,}000} = 0.0163$. The limits $R^*(\alpha') - d_\delta(M)$ and $R^*(\alpha') + d_\delta(M)$ are referred to as *confidence limits*, and when they are known the empirical reliability function can be used for design purposes. To be on the safe side, the criterion

$$R(\alpha') \geqslant r \tag{11.98}$$

where r is the required reliability, is replaced by

$$R^*(\alpha') - d_\delta(M) \geqslant r \tag{11.99}$$

The maximum nondimensional load α_{allow} that satisfies the equation

$$R^*(\alpha_{\text{allow}}) - d_\delta(M) = r \tag{11.100}$$

is then the allowable strength for the entire ensemble of bars. For example, for $A = 0.01$, $\alpha_{\text{allow}} = 0.75$, as can be seen from Fig. 11.12.

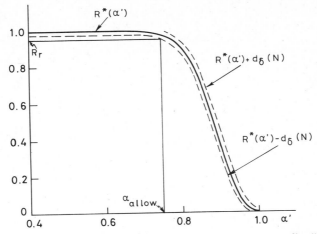

Fig. 11.12. Determination of nondimensional allowable load α_{allow} at 95% reliability.

It is interesting to compare the prediction of the maximum absolute difference $\max|R^*(\alpha') - R(\alpha')|$ for a case capable of exact solution, such as is given by the single-term Galerkin method in Eq. (11.91). The results are in such excellent agreement with those of the Monte Carlo method that the latter are indistinguishable from them and cannot be plotted separately (curve $B = 1$, Fig. 11.6). The maximum difference between $R(\alpha')$ and $R^*(\alpha')$ is 0.005, much smaller than the value $d_{0.01}(10,000) = 0.0163$. The implication is that the predicted maximum absolute difference, $\max|R^*(\alpha') - R(\alpha')|$, according to the Kolmogorov-Smirnov test for goodness of fit may be conservative, and the actual difference in real life may fortunately be much smaller.

In this section, the reliability of the model imperfection-sensitive structure was considered. Applications of the expounded ideas to shells with axisymmetric or general nonsymmetric random imperfections have been given in papers by Elishakoff and Arbocz. The first step in such an application consists in compiling extensive experimental information on shells produced by a specific manufacturing process, with a view to estimating their mean and autocovariance functions or, alternatively, via equations of type (11.84) and (11.87), the mean vector and variance-covariance matrix. The contents of existing initial-imperfection data banks can be incorporated directly for estimation of these characteristics. Then, using the latest deterministic theoretical findings and powerful numerical codes, the Monte Carlo method is applied. Here, clearly, besides the random initial imperfections, remaining random parameters (such as Young's modulus, length, axial load, characteristics representing the boundary conditions, etc.) have to be simulated in a suitable manner. The final product of such an analysis is the reliability function associated with the specific manufacturing process. Thus the imperfection-sensitivity concept can be introduced into engineering design, and the existing gap between theory and practice will be narrowed.

PROBLEMS

11.1. Simulate a random variable with Weibull distribution, using the inverse function method.

11.2. Solve Problem 5.13 using the Monte Carlo method, and compare with the exact solution.

11.3. Consider the random vector $\{Y\} = (Y_1, Y_2)$ with joint probability density $f(y_1, y_2) = 5y_1$, for $0 < y_1 < 1$, $0 < y_2 < 1$. Simulate this random vector.

11.4. Perform the Cholesky decomposition of the variance-covariance matrix $v_{jk} = \sigma^{|j-k|}$, where σ is a positive number. Simulate a normal random vector with zero mathematical expectation and this variance-covariance matrix.

11.5. A beam simply supported at its ends is subjected at sections $x = l/3$ and $x = 2l/3$ (where l is the span of the beam) to two concentrated loads with identically vanishing mean values and the variance-covariance matrix

$$[V] = \sigma^2 \begin{bmatrix} 1 & 1 \\ 1 & 4 \end{bmatrix}$$

Show that the matrix $[C]$ is

$$[C] = \sigma \begin{bmatrix} 1 & 0 \\ 1 & \sqrt{3} \end{bmatrix}$$

Determine the reliability of the beam by the Monte Carlo method, given that the allowable stress is also a random variable, uniformly distributed in the interval (assuming actual values to l, σ, c, and the section modulus S of the beam).

11.6. What is the reliability in Problem 11.5, if the variance-covariance matrix is

$$[V] = \sigma^2 \begin{bmatrix} 1 & 2 - \varepsilon \\ 2 - \varepsilon & 4 \end{bmatrix}$$

Investigate the variation of the reliability in the interval $0 < \varepsilon \leqslant 1$, checking the case $\varepsilon = 0$ separately.

11.7. A beam simply supported at its ends is subjected to two concentrated loads Y_1 and Y_2 acting at sections $x = l/4$ and $x = 3l/4$, respectively; $l = 1000$ mm, $S = 1800$ mm^3, $E(Y_1) = 1500$ N, $E(Y_2) = 1500$ N, $\sigma_{\text{allow}} = 250$ N/mm^2. The variance-covariance matrix is

$$[V] = 300^2 \begin{bmatrix} 1 & \frac{1}{2} \\ \frac{1}{2} & \frac{1}{3} \end{bmatrix} N^2$$

Find the reliability of the beam both exactly and by the Monte Carlo method.

11.8. Consider a bar on a mixed quadratic-cubic elastic foundation, with its displacement described by the differential equation

$$EI\frac{d^4w}{dx^4} + P\frac{d^2w}{dx^2} + k_1w - k_2w^2 - k_3w^3 = -P\frac{d^2\bar{w}}{dx^2}$$

where the constants k_1, k_2, and k_3 are the foundation parameters. Using the Galerkin method in a single-term approximation, derive an equation analogous to Eq. (11.69) and draw a parallel to the model structure considered in Sec. 5.5. Show also that the relation between the nondimensional buckling load α^* and the initial imperfection $\bar{\xi}_m$ is in full agreement with Eq. (5.50). Finally, derive the reliability function and the allowable nondimensional load. (Multiterm analysis of this problem is given by Elishakoff, 1981.)

CITED REFERENCES

Amazigo, J. C., "Buckling of Stochastically Imperfect Structures," in B. Budiansky, Ed. *Buckling of Structures* (IUTAM Symp., Cambridge, MA, 1974), Springer-Verlag, Berlin, 1976, pp. 172–182.

Bolotin, V. V., "Statistical Methods in the Nonlinear Theory of Elastic Shells," *Izv. Akad. Nauk SSSR, Otdel. Tech. Nauk*, **3** (1958); English transl., NASA TTF-85, 1962, pp. 1–16.

Boyce, W. E., "Buckling of a Column with Random Initial Displacements," *J. Aerospace Sci.*, **28**, 308–320 (1961).

Budiansky, B., "Theory of Buckling and Post-Buckling Behavior of Elastic Structures," *Advan. Appl. Mech.*, **14**, 1–65 (1974).

Davidenko, D. F., "On a New Method of Numerical Solution of the Systems of Nonlinear Equations," *Dokl. Akad. Nauk SSSR (Proc. Acad. Sci. USSR)*, **88**(4), 68–78 (1951).

Elishakoff, I., "Simulation of Space-random Fields for Solution of Stochastic Boundary-Value Problems," *J. Acoust. Soc. Am.*, **65**(2), 399–403 (1979a).

Elishakoff, I., "Buckling of a Stochastically Imperfect Finite Column on a Nonlinear Elastic Foundation—A Reliability Study," *ASME J. Appl. Mech.*, **46**, 411–416 (1979b).

Elishakoff, I., "Reliability Approach to the Random Imperfection Sensitivity of Columns," Dept. of Aeronaut. Eng., Technion-I.I.T., *TAE Rep.* **441**, March 1981.

Elishakoff, I., and Arbocz, J., "Reliability of Axially Compressed Cylindrical Shells with Random Axisymmetric Imperfections," *Intern. J. Solids Structures*, **18**, 563–585 (1982).

Elishakoff, I., and Arbocz, J., "Stochastic Buckling of Shells with General Imperfections," in *Stability in the Mechanics of Continua*, F. M. Schroeder, Ed., Springer-Verlag, Berlin, 1982, pp. 306–217.

Frazer, W. B., "Buckling of a Structure with Random Imperfections," Ph.D. thesis, Div. Eng. and Appl. Phys., Harvard Univ., Cambridge, MA, May 1965.

Hammersley, J. M., and Handscomb, D. C., *Monte Carlo Methods*, Methuen, London, 1964.

Hutchinson, J. W., and Koiter, W. T., "Postbuckling Theory," *Appl. Mech. Rev.*, **1970**, 1353–1366.

Jacquot, R. G., "Nonstationary Random Column Buckling Problem," *J. Eng. Mech. Div.*, *Proc. of ASCE*, **1972**(EM5), 1173–1182.

Lehmer, D. H., "Mathematical Methods in Large-scale Computing Units," *Ann. Comp. Lab. Harvard Univ.*, **26**, (1951), 141–146.

Massey, F. J., Jr., "The Kolmogorov-Smirnov Test for Goodness of Fit." *J. Amer. Stat. Assoc.*, **46**, (253), 68–78 (1951).

McCracken, D. D., "The Monte Carlo Method," *Sci. Am.*, **1955**(5), 90–96.

Metropolis, N., and Ulam, S., "The Monte Carlo Method," *J. Am. Stat. Assoc.*, **44**(247), 335–341 (1949).

NASA, "Buckling of Thin-Walled Circular Cylinders," NASA SP-8007, August 1968.

Qiria, V. S., "Motion of the Bodies in Resisting Media," *Proc. Tbilisi State Univ.*, **44**, 1–20 (1951).

Rubinstein, R. Y., *Simulation and the Monte Carlo Method*, John Wiley & Sons, New York, 1981.

Sobol', I. M., *Numerical Monte Carlo Methods*, "Nauka" Publ. House, Moscow, 1973, pp. 11–12.

Timoshenko, S. P., and Gere J. M., *Theory of Elastic Stability*, 2nd ed., McGraw-Hill, New York, 1961, pp. 94–98.

Von Neumann, J., "Various Techniques used in Connection with Random Digits," *Nat. Bur. Stand.*, *Appl. Math. Ser.*, **12**, 36–38 (1951).

RECOMMENDED FURTHER READING

Allen, A. O., *Probability, Statistics, and Queueing Theory with Computer Science Applications*, Academic Press, New York, 1978. Chap. 7: Estimation, pp. 273–291; Chap. 8: Hypothesis Testing, pp. 292–321.

Bykov, V. V., *Digital Modeling in Statistical Radio-Engineering*, Soviet Radio, Moscow, 1971.

Edlund, B. O., and Leopoldson, U. L. C., "Computer Simulation of the Scatter in Steel Member Strength," *Int. J. Computers and Structures*, **5** 209–224 (1975).

Elishakoff, I., "Impact Buckling of Thin Bar via Monte Carlo Method," *ASME J. Appl. Mech.*, **45**, pp. 586–590 (1978).

Elishakoff, I., "Hoff's Problem in a Probabilistic Setting," *ASME J. Appl. Mech.*, **47**, 403–408 (1980).

Ermakov, S. M., *Monte Carlo Method and Related Problems*, "Nauka" Publ. House, Moscow, 1975.

Ermakov, S. M., and Mikhailov, G. A., *Course in Statistical Simulation*, "Nauka" Publ. House, Moscow, 1976.

Holton, I. H., "A Retrospective and Prospective Survey of the Monte Carlo Method," *SIAM Rev.*, **12**(1), 1–63 (1970).

Hoshia, M., and Spence, S., "Reliability Analysis of a Tainter Gate", *Proc. YSCE*, **183**, 111–122 (1970).

Mikhailov, G. A., *Some Problems in the Theory of the Monte Carlo Method*, "Nauka" Publ. House, Novosibirsk, 1974.

Pollyak, Yu. G., *Probabilistic Modeling on Digital Computers*, Soviet Radio, Moscow, 1971.

Shinozuka, M., "Simulation of Multivariate and Multidimensional Random Processes," *J. Acoust. Soc. Am.*, **49**, 357–367 (1971).

Shinozuka, M., and Astil, C. J., "Random Eigenvalue Problems in Structural Analysis," *AIAA J.* **10**, 456–462 (1972).

Shinozuka, M., and Wen, Y. K., "Monte Carlo Solution of Nonlinear Vibrations," *AIAA J.*, **10**, 37–40 (1972).

Shreider, Yu. A., Ed., *The Monte Carlo Method*, Pergamon Press, Oxford, 1966.

Vaicaitis, R., Dowell, E. H., and Ventres, C. S., "Nonlinear Panel Response by a Monte Carlo Approach," *AIAA J.*, **12**, 685–691 (1974).

Appendix

A EVALUATION OF INTEGRALS (4.15) AND (4.22)

Evaluate first the integral

$$I_1 = \int_{-\infty}^{\infty} e^{-(1/2)\xi^2} \, d\xi$$

We note first that $I_1 > 0$, since the integrand is positive. Next, we construct its square and rewrite it as a double integral:

$$I_1^2 = \left[\int_{-\infty}^{\infty} e^{-(1/2)\xi^2} \, d\xi \right] \left[\int_{-\infty}^{\infty} e^{-(1/2)\eta^2} \, d\eta \right]$$

$$= \int_{-\infty}^{\infty} \int_{-\infty}^{\infty} \exp\{ -\tfrac{1}{2}(\xi^2 + \eta^2) \} \, d\xi \, d\eta$$

Finally, we transform it into polar coordinates

$$\xi = \rho \sin \theta, \quad \eta = \rho \cos \theta$$

which yield

$$I_1^2 = \int_0^{\infty} \int_0^{2\pi} \rho e^{-(1/2)\rho^2} \, d\rho = 2\pi$$

which in turn yields

$$I_1 = \int_{-\infty}^{\infty} e^{-(1/2)\xi^2} \, d\xi = \sqrt{2\pi} \tag{A.1}$$

Equation (4.22) follows from the more general integral

$$I_2 = \int_{-\infty}^{\infty} \exp\left[-\tfrac{1}{2}(At^2 + 2Bt + C)\right] dt = \left(\frac{2\pi}{A}\right)^{1/2} \exp\left(-\frac{AC - B^2}{2A}\right)$$

$$(A.2)$$

where $A > 0$. Substituting $t = \eta/\sqrt{A}$, we find

$$I_2 = \frac{1}{\sqrt{A}} \exp\left(-\frac{AC - B^2}{2A}\right) I_3\left(\frac{B}{\sqrt{A}}\right) \qquad (A.3)$$

where

$$I_3\left(\frac{B}{\sqrt{A}}\right) = \int_{-\infty}^{\infty} \exp\left[-\tfrac{1}{2}\left(\eta + \frac{B}{\sqrt{A}}\right)^2\right] d\eta$$

In order to show that I_3 is invariant with respect to $B/\sqrt{A}$, we differentiate it with respect to latter (denoted for convenience by κ):

$$\frac{dI_3(\kappa)}{d\kappa} = \frac{d}{d\kappa}\left(\int_{-\infty}^{\infty} \exp\left[-\tfrac{1}{2}(\eta + \kappa)^2\right] d\eta\right)$$

$$= -\int_{-\infty}^{\infty} (\eta + \kappa)\exp\left[-\tfrac{1}{2}(\eta + \kappa)^2\right] = \left(\exp\left[-\tfrac{1}{2}(\eta + \kappa)\right]\right)_{-\infty}^{+\infty} = 0$$

and see that

$$I_3(\kappa) = \text{const} = I_3(0) = \int_{-\infty}^{\infty} e^{-(1/2)\eta^2} d\eta = I_1 = \sqrt{2\pi}$$

So that, substituting I_1 for $I_3(B/\sqrt{A})$, we arrive at the desired result (A.2), and letting $A = 2\beta^2$, $B = -\alpha$, $C = 0$, we arrive at Eq. (4.22). Letting $A = 2\alpha$, $B = C = 0$, we arrive at Eq. (4.15).

B   TABLE OF THE ERROR FUNCTION

$$\text{erf}(x) = \frac{1}{\sqrt{2\pi}} \int_0^x e^{-y^2/2}\, dy$$

x	0.00	0.01	0.02	0.03	0.04	0.05	0.06	0.07	0.08	0.09
0.0	0.00000	0.00399	0.00798	0.01197	0.01595	0.01994	0.02392	0.02790	0.03188	0.03586
0.1	0.03983	0.04380	0.04776	0.05172	0.05567	0.05962	0.06356	0.06749	0.07142	0.07535
0.2	0.07926	0.08317	0.08706	0.09095	0.09483	0.09871	0.10257	0.10642	0.11026	0.11409
0.3	0.11791	0.12172	0.12552	0.12930	0.13307	0.13683	0.14058	0.14431	0.14803	0.15173
0.4	0.15542	0.15910	0.16276	0.16640	0.17003	0.17364	0.17724	0.18082	0.18439	0.18793
0.5	0.19146	0.19497	0.19847	0.20194	0.20540	0.20884	0.21226	0.21566	0.21904	0.22240
0.6	0.22575	0.22907	0.23237	0.23565	0.23891	0.24215	0.24537	0.24857	0.25175	0.25490
0.7	0.25804	0.26115	0.26424	0.26730	0.27035	0.27337	0.27637	0.27935	0.28230	0.28524
0.8	0.28814	0.29103	0.29389	0.29673	0.29955	0.30234	0.30511	0.30785	0.31057	0.31327
0.9	0.31594	0.31859	0.32121	0.32381	0.32639	0.32894	0.33147	0.33398	0.33646	0.33891
1.0	0.34134	0.34375	0.34614	0.34849	0.35083	0.35314	0.35543	0.35769	0.35993	0.36214
1.1	0.36433	0.36650	0.36864	0.37076	0.37286	0.37493	0.37698	0.37900	0.38100	0.38298
1.2	0.39493	0.38686	0.38877	0.39065	0.39251	0.39435	0.39617	0.39796	0.39973	0.40147
1.3	0.40320	0.40490	0.40658	0.40824	0.40988	0.41149	0.41308	0.41466	0.41621	0.41774
1.4	0.41924	0.42073	0.42220	0.42364	0.42507	0.42647	0.42785	0.42922	0.43056	0.43189
1.5	0.43319	0.43448	0.43574	0.43699	0.43822	0.43943	0.44062	0.44179	0.44295	0.44408
1.6	0.44520	0.44630	0.44738	0.44845	0.44950	0.45053	0.45154	0.45254	0.45352	0.45499
1.7	0.45543	0.45637	0.45728	0.45818	0.45907	0.45994	0.46080	0.46164	0.46246	0.46327
1.8	0.46407	0.46485	0.46562	0.46638	0.46712	0.46784	0.46856	0.46926	0.46995	0.47062
1.9	0.47128	0.47193	0.47257	0.47320	0.47381	0.47441	0.47500	0.47558	0.47615	0.47670
2.0	0.47725	0.47778	0.47831	0.47882	0.47932	0.47892	0.48030	0.48077	0.48124	0.48169
2.1	0.48214	0.48257	0.48300	0.48341	0.48382	0.48422	0.48461	0.48500	0.48537	0.48574
2.2	0.48610	0.48645	0.48679	0.48713	0.48745	0.48778	0.48809	0.48840	0.48870	0.48899
2.3	0.48928	0.48956	0.48983	0.49010	0.49036	0.49061	0.49086	0.49111	0.49134	0.49158
2.4	0.49180	0.49202	0.49224	0.49245	0.49266	0.49286	0.49305	0.49324	0.49343	0.49361
2.5	0.49379	0.49396	0.49413	0.49430	0.49446	0.49461	0.49477	0.49492	0.49506	0.49520
2.6	0.49534	0.49547	0.49560	0.49573	0.49585	0.49598	0.49609	0.49621	0.49632	0.49643
2.7	0.49653	0.49664	0.49674	0.49683	0.49693	0.49702	0.49711	0.49720	0.49728	0.49736
2.8	0.49744	0.49752	0.49760	0.49767	0.49774	0.49781	0.49788	0.49795	0.49801	0.49807
2.9	0.49813	0.49819	0.49825	0.49831	0.49836	0.49841	0.49846	0.49851	0.49856	0.49861
3.0	0.49865	0.49869	0.49874	0.49878	0.49882	0.49886	0.49889	0.49893	0.49896	0.49900
3.1	0.49903	0.49906	0.49910	0.49913	0.49916	0.49918	0.49921	0.49924	0.49926	0.49929
3.2	0.49931	0.49934	0.49936	0.49938	0.49940	0.49942	0.49944	0.49946	0.49948	0.49950
3.3	0.49952	0.49953	0.49955	0.49957	0.49958	0.49960	0.49961	0.49961	0.49964	0.49965
3.4	0.49966	0.49968	0.49969	0.49970	0.49971	0.49972	0.49973	0.49974	0.49975	0.49976
3.5	0.49977	0.49978	0.49978	0.49979	0.49980	0.49981	0.49981	0.49982	0.49983	0.49983
3.6	0.49984	0.49985	0.49985	0.49986	0.49986	0.49987	0.49987	0.49988	0.49988	0.49989
3.7	0.49989	0.49990	0.49990	0.49990	0.49991	0.49991	0.49992	0.49992	0.49992	0.49992
3.8	0.49993	0.49993	0.49993	0.49994	0.49994	0.49994	0.49994	9.49995	0.49995	0.49995

C CALCULATION OF THE MEAN SQUARE
RESPONSE OF A CLASS OF LINEAR SYSTEMS

A necessary condition for the integral (9.47), denoted here as I_n,

$$I_n = \int_{-\infty}^{\infty} \frac{S_F(\omega)\, d\omega}{L_n(i\omega)L_n(-i\omega)} \tag{C.1}$$

(where $S_F(\omega)$ is a polynomial in ω) to be finite, is for $S_F(\omega)$ not to contain orders higher than $2n - 2$. Assume, accordingly, that

$$S_F(\omega) = b_0(i\omega)^{2n-2} + b_1(i\omega)^{2n-4} + \cdots + b_{n-1} \tag{C.2}$$

We also assume that the system is stable, that is, that all the roots of the auxiliary equation (9.9) have negative real parts.

The integral (C.1) has then the following values

$$I_1 = \frac{\pi b_0}{a_0 a_1} \tag{C.3}$$

$$I_2 = \frac{\pi}{a_0 a_1}\left(-b_0 + \frac{a_0 b_1}{a_2}\right) \tag{C.4}$$

$$I_3 = \frac{\pi}{a_0(a_0 a_3 - a_1 a_2)}\left(-a_2 b_0 + a_0 b_1 - \frac{a_0 a_1 b_2}{a_3}\right) \tag{C.5}$$

$$I_4 = \frac{\pi}{a_0(a_0 a_3^2 + a_1^2 a_4 - a_1 a_2 a_3)}$$

$$\times \left[b_0(-a_1 a_4 + a_2 a_3) - a_0 a_3 b_1 + a_0 a_1 b_2 + \frac{a_0 b_3}{a_4}(a_0 a_3 - a_1 a_2) \right] \tag{C.6}$$

$$I_5 = \frac{\pi M_5}{a_0 N_5} \tag{C.7}$$

$$M_5 = b_0(-a_0 a_4 a_5 + a_1 a_4^2 + a_2^2 a_5 - a_2 a_3 a_4) + a_0 b_1(-a_2 a_5 + a_3 a_4)$$

$$+ a_0 b_2(a_0 a_5 - a_1 a_4) + a_0 b_3(-a_0 a_3 + a_1 a_2)$$

$$+ \frac{a_0 b_4}{a_5}(-a_0 a_1 a_5 - a_0 a_3^2 + a_1^2 a_4 - a_1 a_2 a_3)$$

$$N_5 = a_0^2 a_5^2 - 2a_0 a_1 a_4 a_5 - a_0 a_2 a_3 a_5 + a_0 a_3^2 a_4 + a_1^2 a_4^2 + a_1 a_2^2 a_5 - a_1 a_2 a_3 a_4$$

$$I_6 = \frac{\pi M_6}{a_0 N_6} \tag{C.8}$$

$$
\begin{aligned}
M_6 = {} & b_0 \left(-a_0 a_3 a_5 a_6 + a_0 a_4 a_5^2 - a_1^2 a_6^2 + 2a_1 a_2 a_5 a_6 + a_1 a_3 a_4 a_6 \right. \\
& \left. - a_1 a_4^2 a_5 - a_2^2 a_5^2 - a_2 a_3^2 a_6 + a_2 a_3 a_4 a_5 \right) \\
& + a_0 b_1 \left(-a_1 a_5 a_6 + a_2 a_5^2 + a_3^2 a_6 - a_3 a_4 a_5 \right) \\
& + a_0 b_2 \left(-a_0 a_5^2 - a_1 a_3 a_6 + a_1 a_4 a_5 \right) \\
& + a_0 b_3 \left(a_0 a_3 a_5 + a_1^2 a_6 - a_1 a_2 a_5 \right) \\
& + a_0 b_4 \left(a_0 a_1 a_5 - a_0 a_3^2 + a_1^2 a_4 + a_1 a_2 a_3 \right) \\
& + \frac{a_0 b_5}{a_6} \left(a_0^2 a_5^2 + a_0 a_1 a_3 a_6 - 2a_0 a_1 a_4 a_5 - a_0 a_2 a_3 a_5 \right. \\
& \left. + a_0 a_3^2 a_4 - a_1^2 a_2 a_6 + a_1^2 a_4^2 + a_1 a_2^2 a_5 - a_1 a_2 a_3 a_4 \right)
\end{aligned}
$$

$$
\begin{aligned}
N_6 = {} & a_0^2 a_5^3 + 3a_0 a_1 a_3 a_5 a_6 - 2a_0 a_1 a_4 a_5^2 - a_0 a_2 a_3 a_5^2 - a_0 a_3^2 a_6 \\
& + a_0 a_3^2 a_4 a_5 + a_1^3 a_6^2 - 2a_1^2 a_2 a_5 a_6 - a_1^2 a_3 a_4 a_6 + a_1^2 a_4^2 a_5 + a_1 a_2^2 a_5^2 \\
& + a_1 a_2 a_3^2 a_6 - a_1 a_2 a_3 a_4 a_5
\end{aligned}
$$

Expressions for $I_7 - I_{10}$ are listed in the book by Newton et al. The general formula is (see James et al.)

$$I_n = \frac{\pi (-1)^{n+1}}{a_0} \frac{M_n}{H_n} \tag{C.9}$$

where M_n and H_n are the following determinants:

$$
M_n = \begin{vmatrix}
b_0 & h_{12} & h_{13} & \cdots & h_{1n} \\
b_1 & h_{22} & h_{23} & \cdots & h_{2n} \\
b_2 & h_{32} & h_{33} & \cdots & h_{3n} \\
\vdots & & & & \\
b_{n-1} & h_{n2} & h_{n3} & \cdots & h_{nn}
\end{vmatrix} \tag{C.10}
$$

$$H_n = \begin{vmatrix} h_{11} & h_{12} & h_{13} & \cdots & h_{1n} \\ h_{21} & h_{22} & h_{23} & \cdots & h_{2n} \\ h_{31} & h_{32} & h_{33} & \cdots & h_{3n} \\ \vdots & & & & \\ h_{n1} & h_{n2} & h_{n3} & \cdots & h_{nn} \end{vmatrix} \qquad \text{(C.11)}$$

where

$$h_{jk} = a_{2j-k}$$

$$a_j = 0, \quad j < 0 \text{ or } j > n \qquad \text{(C.12)}$$

Note that M_n is obtained from H_n by replacing the first column with $b_0, b_1, \ldots, b_{n-1}$. Note, also, that H_n is the Hurwitz determinant for the aux-iliary equation (9.9), the roots of which have negative real parts.

Example C.1

We calculate the mean-square response of the single-degree-of-freedom system to ideal white noise excitation. According to Eq. (9.64), we have

$$E(X^2) = \frac{S_0}{m^2} \int_{-\infty}^{\infty} \frac{d\omega}{|(i\omega)^2 + 2\zeta\omega_0(i\omega) + \omega_0^2|^2} \qquad \text{(C.13)}$$

Comparison of Eqs. (9.2), (9.48) and (C.1) yields

$$a_0 = 1 \qquad a_1 = 2\zeta\omega_0 \qquad a_2 = \omega_0^2 \qquad b_0 = 0 \qquad b_1 = 1 \qquad \text{(C.14)}$$

Equation (C.5) then results in

$$E(X^2) = I_2 = \frac{S_0\pi}{2\zeta\omega_0^3 m^2} \qquad \text{(C.15)}$$

which coincides with Eq. (9.42).

Example C.2

Consider a system governed by the differential equation

$$L_n\left(\frac{d}{dt}\right)X = Q_m\left(\frac{d}{dt}\right)Y \qquad \text{(C.16)}$$

where $L_n(d/dt)$ denotes the expression in Eq. (9.2), and

$$Q_m\left(\frac{d}{dt}\right) = b_0\frac{d^m}{dt^m} + b_1\frac{d^{m-1}}{dt^{m-1}} + \cdots + b_{m-1}\frac{d}{dt} + b_m, \quad b_0 \neq 0 \qquad \text{(C.17)}$$

We denote

$$L_n\left(\frac{d}{dt}\right)X = F(t) \qquad Q_m\left(\frac{d}{dt}\right)Y = F(t) \qquad \text{(C.18)}$$

and require that both systems be stable. Then we have

$$S_X(\omega) = \frac{S_F(\omega)}{L_n(i\omega)L_n(-i\omega)} \qquad S_Y(\omega) = \frac{S_F(\omega)}{Q_m(i\omega)Q_m(-i\omega)}$$

and eliminating $S_F(\omega)$, we obtain

$$S_X(\omega) = \frac{Q_m(i\omega)Q_m(-i\omega)}{L_n(i\omega)L_n(-i\omega)}S_Y(\omega) \qquad \text{(C.19)}$$

$X(t)$ has a finite mean square if

$$\int_{-\infty}^{\infty} \frac{Q_m(i\omega)Q_m(-i\omega)}{L_n(i\omega)L_n(-i\omega)}S_Y(\omega)\,d\omega < \infty$$

which is the case when we impose $m < n$. Then the integral

$$E(X^2) = \int_{-\infty}^{\infty} \frac{Q_m(i\omega)Q_m(-i\omega)}{L_n(i\omega)L_n(-i\omega)}S_Y(\omega)\,d\omega \qquad \text{(C.20)}$$

is readily evaluated via Eq. (C.1).
    Consider, as an example,

$$\ddot{X} + 2\zeta\omega_0\dot{X} + \omega_0^2 X = \frac{1}{m}(\alpha\dot{Y} + \beta Y) \qquad \text{(C.21)}$$

where Y is an ideal white noise, $S_Y = S_0$:

$$S_X(\omega) = \frac{S_0}{m^2}\frac{\alpha^2\omega^2 + \beta^2}{\left(\omega_0^2 - \omega^2\right)^2 + 4\zeta^2\omega_0^2\omega^2} \qquad \text{(C.22)}$$

To determine

$$E(X^2) = \int_{-\infty}^{\infty} S_X(\omega)\,d\omega$$

we substitute in (C.1)

$$a_0 = 1 \qquad a_1 = 2\zeta\omega_0 \qquad a_2 = \omega_0^2 \qquad b_0 = -\alpha^2 \qquad b_1 = \beta^2 \quad \text{(C.23)}$$

and

$$E(X^2) = I_2 = \frac{S_0 \pi \left(\alpha \omega_0^2 + \beta^2 \right)}{2 \zeta \omega_0^3 m^2}$$

and denoting $\alpha \dot{Y} + \beta Y = F(t)$, arrive at

$$E(X^2) = \frac{S_F(\omega_0) \pi}{2 \zeta \omega_0^3 m^2} \tag{C.24}$$

For $\alpha = 0, \beta = 1$, we return to Eqs. (9.42) and (C.15).

Several authors used expressions of type (C.22) to approximate the spectral density of earthquake acceleration, with ω_0 representing a characteristic ground frequency, α and β real constants, and ζ a positive constant.

Note that an alternative form of Eqs. (C.3)–(C.6) is given by Crandall and Mark.

CITED REFERENCES

Crandall, S. H., and Mark, W. D., *Random Vibration in Mechanical Systems*, Academic Press, New York, 1963, p. 72.

James, H. M., Nichols, N. B., and Phillips, R. S., *Theory of Servomechanisms*, McGraw-Hill, New York, 1947, p. 369.

Newton, G. C., Jr., Gould, L. A., and Kaiser, J. F., *Analytical Design of Linear Feedback Controls*, John Wiley & Sons, New York, 1964, pp. 471–481.

D SOME AUTOCORRELATION FUNCTIONS AND ASSOCIATED SPECTRAL DENSITIES

$R_X(\tau)$	$S_X(\omega)$	Existing Derivatives
$2\pi\delta(\tau)$ Ideal white noise	S_0	None
$\dfrac{2\sin\omega_c\tau}{\tau},\quad \omega_c > 0$ Band-limited white noise	$U\left(1 - \dfrac{\|\omega\|}{\omega_c}\right)$	Any number
$e^{-\alpha\|\tau\|},\quad \alpha > 0$ Exponential correlation	$\dfrac{\alpha}{\pi}\left(\dfrac{1}{\alpha^2 + \omega^2}\right)$	None
$e^{-\alpha\|\tau\|}\cos\Omega\tau,\ (\alpha > 0, \Omega > 0)$ Damped exponential correlation	$\dfrac{\alpha}{\pi}\left(\dfrac{\omega^2 + \alpha^2 + \Omega^2}{\omega^4 + 2\omega^2(\alpha^2 - \Omega^2) + (\alpha^2 + \Omega^2)^2}\right)$	None
$e^{-\alpha\|\tau\|}(\cos\Omega\tau + \gamma\sin\Omega\|\tau\|),$ $(\alpha > 0, \Omega > 0, \gamma \leqslant \alpha/\Omega)$	$\dfrac{1}{\pi}\left(\dfrac{(\alpha + \gamma\Omega)(\alpha^2 + \Omega^2) + (\alpha - \gamma\Omega)\omega^2}{(\alpha^2 + \Omega^2)^2 + 2(\alpha^2 - \Omega^2)\omega^2 + \omega^4}\right)$	One
$e^{-\alpha\|\tau\|}\left(\cosh\Omega\tau + \dfrac{\alpha}{\Omega}\sinh\Omega\|\tau\|\right),$ $\alpha > 0, \Omega > 0$	$\dfrac{\alpha^2 + \Omega^2}{\pi}\left(\dfrac{2\alpha}{\left[(\alpha - \Omega)^2 + \omega^2\right]\left[(\alpha + \Omega)^2 + \omega^2\right]}\right)$	One
$e^{-\alpha^2\tau^2},\ \alpha > 0$	$\dfrac{1}{2\alpha\sqrt{\pi}}\exp\left[-\left(\dfrac{\omega}{2\alpha}\right)^2\right]$	Any number
$e^{-\alpha^2\tau^2}\cos\Omega\tau,\quad \alpha > 0, \Omega > 0$	$\dfrac{1}{2\alpha\sqrt{\pi}}\exp\left\{-\dfrac{\omega^2 + \Omega^2}{4\alpha^2}\right\}\cosh\left(\dfrac{\omega\Omega}{2\alpha^2}\right)$	Any number
$\cos\Omega\tau,\quad \Omega > 0$	$\tfrac{1}{2}[\delta(\omega - \omega_0) + \delta(\omega + \omega_0)]$	Any number

E GALERKIN METHOD

Given a differential equation (supplemented by a set of the boundary conditions at $x = 0$ and $x = l$)

$$L(w) = 0 \tag{E.1}$$

where $w = w(x)$ is the function of x we seek, and the operator L is not necessarily homogeneous.

The Galerkin method consists in approximating the solution in the form

$$w_N(x) = \sum_{j=1}^{N} a_j \psi_j(x) \tag{E.2}$$

where a_j are constants, N the number of terms taken into account, and $\psi_j(x)$ the *coordinate functions* assumed to satisfy the same boundary conditions as the exact solution $w(x)$ and to form a *complete set* in $x \in [0, l]$, or, in other words, a piecewise continuous function $f(x)$ can be approximated by the combination $\sum_{j=1}^{N} c_j \psi_j(x)$, such that

$$\int_0^l \left[f(x) - \sum_{j=1}^{N} c_j \psi_j(x) \right]^2 dx$$

can be as small as necessary.

If (E.2) happens to satisfy not only the boundary conditions but also the differential equation (E.1), it means that we have hit upon the exact solution and the problem is solved. Ordinarily, however, this is not the case, and the result of substituting (E.2) into (E.1) does not vanish identically but yields some error

$$L(w_N) = L\left(\sum_{j=1}^{N} a_j \psi_j \right) = \varepsilon_N(x), \quad \varepsilon_N(x) \not\equiv 0 \text{ in } x \in [0, l] \tag{E.3}$$

Further, we require that this error be orthogonal to each individual coordinate function

$$\int_0^l L\left(\sum_{j=1}^{N} a_j \psi_j \right) \psi_1(x) \, dx = 0$$

$$\int_0^l L\left(\sum_{j=1}^{N} a_j \psi_j \right) \psi_2(x) \, dx = 0$$

$$\vdots \tag{E.4}$$

$$\int_0^l \left(\sum_{j=1}^{N} a_j \psi_j \right) \psi_N(x) \, dx = 0$$

Determining from this set of equations the constants $a_1, a_2, \ldots, a_N$, we substitute them in Eq. (E.2) to yield the approximate solution of Eq. (E.1). Note that if Eq. (E.2) represents the exact solution, Eqs. (E.4) are satisfied identically, since a function identically equal to zero is orthogonal to all coordinate functions. If Eq. (E.2) does not represent the exact solution, Eqs. (E.4) are used to yield the approximation.

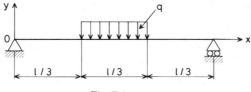

Fig. E.1.

Example E.1

Consider a beam, shown in Fig. E1, under distributed loading over the segment $l/3 \leqslant x \leqslant 2l/3$. We are interested in the displacement at the cross section $x = l/2$. The differential equation describing the equilibrium of the beam is

$$EI\frac{d^4w}{dx^4} = 0, \quad 0 \leqslant x \leqslant \frac{l}{3}, \quad \frac{2l}{3} \leqslant x \leqslant l \quad (E.5)$$

$$EI\frac{d^4w}{dx^4} = -q, \quad \frac{l}{3} < x < \frac{2l}{3} \quad (E.6)$$

The boundary conditions are

$$w = 0 \quad \frac{d^2w}{dx^2} = 0, \quad x = 0, x = l \quad (E.7)$$

Using the Galerkin method in a single-term approximation, we choose the following expression for the coordinate function:

$$\psi_1(x) = b_0 + b_1x + b_2x^2 + b_3x^3 + b_4x^4 \quad (E.8)$$

For the boundary condition $w = 0$ at $x = 0$ to be satisfied, b_0 must equal zero. The requirements that $w'' = 0$ at $x = 0$ yields $b_2 = 0$ and the boundary conditions $w = 0$ and $w'' = 0$ at $x = l$ yield, respectively,

$$b_1l + b_3l^3 + b_4l^4 = 0$$

$$6b_3l + 12b_4l^2 = 0$$

where $l = 3a$, so that putting $b_3 = 1$ gives us

$$\psi_1(x) = x^3 - \frac{x^4}{2l} - \frac{l^2}{2}x$$

The expression for $w_1(x)$ in Eq. (E.2) then reads

$$w_1(x) = a_1\psi_1(x) = a_1\left(x^3 - \frac{x^4}{2l} - \frac{l^2}{2}x\right) \qquad \text{(E.9)}$$

The expression of the displacement at $x = l/2$ we seek will be

$$w\left(\frac{l}{2}\right) = -\tfrac{5}{32}a_1l^3 \qquad \text{(E.10)}$$

The fourth derivative is

$$\frac{d^4w}{dx^4} = -\frac{12a_1}{l}$$

and Eq. (E.4.1) becomes

$$\int_0^{l/3} EI\left(-\frac{12a_1}{l}\right)\left(x^3 - \frac{x^4}{2l} - \frac{l^2}{2}x\right)dx$$

$$+ \int_{l/3}^{2l/3}\left[EI\left(-\frac{12a_1}{l}\right) - q\right]\left(x^3 - \frac{x^4}{2l} - \frac{l^2}{2}x\right)dx$$

$$+ \int_{2l/3}^{l} EI\left(-\frac{12a_1}{l}\right)\left(x^3 - \frac{x^4}{2l} - \frac{l^2}{2}x\right)dx = 0 \qquad \text{(E.11)}$$

After some algebra,

$$a_1 = -\tfrac{121}{2916}\left(\frac{ql}{EI}\right)$$

and

$$w\left(\frac{l}{2}\right) = \tfrac{605}{93,312}\left(\frac{ql^4}{EI}\right) \approx 0.006484\left(\frac{ql^4}{EI}\right) \qquad \text{(E.12)}$$

To check the accuracy of the method, we consider the exact solution. Using the singular function

$$\left\langle x - \frac{l}{3}\right\rangle = \begin{cases} 0, & x < \dfrac{l}{3} \\ x - \dfrac{l}{3}, & x \geqslant \dfrac{l}{3} \end{cases}$$

we rewrite the differential equations (E.5) and (E.6) as follows:

$$EI\frac{d^2w}{dx^2} = M_z(x) = -Rx + \frac{q\langle x - l/3\rangle^2}{2}, \quad 0 \leqslant x \leqslant \frac{l}{2}$$

the reaction R at the left-hand support being $ql/6$. Integration yields

$$EI\frac{dw}{dx} = -\frac{ql}{6}\frac{x^2}{2} + \frac{q\langle x - l/2\rangle^3}{6} + C \qquad\qquad \text{(E.13)}$$

where C is an integration constant and (since the slope dw/dx vanishes at midspan) equals

$$C = \tfrac{13}{648}ql^3 \qquad\qquad \text{(E.14)}$$

Further integration yields

$$EIw = -\frac{ql}{6}\left(\frac{x^3}{6}\right) + \frac{q\langle x - l/3\rangle^4}{24} + Cx + D$$

For the boundary condition $w = 0$ at $x = 0$ to be satisfied, D must equal zero, hence

$$w(x) = \frac{1}{EI}\left[-\frac{ql}{36}x^3 + \frac{q\langle x - l/3\rangle^4}{24} + \tfrac{13}{648}ql^3\right] \qquad\qquad \text{(E.15)}$$

and the displacement at midspan equals

$$w\left(\frac{l}{2}\right) = \tfrac{615}{93,312}\left(\frac{ql^4}{EI}\right) \simeq 0.006591\left(\frac{ql^4}{EI}\right) \qquad\qquad \text{(E.16)}$$

The percentage difference between the approximate solution (E.12) and the exact one (E.16) is seen to be 1.6% in favor of the latter. This is because the very fact of approximation of the exact solution is equivalent to stiffening of the structure.

In this example, a single-term approximation turned out to be very satisfactory. As we have seen in Secs. 10.7 and 11.8, a multiterm analysis is often called for. The convergence of the method is discussed in Mikhlin's book.

From 1915 on, Galerkin realized numerous actual applications of the method, demonstrating its unusual potential, although the original idea apparently dates back to Boobnov in 1913. For the interrelation between the Galerkin method and the other approximate techniques, consult, for example, Singer, Leipholz, and Vorovich.

CITED REFERENCES

Leipholz, H. H. E., "Use of Galerkin's Method for Vibration Problems," *Shock Vibration Problems*, **8**, 3–18 (1976).

Mikhlin, S. G., *Variational Methods in Mathematical Physics*, Pergamon, Oxford, 1964.

Singer, J., "On the Equivalence of the Galerkin and Rayleigh-Ritz Methods," *J. Roy. Aeronaut. Soc.*, Sept. 1962.

Vorovich, I. I., "Boobnov-Galerkin Method, its Development and Role in Applied Mathematics," in A. Ju. Ishlinskii, Ed., *Advances of Mechanics of Deformable Continua* (Dedicated to 100th Anniversary of the Birth of B. G. Galerkin), "Nauka" Pub. House, Moscow, 1975, pp. 121–133.

Index